The Organism-En

Vienna Series in Theoretical Biology

Gerd B. Müller, editor-in-chief

Thomas Pradeu and Katrin Schäfer, associate editors

Transformations of Lamarckism, edited by Snait B. Gissis and Eva Jablonka, 2011

Convergent Evolution: Limited Forms Most Beautiful, by George McGhee, 2011

From Groups to Individuals, edited by Frédéric Bouchard and Philippe Huneman, 2013

Developing Scaffolds in Evolution, Culture, and Cognition, edited by Linnda R. Caporael, James Griesemer, and William C. Wimsatt, 2013

Multicellularity: Origins and Evolution, edited by Karl J. Niklas and Stuart A. Newman, 2016

Vivarium: Experimental, Quantitative, and Theoretical Biology at Vienna's Biologische Versuchsanstalt, edited by Gerd B. Müller, 2017

Landscapes of Collectivity in the Life Sciences, edited by Snait B. Gissis, Ehud Lamm, and Ayelet Shavit, 2017

Rethinking Human Evolution, edited by Jeffrey H. Schwartz, 2018

Convergent Evolution in Stone-Tool Technology, edited by Michael J. O'Brien, Briggs Buchanan, and Metin I. Erin, 2018

Evolutionary Causation: Biological and Philosophical Reflections, edited by Tobias Uller and Kevin N. Lala, 2019

Convergent Evolution on Earth: Lessons for the Search for Extraterrestrial Life, by George McGhee, 2019

Contingency and Convergence: Toward a Cosmic Biology of Body and Mind, by Russell Powell, 2020

How Molecular Forces and Rotating Planets Create Life, by Jan Spitzer, 2021

Rethinking Cancer: A New Understanding for the Post-Genomics Era, edited by Bernhard Strauss, Marta Bertolaso, Ingemar Ernberg, and Mina J. Bissell, 2021

Levels of Organization in the Biological Sciences, edited by Daniel S. Brooks, James DiFrisco, and William C. Wimsatt, 2021

The Convergent Evolution of Agriculture in Humans and Insects, edited by Ted R. Schultz, Richard Gawne, and Peter N. Peregrine, 2022

Evolvability: A Unifying Concept in Evolutionary Biology?, edited by Thomas F. Hansen, David Houle, Mihaela Pavlicev, and Christophe Pélabon, 2023

Evolution "On Purpose": Teleonomy in Living systems, edited by Peter A. Corning, Stuart A. Kauffman, Denis Noble, James A. Shapiro, Richard I. Vane-Wright, and Addy Pross, 2023

Properties of Life: Toward a Theory of Organismic Biology, by Bernd Rosslenbroich, 2023

The Evolution of Techniques: Rigidity and Flexibility in Use, Transmission, and Innovation, edited by Martin Charbonneau, 2024

Cognitive Biology, edited by Luca Tommasi, Mary A. Peterson, and Lynn Nadel

Convergent Evolution in Stone-Tool Technology, edited by Michael J. O'Brien, Briggs Buchanan, and Metin I. Eren

Darwinizing Gaia: Natural Selection and Multispecies Community Evolution, by W. Ford Doolittle, 2024

The Organism-Environment Pairing: A Historical and Philosophical Reappraisal, by Alejandro Fábregas-Tejeda, 2026

See a complete list of books in the Vienna Series at
https://mitpress.mit.edu/series/vienna-series-in-theoretical-biology/.

The Organism-Environment Pairing

A Historical and Philosophical Reappraisal

Alejandro Fábregas-Tejeda

The MIT Press
Cambridge, Massachusetts
London, England

The MIT Press
Massachusetts Institute of Technology
77 Massachusetts Avenue
Cambridge, MA 02139
mitpress.mit.edu

The MIT Press would like to thank the anonymous peer reviewers who provided comments on drafts of this book. The generous work of academic experts is essential for establishing the authority and quality of our publications. We acknowledge with gratitude the contributions of these otherwise uncredited readers.

This book was set in Stone Serif and Stone Sans by Westchester Publishing Services. Printed and bound in the United States of America.

Library of Congress Cataloging-in-Publication Data

Names: Fábregas-Tejeda, Alejandro author
Title: The organism-environment pairing : a historical and philosophical reappraisal / Alejandro Fábregas-Tejeda.
Description: Cambridge, Massachusetts : The MIT Press, [2026] | Series: Vienna series in theoretical biology | Includes bibliographical references and index.
Identifiers: LCCN 2025028449 (print) | LCCN 2025028450 (ebook) | ISBN 9780262052825 paperback | ISBN 9780262052832 pdf | ISBN 9780262052849 epub
Subjects: LCSH: Ecology—Philosophy | Evolutionary developmental biology—Philosophy | Niche (Ecology) | Adaptation (Biology)
Classification: LCC QH540.5 .F237 2026 (print) | LCC QH540.5 (ebook)
LC record available at https://lccn.loc.gov/2025028449
LC ebook record available at https://lccn.loc.gov/2025028450

EU Authorised Representative: Easy Access System Europe, Mustamäe tee 50, 10621 Tallinn, Estonia | Email: gpsr.requests@easproject.com

For Fátima Sofía,
My nurturing environment

The antithesis between organism and environment is accepted as a genuine and irremoveable one. An organism without an environment is never an object of experience and to a biologist it is unthinkable in the sense of unmeaning. Thus if we begin, as good empirical biology should begin, with what we observe, the organism in its environment constitutes the complex with which biology has to deal. . . . The relation between organism and environment does not appear to be so simple as is sometimes supposed.

—British theoretical biologist Joseph Henry Woodger (1929, 332)

Contents

Series Foreword

Biology is a leading science in this century. As in all other sciences, progress in biology depends on the interrelations between empirical research, theory building, modeling, and societal context. But whereas molecular and experimental biology have evolved dramatically in recent years, generating a flood of highly detailed data, the integration of these results into useful theoretical frameworks has lagged behind. Driven largely by pragmatic and technical considerations, research in biology continues to be less guided by theory than seems indicated. By promoting the formulation and discussion of new theoretical concepts in the biosciences, this series intends to help fill important gaps in our understanding of some of the major open questions of biology, such as the origin and organization of organismal form, the relationship between development and evolution, and the biological bases of cognition and mind. Theoretical biology has important roots in the experimental tradition of early-twentieth-century Vienna. Paul Weiss and Ludwig von Bertalanffy were among the first to use the term *theoretical biology* in its modern sense. In their understanding the subject was not limited to mathematical formalization, as is often the case today, but extended to the conceptual foundations of biology. It is this commitment to a comprehensive and cross-disciplinary integration of theoretical concepts that the Vienna Series intends to emphasize. Today, theoretical biology has genetic, developmental, and evolutionary components, the central connective themes in modern biology, but it also includes relevant aspects of computational or systems biology and extends to the naturalistic philosophy of sciences. The Vienna Series grew out of theory-oriented workshops organized by the KLI, an international institute for the advanced study of natural complex systems. The KLI fosters research projects, workshops, book projects, and the journal *Biological Theory*, all devoted to aspects of theoretical biology, with an emphasis on—but not restriction to—integrating the developmental, evolutionary, and cognitive sciences. The series editors welcome suggestions for book projects in these domains.

Gerd B. Müller, Thomas Pradeu, Katrin Schäfer

Acknowledgments

In the course of crafting and thinking through this book, I have come to the realization that academic work is, inescapably, a collaborative endeavor, even if sometimes it involves finding oneself alone before a blank page, with the world momentarily muffled, bracketed, outside. This awareness remains as I type these words to express my gratitude to those who have supported and cared for me, both as an individual (organism!) and as a scholar, in the past few years dedicated to this project.

First and foremost, this book would not have been possible without the mentorship, input, support, and friendship of Jan Baedke. Not in a thousand counterfactually variable worlds, could I have wished for a more caring and awesome *Doktorvater*. Thank you, Jan, for your timely advice and insights; for steering me in the right direction at every turn of the way; for making me passionate about organisms, environments, organicism, and *salsas*, and for instilling the belief in me that the integrated history and philosophy of science is worth pursuing.

During my years spent at Ruhr University Bochum (RUB), I had the privilege of crossing paths with Helmut Pulte. He epitomizes the type of scholar I aspire to become some day—a person who is not only profoundly knowledgeable and incisive but also kind and thoughtful. I wish to extend my heartfelt gratitude to him for the invaluable lessons I have gleaned from our interactions.

While at RUB, even amid the challenges posed by the COVID-19 pandemic and prolonged confinement periods, I was fortunate enough to get to know a lot of wonderful people, including Enno Fischer, Matthew Sims, Anna Pavani, Stefan Reiners-Selbach, and Rudolf Meer. Enno deserves a shout-out for his support during the last stretch of composing the first full draft of this book. I also thank Matt for our friendship and for our inspiring discussions. Additionally, thanks are due to Tanja Markner for her administrative assistance for several years. I am also indebted to the Deutsche Forschungsgemeinschaft (DFG) for their financial support throughout my doctoral studies (project no. BA 5808/2-1).

Of course, my time at RUB is perhaps better encapsulated with another acronym: ROTO. Without a doubt, being part of the DFG-funded Emmy Noether research group, "The Return of the Organism in the Biosciences: Theoretical, Historical, and Social Dimensions," has been a game changer for me. Doing research as a member of this diverse and, let's admit it, eccentric community of thinkers—now all cherished friends—has been immensely gratifying, intellectually fulfilling, and so much fun. Thanks to Jan, Guido I. Prieto, Abigail Nieves Delgado, Azita Chellappoo, Saana Jukola, Daniel S. Brooks, Federico Boem, Vera Straetmanns, Ann-Christin Fischer, Alejandra Petino Zappala, Alexander Böhm, Yağmur Metin, Friederike Andrae, Lydia Pryba, and Sergei Shevchenko.

Above all, the presence and philosophical nourishment offered by my esteemed PhD companion, Guido Prieto—a true Argentinian luminary—felt like a heavenly gift, if indeed miracles were to exist. Thank you, Guido, for so many laughs, tête-à-têtes, and insights; your footprint lives in so many pages that follow.

Additionally, I want to acknowledge all the visiting researchers who enlarged the ROTO family and infused it with greater human vitality: Andrea Olmo Viola, Elisabeth Muchka, Jalal Soltani, Mariano Martín-Villuendas, Alberto Monterde Fuertes, Mark Krenning, Joan Camarena Bononad, Daniel Nicholson, and Tatjana Buklijas. I reserve special thanks for Mariano, with whom I had the pleasure of coediting a bilingual special issue on the philosophy of organismal biology in *ArtefaCToS*. In a short time, he almost convinced me to become a (Deweyan) pragmatist, and not always over beers—imagine what would have happened were he to stay longer. ROTOs forever!

Moving forward, I want to state my appreciation to the participants of the ROTO online journal club for their perceptive opinions and discussions: Ozan Altan Altınok, Robin Bruce, Tim Elmo Feiten, Joana Formosinho, Andrea Gambarotto, Rosine Kelz, Samar Khan, Gregory Kohn, Markus

Kunze, Ludmila Litvin, Auguste Nahas, Frank Paris, Tamar Schneider, Alok Srivastava, and Francisco Vergara-Silva, among others.

This project and my ideas about the organism-environment relationship have been significantly enriched through interactions with numerous colleagues. I am especially grateful to Bendik Aaby, Lynn Chiu, Arantza Etxeberria, Natalie Feiner, Elis Jones, Daniel Liu, Blake Matthews, Daniel Nicholson, Gregory Radick, Adrian Stencel, Javier Suárez, Sonia Sultan, Marco Tamborini, Rose Trappes, Tobias Uller, Charles Wolfe, and Özlem Yılmaz. Charles Wolfe provided valuable feedback on a full version of chapter 2, and I really appreciate his observations and critical stance. My perspectives on organismal agency have notably benefited from sustained dialogues with Denis Walsh and Fermín Fulda and with two tremendously brilliant friends, Auguste Nahas and David Harrison. However, none of them are to blame for what I argue in the pages that follow, and all glaring errors and omissions remain my own.

I am also thankful for the constructive feedback I received when presenting various aspects and sections of my work in seminars, workshops, summer schools, and international meetings. As my memory will surely falter in listing every name, I refrain from doing it here, but please rest assured that your contributions have been greatly appreciated and incorporated to the best of my capabilities.

From April to June 2022, I was a visiting student at the Department of History and Philosophy of Science, University of Cambridge. I want to thank Nick Hopwood for sponsoring my research stay. Tim Lewens invited me to his group's seminars, during which I presented portions of my book (now components of chapter 4). I am thankful to Tim, Azita Chellappoo, Andrew Buskell, Arsham Nejad Kourki, Raphael Scholl, Philipp Spillmann, and Olesya Bondarenko for their comments and feedback. My time at Cambridge also allowed me to reconnect with my dear friend and notable historian Daniela Sclavo. Thank you for the inspiration, Dani! Thanks as well to Amelia Urry for the encouraging words during my writing periods at the Whipple Library—I have not forgotten them.

From October 2022 to April 2023, I was granted a Writing-Up Fellowship at the Konrad Lorenz Institute for Evolution and Cognition Research (KLI) in Klosterneuburg, Austria. The KLI affords an unparalleled place to think and write unencumbered. The popular saying that "a place is only as good as the people in it" is definitely true for the KLI: The fellows, the former scientific director (Guido Caniglia), the executive manager (Isabella Sarto-Jackson), and the former president (Gerd B. Müller) have kept the KLI the remarkable place it is. Thanks to Isabella for all the help during my fellowship and thanks to Gerd for his pointy scientific reflections. Thank you, Guido, for always challenging me to think further and for making me wonder about what kind of scholar I want to embody in the future. I sincerely hope I follow some of your footsteps. My Writing-Up Fellow cohort deserves all the gratitude and admiration in the world: David Harrison, Lauren Lambert, Jonatan Palmblad, and Ilvanna Salas León. Thank you for everything, folks! Thanks as well to the postdoctoral fellows and the visiting researchers I met for their support and useful comments and the bonds we created: Marina Knickel, Corey Bunce, Laura Menatti, Leonardo Bich, Hari Sridhar, Joyshree Chanam, Luis Alejandro Villanueva Hernández, Marco Vianna Franco, Sonia Sultan, and Kendall Baker. Laura and Leo merit special recognition for their meticulous reading and thoughtful observations on an early draft of chapter 3. I also want to thank Sonia and Kendall for the invaluable lessons they bequeathed to me. Their influence and theoretical insights have left an indelible mark on my thinking and will undoubtedly follow me wherever I go.

In March 2023, I had the privilege of visiting the BioKoinos group at the Complutense University of Madrid in what turned out to be one of the most intellectually invigorating experiences of my career thus far. During this visit, I received precious feedback on a complete draft of chapter 3 from the members of the group, including Laura Nuño de la Rosa, María Cerezo, and Vanessa Triviño. I am extremely grateful to Laura for her warm hospitality and kindness during my stay.

More recently, I have benefited enormously from the support, advice, and knowledge of Grant Ramsey, Andra Meneganzin, and Gianmaria Dani while working as a postdoctoral research fellow at the Centre for Logic and Philosophy of Science, Institute of Philosophy, KU Leuven. I am so proud and happy to be a member of the Ramsey Lab! Grant has been an incredible mentor and collaborator and helped me prepare a stronger book proposal when the time came. I am also immensely appreciative of the financial support I received from the Research Foundation Flanders during the preparation of the revised version of this monograph—in particular, for grant G070122N.

In the editorial terrain, I am infinitely grateful to Gerd B. Müller for believing in the value of this project and to Katrin Schäfer and Thomas Pradeu, the associate editors of the Vienna Series in Theoretical Biology, for their vote of confidence. Working with the MIT Press has been an absolute delight—thank you so much to the entire editorial team that brought this book to life and especially my editor, Anne-Marie Bonno, for her support, advice, patience, and openness. I would also like to thank the two reviewers of the full book manuscript for their useful and detailed comments and suggestions.

Ever since I was a student back in Mexico, my academic pursuits have been profoundly shaped by Francisco Vergara-Silva, who continues to be a cornerstone of my scientific, historical, and philosophical thinking. Paco, my earnest thanks once more, and I look forward to our future collaborations and encounters with great anticipation.

Despite the geographical distance that separates us, my dear friends Brito, Ricardo, and Mario have been an endless source of solace, respite, and comfort throughout the years devoted to writing this book. Your friendship stands as one of the proudest achievements of my life.

I want to thank my parents and my sister for their support, boundless love, and untiring generosity. Though I deeply miss you, please know how grateful I am to you for fostering my personal and professional growth, now and always. Without you, I would not be here, penning these words. Thank you for everything.

I can't forget to express my deepest gratitude to my network of support back in Mexico: Wiky, Diana, Belén, Oscar, and Helvita. Thank you for your help and trust.

The final acknowledgment is the most special of all to me. Fátima, I still cannot find the words to convey what you mean to me, how much richer my life is, in so many dimensions, by your side, how you have sustained me, with your love, intelligence, and kindness, throughout the years. Thank you, thank you. This is for you.

1 Organisms and Their Environing Worlds

A *relationship* between two things (organism, environment) is not to be confused with either one of those things nor usefully discussed in terms of only one of them.

—American paleontologist George Gaylord Simpson (1958, 520; emphasis in original)

1.1 Enter the Juxtaposition: An Organism and Its Environment

Snow flies, belonging to the *Chionea* genus of flightless dipterans, exhibit remarkable resilience in the face of the harsh winter conditions prevalent in boreal and alpine settings of the northern hemisphere. Every individual organism encounters the peril of freezing-induced mortality during this season, and yet they remain behaviorally active throughout their lives (figure 1.1). If one ventures to visit their dwelling sites, it is possible to see them navigating snowy landscapes in spite of being ectotherms. They have developed a strategy to endure subzero temperatures by resorting to a peculiar and extreme behavior—the rapid amputation of their own legs. This serves as a measure to prevent the propagation of ice crystals to their inner organs and prolong their survival in the unforgiving cold (Golding et al. 2023).

In contrast, ticks encounter different environing conditions in which cold is not an issue when they seek out a suitable furry mammal to thrive on. Once they find one, they are drawn to the electrostatically charged fur of their future hosts through forces that span air gaps several times their own body length. As scientists tell us, this interaction seems to rely on the induction of electrical polarization *within* individual ticks, rather than static charges on neighboring electric fields. To ensure that they remain

Figure 1.1
Picture of a snow fly (*Chionea* sp.) in its "environment." *Source*: Photograph captured by Flickr user trasroid (CC BY-NC 2.0), originally in color and reproduced here in black and white.

attractive in response to the electric fields of their potential hosts, ticks exhibit polarization, and their charges shift constantly (England et al. 2023). These cunning ectoparasites, otherwise unable to jump, settle on their hosts, and thus expand their effective range, by capitalizing on their surrounding conditions.

For those not extensively familiar with the weirdness of the biology encapsulated in the two examples above, nonhuman organisms might appear otherworldly, existing in a realm detached from our own. After all, electric fields do not seem to directly intercede in our behavioral choices, and the temperatures we withstand seldom require acute measures to cope with them. However, this anthropocentric perspective distorts biological reality, as all organisms, including us, have evolved to face radically different environments. Organisms are intricately intertwined with and deeply connected to their adjoining conditions of existence. No organism lives outside the world that affects it, but within and through it. To capture the complex, ontologically heterogeneous settings that living beings encounter throughout their life cycles, biologists affirm that *every organism is embedded in its environment.*

Within the epistemic sphere of biology, it seems perfectly natural for scientists to contend that organism and environment are "relata" to one another. (By "relata," I am indicating the entities between which a relationship exists, and accordingly, the singular form, "relatum," designates one of said entities.) Indeed, "organism" and "environment" are two central concepts of the life sciences that are often juxtaposed: The ensuing relationship posited is of paramount importance in manifold developmental, physiological, ecological, and evolutionary explanations. Likewise, this dyad features transversally in model building and scientific theorizing. Furthermore, this coupling significantly influences empirical research: It matters for the segregation of variables (e.g., what is purportedly considered "environmental" and what is considered "organismal") and for the general setup of experiments and fieldwork. For instance, in disciplines like physiology and developmental biology, researchers often strive to maintain the environmental background conditions of organisms constant to prevent confounding effects and examine the specific, counterfactually stable causal roles played by, say, certain tissular or genetic changes in a process of interest. Equally, scientists can reverse the approach, with organismic variables held constant in an effort to comprehend the causal contributions of particular abiotic or biotic factors to a specific scrutinized phenomenon (for discussion, see Fábregas-Tejeda 2024a).

However, despite its prominence in the theory and praxis of biology at large, the organism-environment relationship *in itself* has not been the target of much scholarly work in the history and philosophy of science beyond some of its well-explored facets, such as the purported asymmetry involved in organic adaptation (i.e., the fit of organismal structures to environmental demands brought on by the action of natural selection), or the "form-function" dichotomy.[1] Importantly, these explorations on the organism-environment relationship have predominantly been anchored to the domain of evolutionary processes, often concentrating on adaptive evolution. As such, they have inadvertently overlooked the developmental manifestations of this bond.

For instance, in a classic and influential philosophical treatment, Sober (1984) epistemically circumscribed the organism-environment relationship to a specific explanatory province related to evolution. He famously posited that the theory of natural selection, akin to other purported scientific theories of "forces," should encompass a dual set of laws: on the one hand, *source laws*, serving to delineate the circumstances giving rise to evolutionary forces (e.g., What are the causes of natural selection? In which conditions does drift obtain or is gene flow expedited?), and on the other, *consequence laws*, proposed to expound upon the evolutionary outcomes arising from the presence or absence of said forces (e.g., What happens to populations when selection is absent? How are evolutionary trajectories wrought by high mutation rates?). Within this distinction that renders an alleged epistemic division of labor for evolutionary biology, Sober (2000, 22) explicitly claimed that the organism-environment relationship only pertains to the former. According to him, the significance of this relationship for the derivation of source laws arises from factors such as alterations in the patterns of interactions between organisms and environments that prompt natural selection to favor certain variants over others, the causal influence of environmental elements in instigating mutations within organisms, or the ability of scientists to ascertain "what it is in the environment that induces . . . fitness differences" through "experimentally manipulating the organism/environment relationship" (Sober 2000, 21, 69).

In the years following Sober's work, some authors have surveyed how certain expanding fields and research areas, such as niche construction theory, evolutionary developmental biology (evo-devo), and ecological development (eco-devo), urge new conceptualizations of the organism-environment relationship in evolutionary biology (e.g., Rendón and Folguera 2012 and references therein). Along these lines, Aaby (2021a) has provided one of the most comprehensive examinations of this dyad in the context of evolution. He argued that the active behaviors and activities of organisms, construed through the lens of the organism-environment relationship, can fulfill twofold epistemic roles in evolutionary explanations: as the subjects to be explained (i.e., *explananda*) and as the explanatory factors themselves (i.e., *explanantia*). One of Aaby's central contentions is that the organism-environment relationship constitutes the fundamental ontological unit of biological science. Contrary to viewing organisms and environments as two causally independent systems that merely interact,

Aaby suggests that organisms and environments are inseparable and codependent entities. Neither can exist in isolation; they are tightly interdependent and mutually determined by each other.

Several scholars have indeed recently turned their attention to the epistemological investigation of the organism-environment relationship (e.g., Barker et al. 2014a; Pontarotti et al. 2022; Baedke and Buklijas 2023). Moreover, scholars have contributed historical examinations of the concept(s) of environment that, among other emphases, foreground theorizations concerning the interactions and couplings between organisms and their surroundings, as well as concentrating on the motivations and circumstances of the scientists advancing them (e.g., Sprenger 2019; Benson 2020). In an alternative register, different construals of the organism-environment relationship have been explored in the context of particular philosophical traditions that have been in direct and intense dialogue with the biological sciences. For instance, Pearce (2020) has recounted in detail how biological conceptions of the interaction between organisms and environments featured prominently in American pragmatism throughout several cohorts of its central exponents.

It is more than fair to say that this scholarly landscape yields a rich, variegated picture of conceptualizations about organisms, environments, and their connections, in past and present scientific and philosophical work. The immense value of this scholarship notwithstanding, it is still unclear *what kind of relationship is instantiated when organism and environment are considered relata.* This is the central question that begets this book.

In the famous essay in which the British biologist John Odling-Smee introduced the term "niche construction," he delineated the organism-environment relationship as "the meeting point between these two systems. Simon (1982) describes such a meeting point as the *interface* between the inner organization of the organism and the outer organization of its environment" (Odling-Smee 1988, 93; emphasis in original).[2] But what exactly is this meeting point, this interface, in development and evolution? Are we dealing with a symmetrical relationship or an asymmetrical one? Is it a causal relationship or one of ontological codetermination, as suggested by Aaby (2021a)? Are organism and environment two different systems that come together in the world? Can we separate them in a principled way? What kind of relata are organisms and environments? How should we understand them, both ontically and epistemically? How have scientists reasoned about the organism-environment pairing throughout history? What are the heuristic or explanatory payoffs to presuming that there is such a relationship for biological practices, explanations, and modeling efforts?

Questions related to organism-environment separation, reciprocity, and the ontic and epistemic statutes of these relata have not received systematic treatments. And, as I mentioned before, philosophers and historians of biology, with few exceptions, have exhibited a discernible bias in their explorations of the organism-environment relationship; their perspectives have gravitated toward construing it solely within the contours of evolutionary adaptation. Nonetheless, could there be a more comprehensive understanding yet to be unraveled? What does this relationship entail for *token* developing organisms (i.e., individual instances of an organismal *type*) and their environing conditions, and for the sciences that attempt to study them?

This book aspires not only to contribute to the ongoing philosophical discourse that has focused on the evolutionary side of this relationship but also to interject development into the picture, thereby fostering prospects for forging connections between these two biological planes. At the same time, I intend to shed light on the deep-rooted roles played by the organism-environment relationship *within* the epistemology of biological science. My inquest also explores the underlying rationale behind biologists' conceptualization processes: What prompts them to adopt their particular perspectives on the relation between living beings and their surroundings (e.g., to interpret and juxtapose organism and environment as interacting relata), and what are their epistemological underpinnings? My reflections encompass not only ontological considerations regarding the nature of the organism-environment relationship but also extend to the epistemic dimensions of the scientific handling of this link (e.g., how scientists explain and gain understanding through it). In other words, I am interested in both the *de re*—what the relationship *is*—and the *de dicto*—what the relationship *is said to be* in biological disciplines—readings of the organism-environment pairing. These inquiries collectively constitute the problem space that this book tries to address.

In what follows, I explain why answering the question of what kind of relationship is instantiated when organism and environment are considered relata is relevant in our current juncture. Following this, I present the methodological framework that this book will adopt (i.e., an integrated history and philosophy of science approach) and offer a rationale for its selection (section 1.2). Finally, I provide a concise overview of the book's contents, outlining its scope, structure, and guiding axes (section 1.3).

1.1.1 Organisms and Environments Matter

Within contemporary biology, the problem concerning the organism-environment relationship in and of itself becomes more pressing given the complexified picture of a broad range of organism-environment interactions uncovered through postgenomic, data-intensive studies (e.g., in epigenetics, phenomics, microbiome, and exposome research) and new empirical and conceptual insights coming from, among others, disciplines and research areas such as evo-devo, immunology, ecology, ecophysiology, global change biology, phenotypic plasticity research, niche construction theory, and 4E (embodied, embedded, enacted, and extended) approaches to cognition (e.g., Aguilera et al. 2010; Bollati and Baccarelli 2010; Akbar et al. 2022; Pontarotti and Merlin 2023; Ge et al. 2013; Abouheif et al. 2014: Gilbert 2016; Sultan 2021; Wingfield et al. 2011; Baskett 2012; Hongbo et al. 2015; Pfennig 2021; Matthews et al. 2014).

Take, for instance, the study of phenotypic plasticity. This refers to the capacity of organisms, with an invariant genotype, to alter their development trajectories and therefore their resulting phenotypes in response to different environmental conditions or stimuli (see, e.g., Pigliucci 1996; Sultan 2003; Fusco and Minelli 2010). In this sense, a genotype does not determine organismal phenotypes independently of environmental contributions. A well-known example is the case of *Daphnia* microcrustaceans that only in response to the presence of predators (e.g., hemiptera of the genus *Notonecta*) develop, for instance, pointed helmets and elongated tail spines (for an overview, see Weiss 2019). Other examples of plasticity in metazoans include the seasonal polyphenisms of some butterflies, the seasonal changes in the texture and color of the fur of Arctic wolfs (*Vulpes lagopus*), and the formation of castes in groups of hymenopterans, such as ants (Moczek et al. 2011 and references therein). In plants, there are also countless cases of phenotypic plasticity. For example, the morphology of the leaves of *Ranunculus flabellaris* depends on their immediate environment: submerged shoots develop leaves with elongate narrow lobes, while aerial shoots form leaves with shorter, broader lobes (Bruni et al. 1996). As another well-researched case, annual plants of the genus *Polygonum*, even if derived from genetically identical clonal lines, produce broad, thin leaves when grown under canopy shade or in laboratory conditions that simulate it and have narrower leaves, but with thick cuticle and thick mesophyll, when grown under a high-insolation regime (Sultan 2007).

The importance of plasticity lies not only in the ability of organisms to cope with environmental challenges during their ontogeny. This developmental capacity may constitute a first step in the emergence of evolutionary innovations through processes of *genetic assimilation*. After a certain number of generations, plastic phenotypes can be reconstructed in a manner detached from the contingencies of the environment; that is, their production is "canalized," to use a term pioneered by Conrad Hal Waddington, by acquiring a genetic basis (see, e.g., Waddington 1942, 1953, 1956). As Mary Jane West-Eberhard (2003) has suggested, in some instances, genes can be "followers," and not necessarily initiators, in phenotypic evolution. Against this background, some evolutionary biologists advocate a view in which, contrary to traditional theoretical expectations, plasticity precedes genetic change (i.e., the "plasticity-first view of evolution" or "plasticity-led evolution"). Outside the aseptic rooms of research institutes, there are increasing reports of cases of genetic assimilation in natural populations (e.g., Kulkarni et al. 2017; Vigne et al. 2021; for discussion, see Pigliucci et al. 2006; Ehrenreich and Pfennig 2016). In addition to promoting the emergence of new morphological variants, plasticity enables rapid divergence between populations and has been linked to speciation; in some cases, plasticity has the power to reveal genetic variation previously hidden from natural selection, known as "cryptic variation" (for discussion, see Schlichting and Wund 2014). Phenotypic plasticity research serves as merely one illustrative instance, among numerous others that could be mentioned, of how important organism-environment interactions are for the investigations of contemporary biology.

It is crucial to recognize that the organism-environment relationship has been a subject of contention among biologists in other recent periods as well. For instance, there has been ample disagreement concerning the "directionality" and character of this dyad. Some scholars have depicted this relationship as unidirectional, even amid some of the previously recounted interactions. A famous instance can be found in the influential book *Le hasard et la nécessité*, by Jacques Monod, wherein he proclaims that "through its properties, by the microscopic clockwork function that establishes between DNA and protein, as between organism and medium, *an entirely one-way relationship*, this system obviously defies any 'dialectical' description" (Monod (1970) 1971, 110; emphasis added). In contrast, alternative perspectives have theorized that organisms and their environments compose a dialectical system wherein "one thing cannot exist without the other, that one acquires its properties from its relation to the other, that the properties of both evolve as a consequence of their interpenetration" (Levins and Lewontin 1985, 3). In this sense, the nature of the organism-environment relationship has been a long-standing subject of inquiry for scientists as well, so it is not exclusively a curiosity for philosophers and historians to ponder or a topic that has just gained popularity in the life sciences. On the contrary, it has been an enduring problem in the history of biology, as we will see in the following chapters.

As an example of the renewed scientific focus on organism-environment interactions, a book that serves as a melting pot for several of these strands and disciplinary orientations, enriched by numerous empirical case studies, is *Organism and Environment: Ecological Development, Niche Construction, and Adaptation*, authored by Sonia Sultan. There, Sultan claims,

> A flood of new research insights has revealed as never before the intimate causal relationship in which organisms and their environments are bound together. Development is modulated—in astonishingly complex ways—by induced epigenetic modifications, parental nutrient and stress levels, chemical and physical conditions at the cellular level, and a host of other environmental factors and feedbacks. Just as the environment participates in shaping the organism, the organism's presence, activities, products, and byproducts modify its environment. By virtue of these environmental modifications, organisms influence their own functional and selective circumstances and, through effects on shared habitats, those of cohabiting neighbors. At this point, these phenomena are solidly established. The challenge is to expand our understanding to encompass these complex causal interactions and feedbacks. *This book is a close re-examination of the organism–environment relationship.* (Sultan 2015, xiii; emphasis added)

As Sultan notes, while comprehending the ways in which environments shape organisms (e.g., through phenotypic plasticity) is of paramount importance for biological investigations, an equally weighty epistemic mission is grasping and tracing how organisms, in turn, build their environments, exert influence on and transform them through their activities and actions, and help establish what elements of the world affect them. In this respect, Sultan's book crucially builds on the influential conceptual work of Richard Lewontin in *The Triple Helix: Gene, Organism, and Environment* (Lewontin 2000). If Lewontin's and Sultan's inspiring books represent comprehensive scientific reexaminations of the organism-environment relationship for evolutionary biology, this work attempts to be a close reappraisal of this dyad from a metascientific perspective for biology at large. It is from this vantage point that this project aims to address its prime question (i.e., What kind of relationship is instantiated when organism and environment are considered relata?) and the constellation of *de re* and *de dicto* interconnected issues that I outlined above.

My project is situated within a movement that is recognized in the literature as "the return of the organism" to twenty-first century biosciences. This scientific-philosophical trend emphasizes the importance of organisms as active ontogenetic and evolutionary units and their environmental contexts. The resurgence of organisms, mostly emphasized by historians and philosophers of biology, has intermittently garnered attention over the past thirty-odd years and has progressively gained momentum over the last decade (e.g., Laubichler 2000; Ruiz-Mirazo et al. 2000; Arthur 2004; Bateson 2005; Huneman and Wolfe 2010; Huneman 2010; Nicholson 2014; Walsh 2015; Fábregas-Tejeda and Vergara-Silva 2018a; Baedke 2019; for a recent survey, see Prieto 2024, chapter 1). Philosophers of biology have gone from thinking about the organism as "a highly contestable notion" (Sterelny and Griffiths 1999, 173) to highlighting its importance in current biotheoretical work. In manifold fields and research areas, organisms are being conceptualized as causally efficacious ontogenetic units that transcend the properties of their parts (e.g., their genes) while standing in a deeply entangled relationship with their environments. The organism has become a central unit to understand and interrelate intraorganismic (e.g., genetic), interorganismic (e.g., inheritable, epigenetic), and extra-organismic (e.g., environmental, ecological) processes (for discussion, see Baedke 2025). Such circumstances are not completely novel within the annals of the history of biology. Quite the opposite, the organism concept has exhibited a cyclic ebb and flow and has aptly been referred to as "biology's phoenix" (see Benson 1989; see also Peterson 2016; Shields 2017). Throughout the course of biological science, organisms have recurrently surged onto and receded from the scientific forefront. Along these lines, this book intends to be a contribution to what could be called the "history and philosophy of organismal biology" (see also Fábregas-Tejeda and Martín-Villuendas 2023).

In the next section, I contextualize my chosen approach to address the core questions posed in the foregoing discussion, while also introducing overarching methodological considerations.

1.2 Adopting an Integrated History and Philosophy of Science Approach

Despite the diversity of theorizations on organism-environment interactions in contemporary biology, a clearer conceptual understanding of what organisms and environments are, and how both are interrelated, is still needed. According to Barberousse and colleagues, the concept of "organism" is indeed of central importance for the life sciences; however, there is "a sort of blind spot in today's biology, since only few explicit definitions of this concept are available" (Barberousse et al. 2009, 2).[3] I think the same applies to the environment concept and, in particular, to the very idea of *organismal environment*. In the domain of biological science, various entities, ranging from biomolecules, cells, tumors, and organs to *token* organisms, holobionts, populations, ecological communities, ecosystems, and so on, are regarded as being enveloped by or comprising their respective environments. Biologists postulate and investigate environments across different ranges of scale and organizational complexity. The key desideratum underlying this practice is that a biological environment is always a relational construct: Environments are always environments of *something*; there are no environments if nothing is being environed. Moreover, diverse types of environments are extensively discussed within biology (e.g., developmental environments, selective environments, or ecological environments). Therefore, the diversity of environments encompasses not only the range of different entities that are purportedly environed but also the nature of what surrounds and influences them—in other words, what is "environing" them and how (Fábregas-Tejeda 2024a). In this book, my focus is placed on the organismal environment (i.e., that which environs a token developing organism [its counter relatum]) and not on environments simpliciter or sensu lato.

Explicating the ontological and epistemic statuses of organismal environments constitutes a valuable stride in comprehending the bedrocks of the organism-environment relationship. Barker et al. (2014b) point out that "in biology and philosophy, there is a lack of detailed conceptual models of the organism-environment relationship" (3). This book aims to provide the outlines of such a (conceptual) model to promote further scientific and philosophical investigations. For these reasons, my explorations of organismal environments will extend to how scientists have reasoned about them within biological science.

It is pertinent to note from the outset that the scope of this study of the organism-environment relationship is delimited to the disciplinary domain traditionally classified as "biology." Although future endeavors should extend this restricted focus, this book confines itself to the examination of works within the biological sciences. I acknowledge that a metascientific exploration of the organism-environment relationship holds relevance across a spectrum of scientific disciplines. (One compelling instance, among many, is the disciplines that study cognition.)[4] Maybe interested readers will help me expand the purview and breadth of this research in the future to include other scientific areas outside biology.

In this work, to tackle the organism-environment relationship, and the nature of its composing relata, I focus on particular domains within philosophy of science scholarship, including the philosophy of causation and scientific explanation, as well as the epistemology of abstraction and scientific understanding. My philosophical outlook comprises a palette of methods and approaches that cross the terrain of epistemology and metaphysics, ranging from traditional conceptual analysis and exegetical argument reconstruction to tools from the analytic metaphysics of science. Particularly in the philosophy of science, conceptual analysis becomes a critical metaevaluation of the concepts and ideas wielded by scientists, of their relevance and explanatory power, and of how they are used (and misused) in concrete practices within certain fields of research. In chapter 3 (section 3.2.1), I explore in more detail why metaphysical investigations are relevant for biology and what they entail for understanding organismal environments.

Barker and colleagues have also noticed an additional lacuna within scholarly discussions. They claim that "despite the burgeoning interest in new and more complex accounts of the organism-environment dyad by biologists and philosophers, little attention has been paid in the resulting discussions to the history of these ideas" (Barker et al. 2014b, 2). In this vein, this study aims to address these oversights by delving into the history of reflections concerning the juxtaposition of organisms and environments.

A historical lens is crucial in this book because, in my view, a full understanding of the organism-environment relationship demands an exploration of the historical trajectories of biologists' reasoning on this contested matter. This is why I adopt an integrated history and philosophy of science approach, which I sketch below.

1.2.1 Doing Integrated History and Philosophy of Science: Motivations, Trade-offs, and Analytic Foci

The relation between the history and the philosophy of science has long been debated and was especially contentious in the second half of the twentieth century. Numerous scholars have suggested that achieving genuine integration between these disciplines is an insurmountable task owing to disparities in epistemic standards; differing degrees of abstraction, objectives, and methods; contrasting attitudes on the descriptive-normative divide; incompatible metaphysical frameworks; mismatched success criteria; and other such factors. Additionally, there are competing interpretations of what bringing in history would mean for the philosophy of science, and at the same time, historians tend to wonder what their narratives and explanations even gain by embracing philosophical questioning (for discussion, see Esposito 2018; Martin 2006). Only within the past few decades has the scholarly landscape taken a slightly more optimistic turn, with the consolidation of "integrated history and philosophy of science" (hereafter abbreviated as &HPS).[5]

A foundational idea within the framework of &HPS is the recognition that a comprehensive philosophical examination of a specific scientific domain invariably entails an exploration of its historical progression. Historical trajectories of concepts, theories, practices, and related elements hold insights into their epistemic and methodological roles, ontic referents, and functional significance within the sciences. Conversely, investigating a particular facet of the history of science requires an examination of concepts, theories, phenomena, assumptions, and the like, thereby necessitating the inclusion of rigorous philosophical discussions that can extend the depth of historical analyses (Beisbart et al. 2019, 6; see also Arabatzis 2017). This bidirectional exchange between historical and philosophical contributions underscores their tight connection under the &HPS banner.

Schürch (2019) has stressed that &HPS approaches become particularly relevant when the metascientific issues at stake encompass interconnected philosophical and historiographical questions. In this respect, I think that the problem space pertaining to the organism-environment relationship is one where many philosophical and historical questions intersect. Therefore, it is pertinent to employ an &HPS approach in addressing these matters—for example, to understand how organism and environment have been historically construed as relata, to comprehend the ontological and epistemological underpinnings of their juxtaposition, and to uncover the roles these concepts have played in biological science.

Peering into the organism-environment relationship in biology is no easy enterprise, but there is much to be gained from an &HPS perspective. As it is commonly acknowledged, "there is considerable variation in how scientists conceptualize the organism-environment relationship. . . . For instance, microbiologists may focus on the small-scale biochemical reactions that characterize the immediate ecology of a bacterium, whereas zoologist may draw attention to the evolutionary forces that have shaped how an animal is adapted to its surroundings" (Lockman et al. 2018, 702). This quotation exemplifies variation in two distinct domains of scientists' conceptualizations of the organism-environment relationship: variation based on the specific group of organisms being studied and variation rooted in the disciplinary perspective that is espoused. And I could add a third realm of variation: diachronic change of perspectives on the organism-environment relationship throughout the histories of diverse biological disciplines and, even more so, in local contexts of research. This three-fold variation poses a substantial challenge, as philosophical and historical approaches on their own would take distinct routes to address it. For instance, philosophers might opt for generality and broad strokes, abstracting away from non-shared particularities to uncover how things and viewpoints hinge together in the most wide-ranging sense possible, while historians would probably pick delimited, situated case studies to explore only one of the edges of this extensive spectrum of variation in scientific practice. It is clear that an exhaustive examination of the diverse conceptualizations of the organism-environment relationship, in diverse synchronic and diachronic spheres of conceptual variation in biology, would be out of reach for a single book. What, then, is then to be done with this variation from a circumscribed &HPS approach?

Kuukkanen (2016) contends that the challenge in effectively integrating the history and philosophy of science can be traced back to the incompatible metaphysical perspectives endorsed by these disciplines: Historians commonly embrace a strong historicist stance, taking as a baseline assumption that all subjects of study are (in principle) variable; in contrast, philosophers often gravitate toward essentialism, aiming to uncover fundamental qualities that allow them to subsume as many examples and cases as possible under broad abstract frameworks and theories. From this perspective, trade-offs and compromises will be unavoidable in any &HPS undertaking. In the context of the threefold variation I just mentioned, my goal is to find a balanced middle ground between radical historicism and unwavering, contextless essentialism regarding the organism-environment relationship. In this work, I will not focus on a particular clade or group of organisms that have been extensively studied by biologists but, rather, will aim for generality: I strive for a stratum of discourse that would allow me to talk meaningfully—if yet somewhat abstractly—about the organism-environment relationship with taxonomic projectability. Likewise, to answer the central questions of this project, the peculiar specifics of various biological disciplines, characterized by intricate socio-material contexts and historically channeled practices and standards, may not always find inclusion. This reflects a trade-off that might not be easy to shake off when the main question to be answered, a fairly abstract one, for sure, is "What kind of relationship is instantiated when organism and environment are considered relata?" Ideally, should my efforts succeed, more focused historical investigations could subsequently inspect particular scientific practices and disciplines that hinge on *situated understandings* of the organism-environment relationship (see, e.g., Gerber 2020).

Conversely, I am more concerned with the third compass of historical and contextual variation: diachronic changes in *broad understandings* of the organism-environment relationship in biological science. This is a dimension that I will try to flesh out throughout these pages (especially in chapters 5 and 6 with respect to the issues of organism-environment reciprocity and agency). As a case in point, I will focus on how the conceptualizations of the organism-environment relationship changed in postwar biology compared to interwar biology, and I will also center my attention on recent developments in the life sciences that once again place a significant emphasis on organism-environment interactions. I am interested in uncovering far-reaching shifts in scientific construals of the organism-environment relationship that are not organism-specific or discipline-specific or tethered to ultra-localist research contexts but that bear important consequences for the ontology and epistemology of biology at large.

As is starting to become clear, the overall balance of this book will ultimately lean more toward the P in the &HPS formula. Attaining a perfectly symmetrical &HPS outcome might prove elusive due to the multitude of factors—coupled with varying epistemic standards to uphold—that liken it to navigating between two mythical disciplinary sea monsters, Scylla and Charybdis, situated on opposing sides of the Strait of Messina: Veering away from one implies getting closer to the other, and striking a favorable equilibrium with one might lead to distancing from what is commendable in the other. For instance, given that my driving questions are theoretical and broad in nature (e.g., Does the organism-environment dyad instantiate a symmetrical relationship or an asymmetrical one? Can it be spelled out in causal terms? Is it possible to separate a developing organism from its environment?), my historiographic approach will diverge from recent investigations into the historical construction of environment concepts, as exemplified by Etienne Benson's excellent work on the topic. Mine won't be "materialized intellectual history," as Benson calls his approach, which centers on uncovering how concepts are "embodied in practices, technologies, and social relations" and thus how they can transform the material and social worlds around them (Benson 2020, 11). My historiographic framework will be cemented in conceptual history, which trades in a spectrum of abstraction that allows an alignment with the philosophical approaches that I harness in this work (e.g., conceptual analysis and metaphysical examinations of science). Importantly, "organism" and "environment" can be said to be two "nomadic concepts" (see Surman et al. 2014), not bound to any particular biological discipline but transversally distributed among many of them. Likewise, the organism-environment pairing is portable and not discipline-specific (see section 2.3). This is another justification for espousing a transversal conceptual history vantage point for investigating the organism-environment relationship in biology.

For the reasons stated above, my historical analyses will not explore the cultural and social arenas in which conceptions regarding the organism-environment relationship were conceived or in which they circulated to become prevalent scientific currencies. Some of these matters have already been addressed within existing scholarship (see, among others, Sturdy 1988, 2011; Cross and Albury 1987; Abir-Am 1987; Beyler 1996; Harrington 1996; Coen 2006; Robison 2018, chapter 8; Geroulanos and Meyers 2018, 161–206; Taylan 2018, 2021; Gelderloos 2019; Sprenger 2019; Botz-Bornstein 2020; Schnödl and Sprenger 2021; Baedke and Brandt 2022). Nonetheless, additional work in this area remains to be done, and I hope this book provides an extra incentive to revisit these issues under a new light.

Hasok Chang (2012) has argued that the connection between history and philosophy should be understood as a link between the "concrete" and the "abstract" and that employing an abstract (philosophical) framework is essential for recounting concrete narratives. If historians are unable to locate suitable philosophical constructs to contextualize their narratives, Chang tells us, they should endeavor to forge novel ones. Following Chang's courageous recommendation, I devise my own taxonomies of philosophical positions (especially in chapters 4–6) to better frame the debates and points of contention of early twentieth-century biology on distinct axes of the organism-environment relationship. I then put these taxonomies and frameworks to work to examine contemporary debates in biology.

Research done under the &HPS umbrella enables one to reveal the origins of concepts that now may seem self-evident for scientific practice (e.g., environment), shedding light on the underlying epistemic and ontological assumptions that made scientists postulate them in the first place, all while excavating "archives of lost theoretical alternatives with the potential to open up conceptual solutions to contemporary issues" (Gambarotto 2019, 150). In this book, I mobilize positions that were advanced by historical actors to wage in contemporary debates that remain unresolved (e.g., What is the best construal of organism-environment reciprocity?). While facing plausible charges of anachronism, I believe this exercise holds value due to the heightened awareness among scientists in the first half of the twentieth century regarding the opacity of the concept of environment and the sophisticated theorizations they had on organism-environment interactions. In contrast, biologists in the latter half of the twentieth century, and well into the twenty-first century, largely accepted certain ideas of environments as a given, as additional run-of-the-mill theoretical and pragmatic components of their everyday work, ceasing active inquiries into how to conceptualize the organism-environment relationship and its impact in the biosciences. Particular conceptions of environment, which are metaphysically opaque, as we will see in detail in chapter 3, have become so entrenched in biological practice that scientists and philosophers stopped questioning them for a very long time.

The preceding discussion highlights that there is indeed a *presentist* bent in this work—that feared, sullied word among historians. After all, this book is written *for* the philosophers, historians, and scientists of my time. The historical reconstructions of this work are justified and sought out in part because it matters for contemporary biology and the philosophy of biology to clarify the status of the organism-environment relationship. It is not surprising, then, that my historical explorations into how early-twentieth-century scientists reasoned about the organism-environment pairing (e.g., concerning its boundaries, reciprocity, and asymmetries) are also mined for their untapped theoretical potential. To a certain extent, some form of presentism is inevitable whenever reconstructing the history of science (and especially doing &HPS), but we can be careful to avoid Whiggish excesses. According to Chang (2021), the present self of a historian remains intertwined with the process and outcome of historical writing: Our personal intentions and viewpoints naturally shape the creation of historical narratives, just as they influence any other form of communication and comprehension. As Laurent Loison suggests, for historians of science, the "question is therefore no longer *if* we have to make room for presentism, but rather *how* we should use presentism" (Loison 2016, 29; emphasis in original). Various forms of presentist history exist, shaped by historians' epistemic goals and viewpoints. By critically and empathetically engaging with the past, Chang (2021) argues, we can rediscover forsaken paths and unveil novel directions for inquiry. This is precisely what I want to do by revisiting past theorizations on the organism-environment pairing. I think they are crucial to understanding one of the relationships that has been at the core of biological science for so long.

The burgeoning interest in more complex and nuanced accounts of the organism-environment dyad certainly calls for thorough, interdisciplinary analyses, and here both the history and the philosophy of science have important things to say.

To conclude this introductory chapter, I now provide a chapter-by-chapter overview of the contents of this book.

1.3 Guiding Axes and Content

As I mentioned in previous sections, my central motivation is to understand how the organism-environment relationship has been construed in the history of biological science and why it remains a point of contention in contemporary debates. What kind of relationship is embodied when organism and environment are treated as relata? Is it symmetrical or asymmetrical, or both, depending on the epistemic and ontological lens chosen to peruse it? How was this relationship conceptualized and mobilized in previous debates on ontogeny and evolution across disciplines? Are there any boundaries between organisms and environments, or are they inseparable through and through?

I investigate two important instantiations of discussions apropos the organism-environment relationship: (1) organicist and holistic perspectives of the early twentieth century, especially those that were propelled in interwar biology (e.g., British organicism, German *Ganzheitsbiologie*, and dialectical materialism), and (2) contemporary debates unfolding within evolutionary biology, ecology, and the philosophy of biology. These are two instances in the history of biology and the philosophy of biology when organism-environment interactions have been intensely debated.[6] Contrasting these instances provides insights into the epistemic and ontological nature of the organism-environment relationship and the changing (epistemic, explanatory, heuristic) roles it has played within biological science.

With a specific focus on channeling the study's scope, I dissect the organism-environment relationship through three interconnected axes:

1. *Organism-environment separation*: This axis refers to the theoretical and pragmatic attempts to establish, dissolve, reconsider, and renegotiate the boundary between organisms and their environment.
2. *Organism-environment reciprocity*: This axis revolves around diverse construals of reciprocity, including ontological coconstitution, mutual structural fitting, and reciprocal causation.

3. *Organismal agency*: This axis covers the capacity of organisms to actively mold their experienced and physical surroundings through their activities, driven by intrinsic goals like self-maintenance and reproduction.

These three axes aim to capture distinct facets of the organism-environment relationship: (1) the nature of the relata (reflected on one side, through organismal agency), (2) their limits (organism-environment separation), and (3) the kind of reciprocal relationship established between them and its overall conformity to symmetry (or lack thereof). Discussions on the nature of the relata involved are complemented by an additional chapter devoted to untangling the organismal environment (chapter 3). This book addresses historical questions related to the three axes (1–3) that comprise its scope, such as how biologists have construed this relationship, the extent to which they have advocated or rejected a limit between these entities, and how they have conceptualized the reciprocity between organisms and their environments. Moreover, it seeks to explore the perspectives of past and present scientists on the nature of organisms' activities (e.g., are these goal-directed and agential?) and the environments they are embedded in.

Chapter 2 provides a general historical background to the pairing of organisms and environments in the life sciences. It offers a brief reconstruction of the conceptual trajectories of the "organism" and "environment" concepts until the late nineteenth century, when their juxtaposition became widespread. Moreover, it chronicles the trajectories of the organism-environment pairing in the early decades of the twentieth century and introduces the interwar 'Organicist Movement' as the main contrast case for contemporary debates on the organism-environment relationship.

Chapter 3 tries to get a grip of the ontic and epistemic standing of the organismal environment. I provide a general overview of how biologists construe organismal environments and describe important *desiderata* that scientists express in relation to their characterizations (i.e., causal effectiveness, relationality, and dynamicity). Successively, I employ tools from analytic metaphysics to gain a deeper understanding of the organismal environment. I contend that the biotic and abiotic elements that are circumscribed within a particular environment do not possess any common ontological measure and instead share individual extrinsic connections with a given *token* organism. This raises the question of whether it is metaphysically consistent to treat the environing context of a developing organism as a unitary entity. By reviewing criteria of identification, identity, persistence, and countability, I argue that this is not possible and that it is better to conceptualize an organismal environment as a "class-as-many." I also address the *indexicality* of organismal environments, which recovers the insight that what counts as "environment" depends on particular ontogenetic contexts and the activities an organism performs during its life cycle. Exploring the epistemological implications of this appraisal, I highlight the role of abstraction in how scientists typically conceptualize organismal environments. I underscore the epistemic processes that allow scientists to reach the idea of an "environment" in their theorizations and practices. I argue that, epistemically speaking, an organismal environment corresponds to a higher-order composite entity formed through the aggregation of diverse factors. This view of the environment as an ontic plurality masked by a linguistic singularity is epistemically useful and heuristically fruitful for biologists; thus, I emphasize that we should think about an environment primarily as a *device of understanding*, a conceptual tool that scientists employ to grasp and make intelligible the complex settings of the world in which organismal ontogenies are situated and embedded. All of this arid terminology will be explained in due course—so please bear with me.

The reader might already notice that chapters 2 and 3 adopt distinct perspectives—historical and philosophical, respectively. The intertwinement of history and philosophy through the &HPS approach comes into full focus in chapters 4–6, where the examinations of organism-environment separation, reciprocity, and agency feature both elements. As I see it, chapters 2 and 3 offer the basic historical background and the central philosophical building blocks regarding the organismal environment that will be fundamental for the arguments and reconstructions that take place in future chapters.

In chapter 4, my objectives converge on two fronts with regard to axis 1. Primarily, I seek to illuminate the array of perspectives that have shaped discussions surrounding the problem of organism-environment separation. The historical terrain of these positions is traversed by introducing instances from the Organicist Movement, and afterward I survey ongoing discussions in contemporary biology and the philosophy of biology. A unique contribution to historiography and philosophy also comes in the form of a taxonomical delineation underlining four distinct stances on organism-environment separation that have thus far eluded systematic classification: single boundary seekers, boundary skeptics, boundary renegotiators, and boundary pragmatists. However, I also suggest a novel standpoint, emerging from the amalgamation of some strengths of preceding viewpoints, with the intention of advancing the controversy of organism-environment separation. The stance of "shifting boundaries defenders," introduced in this work, proposes that there is no single organism-environment boundary in development but are several of them synchronically instantiated and, moreover, that these also shift diachronically throughout ontogenetic time.

Chapter 5 inspects and evaluates different construals of organism-environment reciprocity (axis 2), such as ontological coconstitution, mutual structural fitting, and reciprocal causation. To accomplish this, I recount how these concepts were embraced and debated by organicist and holistic biologists, focusing particularly on their interpretations of and disputes on these ideas. I then proceed to shed light on the waning of these accounts in the latter half of the twentieth century (e.g., through the rise of the study of reciprocity of other relata, such as gene-environment, and the increasing importance conceded to the idea of "population environment") followed by their resurgence in recent scientific currents. I contextualize existing debates in biology and philosophy regarding organism-environment reciprocal causation vis-à-vis ontological coconstitution. The chapter concludes by presenting my original philosophical contribution to this ongoing debate: I advocate for the viewpoint that reciprocal causation, grounded in both ontological and epistemic considerations, provides the most fitting perspective on organism-environment reciprocity. Coming back to axis 1, I substantiate how different construals of reciprocity impact how the nature of organism-environment separation is theorized, and I argue that there are good epistemic and ontological reasons for drawing boundaries between these relata, their complex interwovenness notwithstanding. I maintain that a perspective of organism-environment reciprocity that is parsed out in terms of reciprocal causation is apt for contemporary debates, especially for understanding how specific abiotic or biotic factors alter developmental trajectories and how, in turn, organisms shape their surroundings with important developmental and evolutionary consequences.

Chapter 6 concentrates on axis 3 as way to flesh out the asymmetries imbued in the organism-environment relationship. I argue that considering organisms as agents that are capable of performing goal-directed actions in accordance to their own normativity and for their own sake (e.g., in terms of self-preservation) changes how this biological duo should be understood. I look into various historical perspectives on organismal purposiveness and agency within early twentieth-century biology, and I show the varying degrees of importance attributed to environments. In certain cases, organismal environments were backgrounded, merely serving as contextual settings for the exercise of agency. In contrast, some accounts foregrounded the significance of organismal environments, recognizing their profound impact on agential activities. I show how this recurring theme resonates with present-day debates on organismal agency, which tend to foreground the environment. In the historical arc, I recount how, much like the scenario with respect to organism-environment reciprocity (chapter 5), viewpoints concerning organismal agency were marginalized from the mid-twentieth century onward, only to resurface in recent years. With regard to the philosophical upshots of the chapter, I contend that exploring organismal agency is essential for a deeper understanding of the organism-environment relationship. Organismal agency represents a foundational asymmetry within this pairing: As agents, organisms distinctly shape the affordance landscapes available to them, embodying bounded loci of causal influence across a range of actual and counterfactual scenarios. Conversely, organismal environments lack the attributes of agency, such as intrinsic purposiveness and normativity, despite their undeniable (but dispersed) causal effects on developing organisms. While organisms and environments engage in reciprocal causation over developmental time, their interactions lack a state of perfect causal symmetry. Recognizing this ontic asymmetry is critical for spelling out the organism-environment relationship in biology.

Finally, chapter 7 weaves together the historical insights and philosophical theses presented throughout this book. It consolidates my responses to the central questions and the problem space I previously delineated (i.e., the nature of the relata and the plausibility of their separation, reciprocity, directionality, and conformance to symmetry), providing perspectives on the organism-environment relationship in terms of both its ontological nature and its epistemic and heuristic significance within biological science (the *de re* and *de dicto* dimensions, respectively, that I sketched in previous sections).

While it might appear that two chapters (chapters 2 and 6) primarily concentrate on a single relatum of the dyad under scrutiny, I remain guided by George Gaylord Simpson's shrewd remark: "A *relationship* between two things (organism, environment) is not to be confused with either one of those things nor usefully discussed in terms of only one of them" (Simpson 1958, 520; emphasis in original). I think it is impossible to talk about organisms and environments without bringing up the counter relata at some point. There are plenty of reasons why the focus of this book is placed on the relationship between organism and environment instead of on their elements in isolation. These reasons will become evident as readers progress through this work and, hopefully, reach its final pages. While neither organisms nor environments individually constitute the pairing, a good understanding of both relata is needed to grasp the intricate relationship that binds them.

The overarching message of this book, bringing together all its threads and knots, is that the organism-environment relationship instantiates a *causal relationship* that can be fruitfully spelled out in terms of "reciprocal causation," and not one of ontological coconstitution, but it is an *asymmetric one* in the end (e.g., causal reciprocity should not be conflated with causal symmetry): It juxtaposes two radically different relata, a self-individuating, agential concretum (i.e., an organism) that acts as a bounded locus of causation—even though it has shifting, dynamic boundaries throughout its life cycle—and an (ontologically heterogeneous, dynamic) indexical aggregative abstractum (i.e., an organismal environment) with dispersed causal consequences and no shared normativity. With regard to the epistemology of life sciences, the (organismal) environment concept and the organism-environment relationship are two of the most powerful thinking tools of the biologist's toolkit, bestowing intelligibility and explanatory expediency upon their endeavors. Importantly, they furnish scientific understanding of, among other things, one of the central features of life and its evolution: Developing organisms qua agents are embedded in their environing worlds, actively participating in their shaping while being shaped by them.

This bundle of conclusions might sound impervious at this point, so we will need proper unpacking in what follows, starting from the historical threads that brought together the concepts of "organism" and "environment" to biological science.

2 Grounding Biological Reasoning in Organisms and Environments

> The organism, which seemed to be an independent unit, capable of acting on its own, gets embedded into the physical medium, like a figure in a tapestry. It is no longer the organism that moves, but the medium in it.
>
> —Spanish philosopher and essayist José Ortega y Gasset (1918, 206–207; my translation)

> The idea of environment has a history, and its ubiquity today conceals the work of metaphysical abstraction that produced it.
>
> —American philosopher and historian Trevor Pearce (2010, 252)

2.1 Excavating Organisms and Environments

In 1958, the American physiological ecologist George Bartholomew advanced both a methodological cautionary note and an epistemic maxim for biology. These were not foreign to, and indeed, were tacitly shared by, some of his contemporaries and many scientists before his time:

> Since an organism is inseparable from its environment, any person . . . must keep constantly in mind that the item being studied is neither a stuffed skin, a pickled specimen, nor a dot on a map. It is not even the live organism held in the hand, caged in the laboratory, or seen in the field. It is a complex interaction between a self-sustaining physicochemical system and the environment. An obvious corollary is that *to know the organism it is necessary to know its environment*. (Bartholomew 1958, 83; emphasis added)

How did such pronouncements become possible in the first place? How did biologists begin to speak of "organism" and "environment" qua singular entities and to study the interconnectedness of both as a prerequisite for acquiring knowledge about living processes? These are some of the questions that motivate this chapter.

Systematically excavating the history of these concepts constitutes a fascinating, yet daunting, case for scientific conceptual history. Here, I will only offer some brushstrokes on how biological reasoning began to be grounded in the juxtaposition between organism and environment. Accordingly, this chapter centers on offering some general historical context for the pairing of organisms and environments within biological science and some initial attempts to theorize and address this relationship. It presents a concise overview, without any pretension of historiographic comprehensiveness, of the developmental paths of the concepts of organism and environment leading up to the late nineteenth century when their juxtaposition became prevalent (section 2.2). Then, I touch on how the organism-environment pairing was established as a general *framing device* of inquiry for several biological disciplines in the early twentieth century (section 2.3). Lastly, I introduce the Organicist Movement of the interwar period, which swiveled on the efforts of an international community of (philosophically minded) biologists striving to cement an epistemological foundation for their science through the organism concept (section 2.4). The Organicist Movement serves as an interesting historical case study marked by intense debate on the interplay between organisms and their environments. In my integrated history and philosophy of science (&HPS) exploration, it also serves as a point of reference for holding a candle to contemporary perspectives seizing this issue. In this sense, section 2.4 offers the backdrop and context for the main historical locus of analysis on the organism-environment relationship (i.e., the theorizations of early twentieth-century organicist and holist authors) for this book (chapters 4–6).

In the next section, I narrow down the scope of my analysis in two distinct ways: (1) temporally, by providing a brief outline of the historical development of the organism and environment concepts, and their interrelation, prior to the first decades of the twentieth century to acquaint unfamiliar readers with some elements of the history of these concepts, and (2) thematically, by mostly limiting this analysis to biological science. Section 2.3 explores the first decades of the twentieth century, but it is worth noting that in the subsequent chapters (chapters 4–6), I will look into the historical trajectories of biologists' conceptualizations of the organism-environment relationship over the course of that century. Regarding the second restraint, I acknowledge that there exists a rich history surrounding the organism-environment relationship beyond the precincts of what is conventionally circumscribed as "biology." In philosophy and psychology, for instance, discussions on organisms and environments stretch back to the very inception of the dyad in the late nineteenth century, as aptly demonstrated by the work of Pearce (2010, 2014b), and continued to unfurl in those fields and others throughout the twentieth century (Pearce 2020).

In what follows, I begin by briefly discussing the historical profiles of the organism and environment concepts (sections 2.2.1 and 2.2.2, respectively) and how they began to be treated together (section 2.3).

2.2 From Inception to the Late Nineteenth Century: The Circuits of the Organism and Environment Concepts

2.2.1 Early Chronicles of the Organism Term and the Emergence of Its Biological Conception

The term "organism" (οργανισμός) finds its roots in classical Greek, with *organon* (ὄργανον) denoting a tool or instrument, coupled with the suffix *-ismos*, indicating the formation of an abstract noun. Historian Tobias Cheung (2006, 2010) has traced the usage of this term across writings spanning the seventeenth to the nineteenth centuries. His work aimed to uncover both the chronological headway and the fluctuating conceptual distinctions associated with "organism." Cheung has shown that "organism" can be discerned in scant earlier medieval sources, but it gained prominence in the 1680s with its appearance in Georg Ernst Stahl's medico-physiological writings. In them, *Organismus* captures a general principle of order in animated bodies, one that is distinct from a *Mechanismus*. By the turn of the eighteenth century, this term, rendered first in Latin as *organismus*, had disseminated into multiple languages, including French (*organisme*), English (organism), Italian (*organismo*), and later, German (*Organismus*).

According to Cheung's analysis, throughout the eighteenth century, "organism" typically denoted a specific principle or a form of order (e.g., a corporeal order) applicable to various domains, including plants, animals, and even the broader natural world (see also Toepfer 2011a, 777–842). However, as the century drew to a close, when the influence of French natural history and responses to Immanuel Kant's transcendental philosophy prompted discussions within the German intellectual sphere, "organism" underwent a conceptual transformation, sprouting into a generic designation for *individual living entities* (Cheung 2006, 331–332; see also Schmidt 1986; Cheung 2000, chapter 4).[1] In this sense, the notion of "organism" was no longer predicated *of* organized bodies. For instance, in Leibniz's *Essais de Théodicée*, he spoke of the *organisme des animaux*, wherein *organisme* denoted the organization of the animal (see Smith 2011), and Linnaeus employed the term "organism" to capture a principle of order that he considered to be proprietary of organic bodies. Rather, organisms qua individual living entities were considered as *having or being constituted* by organized bodies (Cheung 2006, 326, 328).

According to Cheung, this shift culminated around 1830 when "organism" supplanted earlier expressions like "organic," "living body," and "organized body," and from then onward it was firmly established as a recurrent, procedural term within the burgeoning biological sciences that referred to a whole-part unit determined by outer and inner *milieux*.[2] This second usage of the term "organism" aligns with the semantic domain of the *organism concept* that remains prevalent among biologists to this day and which will be discussed throughout this book.[3]

As Wolfe (2014b) has stressed, historically speaking, the organism concept has constituted an ontological go-between amid metaphysical and empirical levels of discussion. The debates surrounding the organism concept were central to the shaping of the disciplinary identity of biology (Bognon-Küss and Wolfe 2019, 8). Some scholars have argued that biology, focusing on the study of the functioning, structure, development, and transformation of living organisms, emerged as an integrative scientific discipline at the outset of the nineteenth century (see, e.g., Foucault 1970; Nyhart 1995). This emergence was the result of a synthesis of methodological and empirical advancements drawn from various scientific domains, including physiology, embryology, comparative anatomy, botany, zoology, natural history, and medicine, alongside specific institutional structurations and certain ideals of what science should amount to (see, among others, Bognon-Küss and Wolfe 2019, 4; Caron 1988; Rodríguez-Caso 2022; van den Berg and Demarest 2020). Notably, the term "biology" was independently coined by several authors hailing from diverse backgrounds—famously, Xavier Bichat, Jean-Baptiste Lamarck, and Gottfried Reinhold Treviranus—underscoring a widespread commitment to an epistemic transition in the scientific investigations of life (see McLaughlin 2002). Of course, as Wolfe (2011b) has contended, the establishment of the science known as "biology" should not be equated with the advent of controversies and investigations regarding the ontological and epistemic status of living entities, which are not a nineteenth century novelty by any means and had, in fact, been ongoing for centuries.[4] However, as the term "biology" gained prominence as the purported unitary, synthetizing science of life, its focus and concerns underwent a nonnegligible renovation.

For instance, according to Cheung (2014), Treviranus's biological work *Biologie oder Philosophie der lebenden Natur für Naturforscher und Aerzte*, published from 1802 to 1822 in six volumes, was underpinned by what he terms an "agent model" (*Agentenmodell*). This model primarily gave emphasis to the internal coherence and regularity of an organism. Treviranus's idea of organismal uniformity was chiefly oriented outward, focusing on the stabilization of internal order through interactions with what we would now call the "external environment" (see also Cheung 2021). Cheung (2014) also argues that an explicit exploration of organismal boundaries in relation to the external world, and the back-and-forth between the inner and outer dominions of an organism, can be found in the work of authors such as Joachim Dietrich Brandis.

As per Cheung's perspective, the dynamics and interfaces between the organism and its surroundings assumed a central role as an organizational framework for biological reasoning in the nineteenth century. These interactions served a dual purpose, simultaneously defining and questioning the boundaries of the organism itself, and thus influencing how biologists thought about its unity and functional coherence. Cheung introduces the concept of the "mutual inwardness of inner and outer worlds" (*gegenseitigen In-sich-Hineinragens von Innen- und Außenwelt*) as an analytic element to apprehend the working framework of early biologists. This "discursive field," in his view, held the promise of furnishing an explanatory apparatus applicable to a broad spectrum of scientific inquiries. For Cheung (2014), the agent model that pertained to individual organisms and their surrounding worlds bore a logic for carrying out biological research, dividing investigations into two layers—an outer and an inner layer.

Historian William Coleman (1977) famously claimed that, in the nineteenth century, three distinct groups of biologists could be singled out: anatomists, histologists, and embryologists who studied organic form;

physiologists who explored vital processes and functions; and researchers who investigated transformations in the interactions between organisms and their changing surroundings. Adding to that historiographic vantage point, Nyhart (2009), in her examination of the German context, has contended that biological perspectives were also articulated and gained traction during the late nineteenth century in places like museums, schools, and zoos. These perspectives cohesively integrated several facets of organismal life, including their interactions with their physical settings and other organisms, their geographic distribution, the complementarity between organic form and physiological function within the organism, and the extent of an organism's adaptation to its conditions of existence. What these historical studies collectively emphasize is the idea that juxtaposing organisms and their surroundings was one of the fundamental moves for opening the (theoretical and experimental) epistemic space of biology qua the integrated science of life.

But how did the concept of environment originate, and how did it find its way into the lexicon of biologists?

2.2.2 Whence the Environment Concept?

It is only in recent years that scholars, in the fields of both the history of biology and environmental history, have turned their focus toward examining the historical development of the concept(s) of environment (see, e.g., Pearce 2010; Meloni 2017, 2019, chapter 3; Benson 2020; Taylan 2022; Caponi 2022; Baedke and Buklijas 2023; Hildebrand 2023; Sprenger 2019, 2023; see also two pioneering references, Spitzer 1942 and Canguilhem [1952] 2008, 98–120). For instance, within the domain of environmental history, Etienne Benson asked, "How have we changed not merely what we think *about* the environment [but] also what we think an environment *is*? These are questions to which historians concerned with changes in the material environment and in the ways humans have related to that environment . . . have paid surprisingly little attention" (Benson 2020, 6–7; emphases in original). In contrast, more often than not, historians of biology have tended to overlook the historical dimensions of the notion of environment.

It is important to stress that charting the historical development of the concept of environment in biology should not be conflated with the project of tracing the history of the idea of "the (global) environment" as a primary subject of scientific and political (environmentalist) concern, as has been increasingly common since the latter half of the twentieth century and which currently dominates our socio-ethical imaginaries about the biosphere.[5] I am interested in the *organismal environment concept*—namely, the environment of any token organism, not in the environment simpliciter as a common language item or in the environment as a global relatum that affects all humans and life on our planet. The same linguistic tag hides radically different concepts that should be distinguished. Henceforth, "environment" will be a shorthand for "organismal environment" in many instances in this book (unless specified).

Etymologically, the noun "environment" is related to the Old French *environ* and *environer*, referring to such terms as "circuit," "surround," "'enclose," and "circumstances" (Jessop 2012). On this, philosopher Chappell (1997) adds, "The French etymology of the word suggests that our environment is whatever is *environ de nous*: that is, our environment is whatever we see around us in performing a turn or veer (*en virant*) through 360 degrees" (2). Over the course of past centuries, a variety of linguistic expressions pertaining to surrounding conditions, including antecedent terms like "ambiance" and "milieu," have been employed within diverse linguistic and enunciative contexts.[6] Spitzer (1942) has interconnected various terms such as "milieu," "ambiance," "circumstances," "environment," and *Umwelt* within the same philological lineage that traces its origins back to the Greek word περιέχω (*periechon*, i.e., that which surrounds, to encompass) from the fifth century BCE. Nevertheless, while these terms have close historical links, they do not invariably signify the same concept, as some of them delineate the world in distinct ways when applied to organisms (see Sprenger 2019).[7]

Just as Georges Canguilhem famously encouraged philosophers to embark on a comprehensive investigation of the milieu concept, I believe the same imperative holds true for the concept of the organismal environment within biology:

> The notion of milieu is in the process of becoming a universal and obligatory means of registering the experience and existence of living things, and one could almost speak of its constitution as a basic category of contemporary thought. But until now, the historical stages of the formation of the concept, diverse uses, as well as the successive reconfigurations of the relationships in which it takes

> part . . . make it rather difficult to make out a coherent whole. For this reason philosophy must, here, initiate a synoptic study of the meaning and value of the concept. (Canguilhem [1952] 2008, 98)

Interestingly, the historical trajectory of the environment concept starts out from the milieu conception. Canguilhem ([1952] 2008) has contended that the notion of milieu was transplanted from physics—in particular, from mechanics—into biology during the latter half of the eighteenth century. The mechanical connotation, although not the exact term, first surfaced with Newton's work, and in its mechanical context, the term can be traced back to the entry titled "Milieu" in d'Alembert and Diderot's *Encyclopédie*. Michaëlsson (1939) claims that the French translations of Newton's works—in particular, the rendition of his *Opticks* undertaken by Pierre Coste in 1720—brought a "resuscitation of the *ambiant* vocable" and derived ones (96). In the eighteenth century, French mechanists employed the term "milieu" in an attempt to capture what Newton had earlier identified as "fluid (medium)" (Spitzer 1942, 171–173; for discussion, see also Altamirano 2016). According to Canguilhem's reconstruction, a fluid qua medium served as an intermediary connecting two bodies, surrounding them and positioning them within its midst. In this sense, Canguilhem gave prominence to the underlying conceptual connection between medium and milieu:

> According to Newton and the physics of central forces, one can speak of an environment, a milieu, because there exist centers of force. The notion of milieu is an essentially relative one. When we consider separately the body that receives an action transmitted by the milieu, we forget that a milieu is a medium, *in between two centers*, and we retain only its function as a centripetal transmitter, its position as that which surrounds a body. In this way, milieu tends to lose its relative meaning and to take on that of an absolute, a reality in itself. (Canguilhem [1952] 2008, 99–100; emphasis in original)

In the French context, Jean-Baptiste Lamarck, influenced by Buffon, introduced the term "milieu" into biological science, although using it exclusively in its plural form (i.e., "milieux") to designate the "set of material (physicochemical) surroundings necessary to the survival and development of organisms" (Taylan 2022, 4; see also Casetta 2023, 24–26). Ferhat Taylan has argued that scientists embraced the broader Lamarckian emphasis on the milieux as a collection of material surroundings for the survival and progression of organisms without necessarily endorsing the contentious principle of "inheritance of acquired characteristics" that has made Lamarck's name so infamous in the history of biology. In light of this, the central element of his usage of "milieux" seems to be the underlying notion of continuity and harmony that he proposed between an organism and its physicochemical surroundings. This was predicated on the presence of active matter that sustains the intricate, (purportedly) fluid-based organization of living entities (Taylan 2022, 7; see also Hodge 1971). For instance, in his *Recherches sur l'organisation des corps vivants*, Lamarck proclaimed, "Instead of being able to say that everything that surrounds living bodies [*corps vivans*] tends to destroy them, I am about to show that independent on the state and order of things in the parts of living bodies that allow organic movement to take place, this movement can nevertheless only take place insofar as the state of the *milieux environnans* favours it" (Lamarck 1802, 78; my translation; for discussion of Lamarck's views, see Giglioni 2013; Gissis 2024).[8] Remarkably, this perspective was adopted by thinkers like the positivist philosopher Auguste Comte, even if they rejected Lamarck's slant on the inheritance of acquired characteristics (Taylan 2022, 5; for discussion, see Casetta 2023, 31–32).

The eighteenth century witnessed a significant surge in interest regarding the study of the environing contexts of organisms, not least because of the heightened exploration of various geographical regions and attempts to transplant plants and animals across ecological contexts (Jordanova 1984, 58; see also Koerner 1999). The historian Ludmilla Jordanova (1984, 59) has argued that Lamarck's fascination with the milieux of organisms was not at all uncommon in his scientific Zeitgeist.

After Lamarck, Comte and the French naturalist Etienne Geoffroy Saint-Hilaire began employing the term "milieu" in its singular form, which embodied an abstract, coalescing concept (Canguilhem [1952] 2008, 99). Regarding the introduction of "milieu" into biology, Spitzer (1942) mentions that this "surrounding element" does not merely "environ" inert substances as it does in physics. Instead, "*milieu ambiant* represents the element in which an organism lives and upon which it depends for sustenance. . . . Thus once the term passes over into the vocabulary of the biologists its reference becomes necessarily enriched" (Spitzer 1942, 175).

In his fortieth lecture on the unity of biology, Comte redefined the concept of milieu as the total ensemble of surrounding circumstances that shields the development and persistence of a living organism (Chien 2007,

74; see also Pearce 2010, 247–248). It was Comte's intention to launch a novel "science of living bodies" in the third volume of the *Cours de philosophie positive*, published in 1838. In his discussions concerning the life-constituting relationship between organic beings and the "outer milieu," and in line with the shift I introduced in the previous subsection, Comte employed expressions like *organisme* and *organisme vivant* instead of using the compound terms "living body" or "organized being" (Cheung 2006, 337). His interpretation of "milieu" encapsulated the diverse mediums or fluids—like water, air, and light—that had long been believed to imbue vitality into living entities since the era of Newtonian physics in the seventeenth century (Chien 2007, 74; see also McVeigh 2021; Taylan 2022). Please notice the centrality of this conceptual change: As some scholars have remarked, environment-related concepts put forth by natural scientists in the nineteenth century, such as Georges Cuvier's *conditions d'existence*, Jean-Baptiste Lamarck's *milieux ambiants* and *milieux environnans*, and subsequently, Charles Darwin's "conditions of life" or "circumstances," are expressed in plural forms (Taylan 2022, 4).[9] This stands in contrast to the singular notion of "milieu" introduced by Comte in 1830, which was later translated into English as "environment" by Harriet Martineau (Pearce 2010, 247).

Take the case of Darwin. Caponi (2022) has argued that the so-called Darwinian revolution ushered in a more elaborate concept of the environment than previously proposed by naturalists such as Linnaeus, Buffon, Lamarck, Geoffrey Saint-Hilaire, and Cuvier. According to Caponi, prior to Darwin, the environing conditions of organisms had been primarily conceptualized as their "inanimate surroundings," a notion typified by the terminology of the German zoologist Karl Semper (see Semper 1881, 39). Expanding on an insight first suggested by Canguilhem, Caponi (2022) proposed that the publication of Darwin's *On the Origin of Species* marked a substantial shift that transformed the biological understanding of the "environment," with an increasing recognition of the influence of *living surroundings*. These cover the relations established among various organisms, including interactions with conspecifics and predators, as the paramount factor in explaining the distribution, configuration, and evolution of life forms. Taylan adds that "rather than being a *milieu* surrounding the individual, the Darwinian environment is a *place* of coexistence, a given biogeographical habitat of interspecific and interindividual relations" (Taylan 2022, 9; emphases in original). Moreover, Caponi (2022) asserts that, according to Darwin, the living component holds preponderant significance within the organismal environment (33). This emphasis on the biotic aspects arises as a consequence of the concept of the *struggle for existence*, an integral piece of the theory of evolution by natural selection.

I concur with Caponi's assessment, as it underscores the essential role of organism-organism interactions within Darwin's theoretical edifice.[10] However, historiographically speaking, it is important to note that characterizing Darwin as possessing an *environment concept* is somewhat anachronistic. He did not once employ the singular form "environment" in the pages of *Origin*. Instead, Darwin's descriptions consistently framed the settings that surround organisms in their struggle for existence as a collection of various elements and factors, rather than a singular entity juxtaposed against them. He unswervingly employed plural forms, such as "conditions of existence" and "conditions of life."[11]

It is only in the second half of the nineteenth century that the general notion of external conditions affecting the health and features of living beings (e.g., the idea of conditions of life prevalent in Darwin's work or the Cuverian *conditions d'existence*) gave way to the abstract notion of an organism's *environment*. Before that, the idea that an individual organism interacted with another singular, abstract entity—the environment—was virtually unknown in Anglo-Saxon life sciences (Pearce 2010, 241; 2014b, 13). As Pearce remarks, "Before the word 'environment' was coined, English only possessed plural terms like 'circumstances'" (Pearce 2010, 242). Pearce has examined the initial popularization of the term (and biological concept of) "environment" by the influential Herbert Spencer, and he has done careful research into the concurrent emergence of the proposal of "organism-environment interaction" within Spencer's body of work (see also Gissis 2024).

During the 1840s, Spencer was exposed to a wide array of concepts and sources (e.g., the works of Charles Lyell, Lamarck, Alexander von Humboldt, and Robert Chambers) that significantly reshaped his perspective on the back-and-forth between organisms and their mutable surroundings (Pearce 2010, 243–244). It was only later that he encountered and read the work of Auguste Comte (Pearce 2010, 245–246). This exposure transpired through the translations of Comte's ideas by George Henry Lewes and Harriet Martineau (Pearce 2010, 247).[12] Pearce contends that Spencer's notion of organism-environment interaction was directly influenced by Comte and, supporting Canguilhem's historiographic thesis, that "it was Comte who first used *milieu* in the singular to mean an organism's external circumstances more generally, coming close to a 'dialectical conception of the relations between organism and *milieu*'" (Pearce 2010, 247). In one of these translations of Comte that Spencer got acquainted with, we find that "the idea of life supposes the mutual relation of two indispensable elements,—an organism, and a suitable medium or environment" (Comte 1853 quoted in Pearce 2010, 248).

In his *Principles of Psychology*, Spencer maintained that a central feature of life is that if there are changes in the surrounding conditions of organisms, these are followed by some changes *within* organisms. He even said that to discriminate a living being from a nonliving one we can simply manipulate elements in its environment and appreciate "that certain things shrink when touched, or fly away when approached, or start when a noise is made" (Spencer 1855, 366). If we are in doubt whether an animal is alive or not, Spencer comically suggested, we can stir it "with [a] stick" to figure it out (Spencer 1855, 366). For Spencer, in contrast, nonliving things suffer direct alterations from their surrounding conditions, but the transmutations undergone by these objects cease eventually and do not tend to produce secondary alterations in internal states in anticipation of some secondary alterations in their environments. Organisms do not simply endure alterations but change in the direction of adjustment to the changes that affect them; they dynamically respond to match their ever-changing environments (Spencer 1855, 372). Behind this simple, pervasive idea, Spencer thought, lies the principle of the *correspondence between organism and environment*: "There remains to notice the hackneyed truth—the truth rendered so common by infinite repetition that we almost forget its significance—that there is invariably, and necessarily, a certain conformity between the vital functions of any organism, and the conditions in which it is placed between the processes going on inside of it, and the processes going on outside of it" (Spencer 1855, 367).

For Spencer, then, the nature of the organism-environment relationship was one of *correspondence*. "The life of the organism will be short or long, low or high," Spencer estimated, "according to the extent to which changes in the environment, are met by corresponding changes in the organism.

Allowing a margin for perturbations, the life will continue only while the correspondence continues" (Spencer 1855, 376). In fact, according to Spencer, life could be defined as the "*continuous adjustment of internal relations to external relations*" (Spencer 1855, 374; emphasis in original; for discussion, see Gissis 2024, chapter 2). In other words, the organism-environment relationship truly grappled with what life was all about. To survive, developing organisms have to internally track whatever is going on in their environing worlds that is taking a toll on them and continuously adjust to it by generating appropriate secondary alterations (Pearce 2010, 248).

Of course, in the Spencerian outlook, evolution was deeply tied to this relationship of correspondence, of "dynamic equilibrium" to use a more familiar expression, and meant "a process that conforms to the general pattern of development, but in a manner facilitated by a conflict of forces both within organisms themselves and between organisms and their environment. . . . Thanks to ceaseless change, organisms are constantly forced to confront new circumstances; according to Spencer, it is their equilibration to those conditions that defines organic evolution" (Renwick 2009, 39–40). According to Spencer, organisms facing shifts in the balance of forces they encounter can ultimately experience one of two outcomes: either death or the endowment of a novel equilibrium. Natural selection, in his view, primarily involves the negative process of eliminating those organisms that do not have the features that allow them to counteract these changes. However, Spencer also recognized the occurrence of positive and creative adaptation, which he believed most organisms employed to confront environmental trials. This form of adaptation, distinct from the outcomes of natural selection in Spencer's view, involves the acquisition and transmission of characteristics throughout an organism's lifetime, akin to Lamarckian principles (see Bowler 2014). Consequently, both individual organisms and the species to which they belong, following Spencerian ideas, can be understood as progressing toward increased diversity and heterogeneity over evolutionary time (Renwick 2009).

Pearce (2010, 2014b) has argued that Spencer's *Principles of Psychology* marked a turning point in the popularization of ideas concerning the interaction between organisms and their environments. As many people read and debated Spencer's work, and his subsequent *Principles of Biology*, it led them to begin contemplating the relationship between organisms and their surroundings, now framed as unitary "environments," through his influential perspective (see Spencer 1864, especially chapters V and VI; see also Francis 2014, chapter 13). The Scottish theoretical biologist D'Arcy Wentworth Thompson even said of Spencer that "no philosopher of modern times, not Kant himself, has exercised in his lifetime so wide a dominion" (Thompson 1913, 3). Certainly, Spencer held a leading position in the intellectual landscape of the latter half of the nineteenth century, exerting a magnetic influence on diverse fields, including both the social and biological sciences. Some academics have gone so far as to characterize his widespread reach as a form of "Global Spencerism" (see Lightman 2015). Despite this, Spencer's influence gradually receded as the twentieth century unfolded.

We should not lose sight that Spencer, as Pearce (2010) has stressed, interchangeably used the terms "environment" and "circumstances" throughout the *Principles*. The metaphysical ramifications stemming from the transition from "circumstances" to "environment" are relevant primarily to succeeding philosophers and biologists, rather than to Spencer himself (Pearce 2010, 249). The entrenched juxtaposition of two singular entities in biological theorization—namely, organism and environment—is a legacy of Spencer's efforts and impact rather than an explicit cornerstone of his thinking. Pearce expounds further on this point:

> Although Spencer and others often treated words like "environment" and "circumstances" as synonyms, the shift from a plural to a singular term had metaphysical and methodological implications. In terms of metaphysics, the successive transitions from individuated particular factors (e.g., climate), to a general plural term (e.g., "circumstances"), to a general singular term (e.g., "environment"), correspond to a progressive concealment of the different elements that make up the world outside the organism and the relations between these elements. . . . However, the singular term "environment", like "organism", is an important heuristic for biologists, insofar as it gives them a way to talk about general causes without exploring the details of micro-level complexity. . . . Hence, the word "environment" does metaphysical work. (Pearce 2010, 249)

In the next chapter, I will try to unravel the organismal environment from both metaphysical and epistemological standpoints. Nonetheless, as we progress in this narrative, we need to show how the concept of environment gained widespread usage within biology and became firmly integrated into its theories, models, and practices during the early decades of the twentieth century. This is the subject of the next section. Again, as Pearce argued, the adoption of a new term, "environment," led to the coalescence of the idea of *organism-environment interaction*, and this displayed a greater degree of adaptability and applicability in biological science than earlier discussions centered on circumstances or other terms that tried to apprehend the plurality of surrounding conditions of organisms and their influences on them (Pearce 2010, 242).

2.3 The Organism-Environment Pairing: A General Framing Device for Biological Science

Spencer's introduction of a singular term facilitated the conceptual division for biologists between two unified and abstract entities that seized many phenomena they were interested in: the organism and its environment. This abstraction rendered the dichotomy highly adaptable, as it refrained from prescribing specific, exhaustive references to each of its constituent terms.[13] As a result, the organism-environment dyad became a pervasive conceptual framing device employed across biology, psychology, and philosophy. Pearce has claimed that the idea of organism-environment interaction became popular "because it was an abstract, portable dichotomy produced by the conjunction of two singular terms" (Pearce 2010, 242). (This is a subject that will be broached in section 3.3.2.)

In what follows, I concentrate on early postulations and mentions of the organism-environment pairing in biological science during the latter part of the nineteenth century (section 2.3.1) and the initial decades of the twentieth century (section 2.3.2). This narrative does not pretend to be exhaustive, but is merely instructive of the dissemination of the environment concept and the rise of the organism-environment pairing as a general framing device for distinct domains of biological science.

2.3.1 Dissemination of the Environment Concept in the Late Nineteenth Century

> A consideration which must not be left out of the account in any discussion of the life-question is *the potent influence of environment.*
>
> —American physician and chemist George F. Barker (1880, 118; emphasis added), using, for the first time, "environment" in the journal *Science.*

As we saw in the previous section, "environment," through the sway of Spencer, came into the biological vocabulary in the second half of the nineteenth century, although loosely at first and in a fragmentary fashion. For instance, during the 1860s, certain botanists referred to individual plants as being influenced by multiple "environments" throughout their development, adhering to the not-yet-overthrown convention in scientific practice of using plural terms to capture the heterogeneous surrounding conditions of organisms. The Scottish botanist John Gibson Macvicar offers an interesting example of this. "Life, at every epoch in the development of the plant," he said, "is no doubt always co-ordinate with the work which it has to do at the time, always adequate to accomplish the development proper to that epoch, if *the environments of the plant* fulfil their part" (Macvicar 1863, 18; emphasis added). For Macvicar, the parts and organs of individual plants can be said to have their particular environments that granted them special characteristics (Macvicar 1863, 24; see also Brown 1883).

The initial looseness of the uses of the environment, and its infiltration into several biological disciplines, can also be exemplified by its juxtaposition to nonliving systems. In a paper published in *Nature* in 1870, the English physiologist and neurologist Henry Charlton Bastian used "environment" in a larger discussion of *spontaneous generation* (see Bastian 1870, 172). A controversial Darwinian, Bastian was a staunch defender of the doctrine of abiogenesis, and he repeatedly stated that he directly observed the appearance of living entities out of nonliving matter in his laboratory (see Strick 1999, 2000). For Bastian, if life was continuous with, and constantly emerging from, nonlife, then it was not problematic to refer to the medium of colloidal aggregates as its "environment." In a similar vein, other authors even talked about "the environment" of crystals and compared it to the environments of plants and animals (e.g., Frazer 1890).

In spite of these unequal applications of the term "environment," one can discern a trend in the biological literature of the 1870s: the postulation of the environment as a singular, merging entity that condensed manifold biotic and abiotic conditions. For example, the American naturalist John Thomas Gulick used the notion of "environment," even in ostensible quotation marks (see also Fox 1883), to encompass all the external conditions that were affecting organisms at particular geographical locations, in spite of being skeptical that evolutionary change could be ignited by outward circumstances alone:

> If the initiation of change in the organism is through change in the "Environment," by what law is the cessation of change determined? If change continues in the organism long after the essential conditions of the "Environment" have become stationary, how do we know that it is not perpetual? Does the change, whether transitory or continuous, expend itself in producing from each species placed in the new "Environment" just one new species completely fitted to the conditions? (Gulick 1873, 497)

Throughout his career, Gulick studied the geographical, nonadaptive variation of several Hawaiian species of snails belonging to the genus *Achatinella*, offering evidence in support of isolation driving species diversification. What puzzled him was how a "diversity of allied species" could live "within the limits of a single island, and in districts which present essentially the same environment." Gulick became convinced regarding "the improbability that these divergences had been caused by differences in *the environment*," so he was "led to search for some other cause of divergent transformation, the diversity of whose action is not dependent on differences in nature external to the organism" (Gulick 1888, 189; emphasis added; see also Gulick 1872). After revisiting the work of Gulick, philosopher Ron Amundson concluded that, according to Gulick's conception, the environment is "*one of the relata in that organism/environment relationship* the character of which constitutes natural selection" (Amundson 1994, 125; emphasis added; see also Gulick 1890).

By the late 1870s and into the 1880s, environments were increasingly conceptualized as distinct entities posed in relational contexts, with organisms serving as their corresponding counterparts. In other words, organism and environment were becoming scientific relata. Even Darwin eventually adopted the term "environment" in his writings, but only until the second edition of *The Variations of Animals and Plants under Domestication*. There, Darwin penned, "If it profited a plant to inhabit a humid instead of an arid station, a fitting change in its constitution might possibly result from *the direct action of the environment*, though we have no grounds for believing that variations of the right kind would occur more frequently with plants inhabiting a station a little more humid than usual, than with other plants" (Darwin 1875, 281; emphasis added). Notice how Darwin speaks of "the direct action of the environment" on a plant, taking the environment to be some kind of single, unified cause—an *explanans*—which translates well as the relatum of an organism in a causal explanation about changes of phenotypic constitution (for discussion, see Pearce 2010, 249).

Another example of this causal use of the environment in evolutionary contexts comes from the British invertebrate zoologist Edwin Ray Lankester. He suggested splitting the term "homology," one of the foundational concepts of comparative reasoning that denotes the historical relation of identity (i.e., sameness) of a character arising from common descent (e.g., the forelimbs of tetrapods), into "homoplasy" and "homogeny." In Lankester's view, the environment could be construed as a "moulding cause" that acts on organismal parts to yield homoplasy, and this contrasted with homogeny, which simply depends on "the inheritance of a common part" (Lankester 1870, 42).

As another example of this, in his 1884 survey of recent approaches to investigate life, Joseph Janvier Woodward claimed that "when we study . . . the double phenomena of long-continued persistence of type, and of slow variation continually occurring, we shall find that *almost all biologists*, whatever their theory of life, *explain these phenomena* on the one hand by heredity, on the other by the sensibility of the organism to *the influence of the environment*" (Woodward 1884, 254; emphases added; see also "Plants Considered in Relation to Their Environment" 1886). In those years, heredity and environment were thus starting to be posed as causal explanantia of both evolutionary stability and organic form divergence.[14]

These are some examples of how the organism-environment dyad demonstrated its epistemic utility and versatility within late-nineteenth-century biological science. By the 1890s, as Pearce's historiographic perspective suggests, it had firmly established itself as a fundamental *framing device* in distinct domains, actively shaping scientific and philosophical discourse. Within biology, "environment" came to be acknowledged as a causal agent, thereby accentuating inquiries into organismic variation and plasticity (Pearce 2010, 249).

Pearce has shown that the closing years of the nineteenth century bore witness to a fervent debate surrounding the so-called factors of evolution. There, Spencer assumed a leading combative role (see Pearce 2020, chapter 5; for coetaneous breakdowns of this debate, see Le Conte 1891; Osborn 1895). At the heart of this polemic lay the investigation into the environment's potential to act as a causal source of organismal variation in the evolutionary process, or merely as a factor contributing to the preservation of existing variations (Pearce 2014b, 18). Spencer, in opposition to the German biologist August Weismann, postulated that the variations observed in organisms were not haphazard but, rather, influenced by their environment (Pearce 2010, 249). In his 1887 book, *The Factors of Organic Evolution*, Spencer distinguished three general factors: (1) the direct action of the environment, (2) natural selection, and (3) Lamarckian modifications of structure caused by modifications of function (Spencer 1887). In contrast, Weismann (1892) enshrined natural selection by arguing for the imperviousness and sequestration of the hereditary germ-plasm to environmental influences, which also ruled out Lamarckian inheritance of acquired characteristics. This move rendered Spencerian factors 1 and 3 negligible for evolution. (For a contemporaneous reading of Weismann and the role the environment plays in evolution, see Mivart 1889.)

During the late 1880s and 1890s, this subject remained hotly debated (see, e.g., Riley 1888; Bateson 1894, 17; Weldon 1894, 25–26; for a defense of the direct action of the environment in shaping organismal traits, see Henslow 1895). What is more, in the last years of the nineteenth century, the debate over the factors of evolution intersected with the rise of a new, yet highly controversial idea, now recognized as the "Baldwin effect," which emerged independently through the contributions of the American psychologist James Mark Baldwin, the British zoologist and psychologist Conwy Lloyd Morgan, and the American paleontologist Henry Fairfield Osborn, as they responded to the challenge of Weismannian hard inheritance (see Radick 2024; see also Loison 2024). This effect proposed, grosso modo, with different nuances depending on the author advancing it, that environment-induced ontogenetic and behavioral changes in organisms could enhance their prospects of survival until corresponding evolutionary changes materialized, all of this without involving bona fide Lamarckian inheritance of acquired characteristics. Significantly, these scholars accentuated the dealings between an organism's constitution and the environmental factors it encounters as the basis for development, an issue that, they argued, had lasting evolutionary consequences (Pearce 2010, 249; 2014b, 20–23; see also Metcalf 1906).[15]

The factors of evolution debate carried on for a couple of decades, with traces found well into the twentieth century (Pearce 2020, chapter 5). In 1918, the Canadian physiologist Ralph Stayner Lillie, of whom we will learn more in chapter 6 (section 6.3.2), reviewed in *Science* Osborn's evolutionary disquisitions on the factors of evolution and highlighted the interaction between organism and environment:

> Professor Osborn's Hale Lectures . . . raise anew the question: are the factors of organic evolution centripetal, consisting in the direct "moulding" action of environmental agencies upon the organism? or are they centrifugal, the expression of the innate formative and other physiological activities of the germ itself, operating under conditions largely independent of the immediate environment? He perceives, however, that the question can not rightly be put as one of alternatives, but that factors of both kinds necessarily enter. *Organism and environment are in continual interaction; what affects the one inevitably affects the other.* (Lillie 1918, 472; emphasis added)

This is yet another piece of historical evidence of how important the organism-environment dyad had become for biological science. In fact, in *Problems of Biology*, published in 1896, George Sandeman made an explicit recognition of the organism-environment relationship as a central target of biological reflection while singling out the unity and wholeness of the organism as the distinctive object of the science of biology:

> There is more observation in the definition of biology as the theory of the outer relations of the whole [organism]. *Organism and environment*, individual and species, parent and child [heredity], are, indeed, *relations which are subjects for biology*. . . . Adaptation to circumstances, and other matters relating to the whole organism reappear in every theory. And there is thus a warrant for regarding the *unity of the organism* as the object of biological study as distinguished from those sciences which deal with aspects only. (Sandeman 1896, 14–15; last emphasis in original)

As the nineteenth century drew to a close, the concepts of "environment" and "organism-environment interaction" had permeated various biological domains and were pretty much ubiquitous (Pearce 2010, 242). In what follows, I will provide a brief exploration of the continued relevance of the organism-environment relationship in early twentieth-century biology (section 2.3.2). In particular, I will offer a concise overview of certain continuities from nineteenth-century discussions, alongside new debates and emerging topics that surfaced. Following this (section 2.3), the Organicist Movement will be introduced as a case study for examining reflections on organism-environment interactions within the context of the &HPS approach of this book, which will be further detailed in chapters 4–6.

2.3.2 The Organism-Environment Dyad at the Onset of the Twentieth Century

By 1900, some authors explicitly recognized that "environment" was already a technical term within biological science. For instance, the physician James J. Walsh claimed that "phrases as 'the struggle for life,' 'the survival of the fittest,' 'natural and sexual selection,' 'atavism,' . . . 'hereditary transmission,' '*environment*,' 'parallelism,' 'altruism,' 'adaptation,' and the like, though *technical terms of very special significance*, have become current coin of the realm wherever our English tongue is spoken" (J. J. Walsh 1900, 466; emphasis added). The idea of "environment" was so prevalent in the life sciences by the beginning of the twentieth century that Walsh did not even seem to realize that this term had actually been imported to biology a few decades back rather than originating within it.

During the initial years of the twentieth century, evolutionary discussions regarding the organism-environment relationship persisted, largely echoing the terms and perspectives established in the late nineteenth century. For instance, Gulick (1905) tried to distinguish between two different kinds of selection: "active (or endonomic) selection" and "coincident (or organic) selection." The former is allegedly due to the "aptitudes," "habitudes," or "methods" that individuals employ to deal with the same environment. In contrast, coincident selection is due to the "discriminative and other accommodational powers of the individuals, preserving the organism from extinction under the stress of great and sudden change, either in the environment or in the relations of the members to each other, and thus giving time for the production and accumulation of variations that coincide with the accommodation in adapting the organism to the new conditions" (Gulick 1905, 65; for discussion of Gulick's views on organic selection, see Loison 2024). We can observe here the seamless juxtaposition of organisms with their environments in the formulation of evolutionary hypotheses and concepts. The same was the case for biogeographical and classificatory issues (e.g., Buxton 1923; Sumner 1924).

Moreover, there was a mounting recognition among biologists that, if this was indeed a tight pair, both organisms and their respective environments had to undergo continuous changes throughout the courses of ontogeny and organic evolution. On this, David Starr Jordan and Vernon Lyman Kellogg pointed out,

> It is a matter of common observation that organisms change from day to day, and that day by day some alteration in their environment is produced. *It is a conclusion from scientific investigation that these changes are greater than they appear*. . . . No character is permanent, no trait of life without change; and as the living organism and groups of organisms are undergoing alteration, so does change take place in the objects of the physical world about them. (Jordan and Kellogg 1907, 7; emphasis added)

Jordan and Kellogg preferred to use the term "bionomics" instead of "organic evolution," and their followers also stressed the centrality of the organism-environment dyad in their scientific pursuits. As James Herbert Orton expressed, "[The] relation of the organism to the environment, especially to limiting factors in the environment, is a subject of much importance to the student of bionomics" (Orton 1933, 693; for discussion, see Largent 1999).

Alongside this enduring discourse, including the next instantiation of the debate over the factors of evolution, there was a surge of developmental perspectives that highlighted the relations between organisms and environments. As early as 1900, for instance, the British physiologist Horace Middleton Vernon conducted experiments with echinoid *Strongylocentrotus lividus* embryos, subjecting some to atypical temperatures during their development. He found that the impact of temperature on larval growth decreased progressively from the time of fertilization. Vernon (1900) titled his study "The Reaction of Developing Organisms to Environment," in a nontrivial gesture that makes it clear that the organismal environment was also being conceptualized as a developmental relatum. In another paper, Vernon claimed, "The conditions of environment under which an organism develops are known to be of considerable influence in the production of variation" (Vernon 1894, 382). Even though Vernon only probed one environmental variable in the experiments (namely, temperature), he was invoking "the environment" of the individual echinoid embryos he assayed as an explanatory relevant entity to account for certain ontogenetic changes.

In the style of Vernon, many more experiments and studies abounded in the first two decades of the twentieth century exploring the stable or variable environmental contexts of development and reproduction in particular stages of ontogeny or across life cycles (see, e.g., Jost 1907; Klebs 1910;

Merrifield 1905; Pearl 1901, 317; Kammerer 1910; Morgan 1927, 412, 632; Meek 1920; Woodruff 1908). Additionally, some biologists also started to pose environments as explanatory relata to tackle problems of cell differentiation during development (e.g., Danchakoff 1917). Based on their results, certain embryologists hypothesized that "the environment probably affects all the physiological processes and not one alone" (Colton 1908, 447).

In the same vein, many empirical studies were conducted to trace or disprove transgenerational effects of environmental action, for instance, the existence of Lamarckian inheritance of acquired characteristics, but not exclusively of this (see, e.g., Sumner 1910; Agar 1913; for a contemporaneous overview, see Abbott 1911). Among many contributions of the time, the German zoologist Richard Woltereck, who was a student of Weismann, formulated the concept of *Reaktionsnorm* (norm of reaction) based on his experiments with the water flea *Daphnia*, during the early 1900s. The *Reaktionsnorm* concept alludes to the entire spectrum of phenotypic potentialities latent in a single genotype, which could be brought forth by different environmental conditions experienced during an organism's development (Woltereck 1909; for historical analyses, see Harwood 1996; Sarkar 1999). The larger issue at stake in the idea of a norm of reaction is "phenotypic plasticity" (see section 1.1.1), and this was a subject discussed in important scientific journals during the first decades of the twentieth century (see, e.g., Child 1902; Brierley 1921; Tincker 1924; Hora 1930).

On the occasion of the semicentennial of the publication of Darwin's *On the Origin of Species*, the German physiologist Jacques Loeb attested to this growing trend toward experimental studies of organismal environments, particularly noting what were, to him, exciting opportunities for manipulation within laboratory settings:

> What the biologist calls the natural environment of an animal is from a physical point of view a rather rigid combination of definite forces. It is obvious that by a purposeful and systematic variation of these and by the application of other forces in the laboratory, results must be obtainable which do not appear in the natural environment. This is the reasoning underlying *the modern development of the study of the effect of environment upon animal life*. It was perhaps not the least important of Darwin's services to science that the boldness of his conceptions gave to the experimental biologist courage to enter upon the attempt of controlling at will the life phenomena of animals. (Loeb 1909, 247; emphasis added)

Along these lines (see also the discussion in Morgan 1926, 492), I believe that the following declaration made by Child, who will be an important historical actor in chapter 4 (section 4.2.1), captures well the theoretical attitude shared by numerous biologists during the initial years of the twentieth century:

> At this time we are primarily concerned with the individual as a living and reacting system, rather than with the species, which is an abstraction from the individuals of which it consists. As a matter of fact, *a living organism is inconceivable except in relation to environment.* It lives and moves and has its being in relation to an external world. Herbert Spencer defined life as continuous adjustment of internal relations to external relations, and no one up to the present has been able to prove that life is anything else. Experimental biology is making it more and more evident that that organism is in continuous relation and reaction to external conditions of some sort. . . . But *whatever the nature of the environment may be in a particular case, it is always present and essential to life.* (Child 1927, 132; emphases added)

An important thread can be underlined from Child's pronouncements. For biologists dealing with developing organisms, the understanding of their objects of study seemed to be intimately tied to their environments. Despite an environment's ontological diversity and instable nature, comprising both biotic and abiotic components in flux as well as material legacies spanning generations, it was presented as an important resource for shedding light on the workings of development and life. Furthermore, the notion of adapting internal relations to external conditions establishes a clear link to earlier theoretical schemes on the organism-environment pairing (e.g., Spencerian ones; see sections 2.2.1 and 2.2.2).[16]

By the 1920s, the reach of the organism-environment dyad as a framing device for scientific inquiry had extended to many domains of biology (see, e.g., Pearse 1922; Goodrich 1924, 51), not only to the study of development and evolution. In a *Nature* review of Lancelot Alexander Borradaile's book *The Animal and its Environment*, we read,

> EVERY zoologist will grant that animals should be studied in relation to their surroundings. Physiologically, as Huxley said, they are whirlpools in the river; morphologically, they are, in part at least, bundles of adaptations, definitely related to environmental conditions. Embryologically considered, . . . they may develop differently when the circumstances are altered. Etiologically, they must be considered in relation to environmental stimuli and environmental sifting; isolation is often an environmental affair; the pulse of evolution changes its throbs with

> the climate. Then there is the behaviour of the creature, so largely concerned with thrust and parry between organism and environment. In certain cases we are most impressed with the grip of the environment on the organism; in other cases, what strikes us is the dominance of the organism over its environment. ("The Animal and Its Environment: A Text-Book of the Natural History of Animals" 1924, 152)[17]

As I have shown, by this time, the concept of environment had been fully integrated into the biological lexicon, and the organism-environment dyad had become omnipresent in both experimental and theoretical pursuits. Nevertheless, the interwar period signified a turning point in the discussions surrounding the organism-environment nexus, as numerous biologists strove to infuse greater theoretical and conceptual depth into the sweeping assertions I have canvassed in the preceding pages. In particular, an international community of scientists and thinkers, steadfast in their commitment to anchoring biology in the unity of the organism, directed particular attention toward the embeddedness of living beings in their surroundings. They explored the thorny problem of organism-environment separation, scrutinized the reciprocal character of the bond between these relata, and probed for any lingering asymmetries within this pairing (the subjects of chapters 4–6, respectively). In the next section, I briefly introduce the Organicist Movement that took root in the landscape of interwar biology, with the aim of providing a historical and philosophical foil for contemporary biological debates on the organism-environment relationship that will be dealt with in the ensuing chapters. Throughout this book, I hope to show that there are still many valuable insights to be rediscovered about the organism-environment relationship by revisiting what past organicist and holistic biologists theorized about it.

2.4 Introducing the Organicist Movement

2.4.1 Interwar Biology and the International Articulation of Organicist and Holistic Positions

As the twentieth century dawned,[18] several debates on the conceptual, epistemic, and ontological underpinnings of biology burgeoned within Anglo-Saxon and German-speaking scientific communities. According to Laubichler (2017), three key focal points were under the limelight: (1) The disjunction between a rapidly increasing number of empirical findings and experimental results, on the one hand, and the dearth of robust conceptual frameworks to deal with them, on the other. This disconnection led to a "data crisis," reminiscent of contemporary anxieties, that gripped the scientific landscape; (2) the attempt to tell apart genuine biological processes and objects from fiat ones, granting them a solid theoretical justification based on empirical findings; and (3) the critical assessment of the epistemological and methodological foundations that undergirded biological inquiry. These concerns intersected with the question of how biology as a science should be taught or its textbooks written and the question of a "general biology" (*Allgemeine Biologie*) emerged constantly in teaching contexts (Laubichler 2017, 95–96; see also Laubichler 2006, 185–205; Reiß 2022).

The German zoologist Julius Schaxel, for instance, mulled over the state of his science at this time: "[Biology] collects immeasurable individual knowledge [*Einzelwissen*], hypotheses proliferate uncontrolled, self-reflection, methodology remains undone. Contemporary biology is in a state of crisis. A general biology, a science of life as such, exists in name only" (Schaxel 1919, 2; German original). In response to this prevailing sense of crisis, of fragmentation of biological knowledge, scholars reflected on the basic concepts that underpin biology (Baedke 2019; see also Laubichler 2000, 298). Notably, among these concepts, both historically and contemporaneously, is the lasting *organism concept* (see Baedke 2025).

During the interwar period (Baedke and Brandt 2022), a diverse array of biological perspectives, with a focus on the active unit of the organism, blossomed in various local contexts worldwide (for discussion, see, e.g., Hein 1969; Haraway 1976; Nicholson and Gawne 2015; Esposito 2016, 2017; Tamborini 2018; Baedke 2019; Herring and Radick 2019). In recent times, some historians have investigated these perspectives, giving particular attention to four distinct contexts: (1) Great Britain, (2) the United States of America, (3) Germany and Austria, encompassing the tradition of German-speaking holism, and more recently, (4) Italy.[19] Scholars have proposed to merge these diverse streams of thought under the banner of the "Organicist Movement." In their recognition of the existence of this international community, Nyhart and Lidgard mention that "the organicist movement in biology . . . peaked in the 1920's to 1940's," highlighting the centrality of an interwar periodization for understanding the debates it brought on (Nyhart and Lidgard 2011, 405). Supporting this last point, recent analyses embracing a digital humanities approach have shown that, from the interwar period, a big corpus of scientific and philosophical publications exist that explicitly discuss the organism concept and organicist viewpoints (see Gibson and Ermus 2019; Böhm et al. 2022).

Various perspectives within this movement have been interpreted as a departure from the long-standing dichotomy between mechanism and vitalism (Allen 2005; cf. Chen 2024). By combining elements from both of these contrasting positions to grapple with the question of whether life is explainable solely through recourse of physicochemical mechanisms, these viewpoints were presented as viable alternatives that offered means to resolve this abiding biological debate (Beyler 1996, 252; see also Schaxel 1917).[20] "The period is a time of basic crisis in which the age-old dichotomy between mechanism and vitalism was reworked and a fruitful synthetic organicism emerged," judged Donna Haraway (1976) in one of the first historical analyses of this movement, "with far-reaching implications for experimental programs and for our understanding of the structure of organisms. . . . Organicism . . . attempts to comprehend and transcend former dichotomies not by abolishing one of the poles but by linking them" (2, 62). Organicist scholars contended that biological explanations resist full reduction to simplistic mechanistic frameworks. However, their perspective did not preclude the possibility of comprehending the distinct character of organisms through wholly scientific approaches, wherein their attributes of self-organization and self-regulation emerge as consequential (but naturalistic) explanatory factors (Nicholson and Gawne 2015, 358).

Organicism, vitalism, and mechanism were vigorously debated positions in early twentieth-century biology communities, and these discussions also found a platform in major scientific venues, including renowned journals such as *Science* and *Nature* (e.g., Parker 1924; Greenwood 1931; "Entwicklungsbiologie und Ganzheit" 1938). As historian Herbert J. Muller asserted, scholars from the Organicist Movement, in contrast to metaphysical vitalists and staunch mechanists, wanted to recenter biological explanations on the living organism: "The vitalists insisted that some altogether new principle—an entelechy, an *élan vital*—was necessary to explain life; the mechanists insisted that the principles of physics were not only adequate but essential. Both tended to lose sight of the living organism in their logical dispute over explanation" (Muller 1943, 106). For organicists, "the fundamental fact in biology, *the necessary point of departure is the organism. . . .*

Although parts and processes may be isolated for analytic purposes, they cannot be understood without reference to the dynamic, unified whole that is more than their sum" (Muller 1943, 107; emphasis added).

As Nicholson and Gawne (2015) have argued, adopting an organicist standpoint within and for biology meant embracing particular ontological and epistemological commitments: (1) Organicists pointed to the centrality of the concept of the organism in biological explanations and the organism was thought of as the fundamental ontological unit for biology; (2) they emphasized the importance of the level-stratified "organization" of living systems as a theoretical principle; and (3) they defended the autonomy of biology, a singular and irreducible intellectual enterprise that could not be fully subjected to the explanatory yokes of physics and chemistry (see also Woodger 1929, 273; Russell 1930, 166; Beckner 1969, 5, 8–9).

Philosopher Hilde Hein (1969) has argued that what distinguished organicism from mechanism or vitalism were different meta-theoretical commitments that lead to disparate standards for understanding and waging biological evidence and provided different heuristic frameworks for research. Along these lines, organicism, for historians such as Erik Peterson, was indeed a "meta-theoretical commitment" regarding how to conceive organisms and living phenomena (Peterson 2016, 249; for a different, critical view, see Wolfe 2014a).

In her pioneering book, Haraway (1976) employs a Kuhnian framework to propose that figures like embryologist Ross G. Harrison in the United States, biochemist and embryologist Joseph Needham in England, and theoretical biologist Paul Weiss in Austria, among others, shared a common disciplinary matrix. According to Haraway, they were deeply immersed in the same "paradigm" and employed analogous conceptual and metaphorical resources to try to understand developing organisms. As a result, Haraway maintains that organicism constituted an epistemic community transcending national borders. Even if we abjure the Kuhnian undertones, one can still argue that organicism represented a scientific-philosophical movement that found expression in various corners of the world. In this regard, Nicholson and Gawne (2015) contend that organicism featured a core set of themes that engaged an international community of scholars who read, translated, deliberated on, and responded to one another's contributions (370–371; see also Haraway 1976, 4, 26, 130; Müller 2017a, 15), as we will also attest throughout this book.

Joseph Needham credited the French zoologist Yves Delage as being the originator of the term "organicism" in his work *L'Hérédité et les Grands Problèmes de la Biologie*, published in 1903 (Needham 1928a, 77). Delage employed this neologism to describe a particular school of thought that conceptualized "life, the form of the body, the properties and characters of its diverse parts, as resulting from the reciprocal play or struggle of all its elements, cells, fibres, tissues, organs, which act the one on the other, modify one the other, allot themselves each its place and part, and lead all together to the final result, giving thus the appearance of a pre-established harmony" (Delage quoted in Needham 1928b, 29–30; for discussion, see Haraway 1976, 34–36; for additional historical context of Delage's work, see Loison 2024). Needham also claimed that, regardless of Delage's early coinage, organicism was "given an entirely new lease of life in 1917 by J. S. Haldane . . . , who then used it to describe his own views" (Needham 1928a, 77).[21]

In actuality, in contrast to Needham's estimation, the Scottish physiologist John Scott Haldane had made use of the term "organicism" for the first time in his writings in 1916. In a lecture delivered before the Harvey Society in October of that year, later published in *Science*, Haldane presented his personal overview of recent developments in physiological science that put the wholeness of the organism and its interwovenness with its environment at the forefront and communicated the term "organicism" in a footnote: "It has been suggested to me that if a convenient label is needed for the teaching upheld in this letter the word 'organicism' might be employed. This word was formerly used in connection with the somewhat similar teaching of such men as Bichat, von Baer and Claude Bernard. Cf. Delage, L 'Hérédité, Paris, 1903, p. 436" (Haldane 1916, 630). The same footnote would appear in Haldane's 1917 book *Organism and Environment as Illustrated by the Physiology of Breathing* (see Haldane 1917, 3). It was within the pages of this book that Needham first encountered the term "organicism," setting in motion its circulation among biologists.[22]

Nonetheless, in the first decades of the twentieth century, there were also terminological alternatives to organicism. For instance, the American zoologist William Emerson Ritter coined the descriptor "organismalism" (Ritter 1919, 28). Austrian theoretical biologist Ludwig von Bertalanffy introduced the term "organismic biology" (*organismische Biologie*) as a moniker for this perspective (Bertalanffy 1932, 80). In a similar vein, Schaxel championed the adoption of an "organismic basic conception" (*organismische Grundauffassung*) for biology (Schaxel 1919, 125). Moreover, some authors also believed that organicism was synonymous to other so-called third way positions, such as "emergent evolution" (see Morgan 1929; for discussion, see Blitz 1992), that were offered as a way to overcome certain entrenched dichotomies in biological thought (e.g., Wheeler 1926, 433; 1929, 105–106).

This is proof that the fine boundaries of organicism were contested during the interwar period. In a retrospective assessment, the philosopher Giovanni Blandino wrote,

> Organicism today is a rather widespread conception which is not always identical in the works of its various supporters for it presents, in fact, many variations, but always with some common fundamental characteristics. Even its denominations are diverse: *Organicism, Organismal or Organismic Theory, Holism* etc. . . . Organicism is characterized by the affirmation of *unity, wholeness, individuality, organization and novelty* of the living organism. (Blandino 1969, 161; emphases in original)

As we have seen, organicism constituted an international movement characterized by interconnected communities engaged in continuous dialogue (e.g., in international conferences; see Ritter 1933). However, its manifestations bore distinctive nuances contingent upon, among other things, the local research contexts in which it flourished. Historian John Parascandola claimed that "there was no one school of organicism or holism, but rather a number of very different theories which shared a common emphasis on the unity and organization of the whole" (Parascandola 1971, 64).[23] Therefore, historiographically speaking, it is possible to discern various organicist and holistic positions in biological science, some of which will be covered in chapters 4–6.[24]

To give some examples: The city of Cambridge, in the United Kingdom, housed the Theoretical Biology Club (TBC), helmed by people such as Joseph Needham, Joseph Henry Woodger, Conrad Hal Waddington, and other scientists and philosophers that pushed an organicist agenda (see section 5.2.5). Nevertheless, even before the inception of this club, biologists were advancing organicist and holistic positions in Great Britain, including the physiologist John Scott Haldane, who substantiated the theoretical tenets of these positions through his own investigations into the physiology of respiration (see Peterson 2016; see also section 4.3.1).

Meanwhile, across the Atlantic Ocean, distinct strands of organicism were cultivated in the United States of America (see Esposito 2016, chapters 5 and 6). Scientists such as Ritter, Child, the zoologist Charles Otis Whitman, and the brothers Frank Rattray and Ralph Stayner Lillie, among others, spearheaded this American rendition of organicism (for a discussion of Child, see section 4.2.1; for coverage of Ralph Lillie's views, see section 6.3.2). Furthermore, Central European regions were very important for the Organicist Movement as well. The Austrian capital city of Vienna was home to the influential organicists Paul Weiss and Ludwig von Bertalanffy (see Drack et al. 2007) and was also a significant pole of organism-centered biology because of the experimental research conducted at the *Biologische Versuchsanstalt*—commonly known as the Vivarium (see Coen 2006; Müller 2017a, 2017c; see also section 5.2.2)—a research institute directed and managed by zoologist Hans Leo Przibram (see section 2.4.2). Other promoters of organicism in the German-speaking context (see Baedke 2019) included the botanist Emil Ungerer, the zoologist Friedrich Alverdes, the neurologist Kurt Goldstein, and the theoretical biologists Jakob von Uexküll and Adolf Meyer-Abich (for discussion of Uexküll and Meyer-Abich, see sections 4.3.3, 4.4.1, and 5.2.5). In particular, in Germany, Schaxel was interested in building an organicist "critical biology," as his work *Grundzüge der Theoriebildung in der Biologie* testifies (see Reiß 2007; see also Laubichler 2017, 99–100; Baedke 2019, 302). Jan Baedke (2019) has proposed a threefold taxonomy of interwar organism-centered perspectives that does not make categorical distinctions between national states or languages but, rather, focuses on theoretical-epistemic affinities: (1) holistic biology (e.g., *Ganzheitsbiologie*), (2) organicism, and (3) dialectical materialism. Throughout this book, following Baedke's classificatory scheme, I will discuss holistic authors, organicists, and dialectical materialists.

In this chapter, I have concisely recounted the historical evolution of the concepts of organism and environment (sections 2.2.1 and 2.2.2, respectively), as well as their initial juxtaposition in biological science (section 2.3). This exploration extended to the examination of the organism-environment pairing during the late nineteenth century and early twentieth century (sections 2.3.1 and 2.3.2). Furthermore, I introduced the Organicist Movement as a key development during the interwar period, which will serve as the primary historical locus of analysis for the organism-environment relationship in the forthcoming chapters. Revisiting the discernments of past biologists will lay the foundation for a deeper exploration of the organism-environment relationship in the context of contemporary debates within the life sciences.

In concluding this chapter, the following subsection will offer only a glimpse into the significance of the organism-environment relationship within the context of the Organicist Movement. A comprehensive examination of how organicist and holistic authors delineated the separation between organism and environment, as well as the reciprocal dynamics of how organisms both shape and are shaped by their surroundings, will serve as the historical thread of chapters 4–6. Complementing these discussions, the trajectories of the Organicist Movement in the second half of the twentieth century will be recounted in chapter 5 (section 5.3).

2.4.2 Organicism and the Organism-Environment Pairing

Organicist biology not only drew from philosophical musings and existing scientific theories but was complemented by a series of experiments that probed the plasticity, robustness, and entanglement of organisms with their surroundings. Of particular note are studies conducted during the first half of the twentieth century on the environmental responsiveness of developing organisms, including their transgenerational effects, at the *Biologische Versuchsanstalt* in Vienna (Müller 2017a; Nickelsen 2017a, 2017b; Nicoglou 2018; Baedke 2019, 296–298). For instance, aiming at advancing a quantitative framework of relations and dependencies of organismal form and function, many studies of plants and animals were undertaken with controlled and intervenable microenvironments (Coen 2006, 493–523; Logan and Brauckmann 2015; Taschwer et al. 2016; Müller 2017b; Nickelsen 2017, 170–175; see also Müller 2017c; Nicoglou 2018, 107–111).[25] A relative novelty in the landscape of biological science, experimental aquariums were set up at the Vivarium with living specimens sourced from diverse collection sites in an attempt to mimic, as best as possible, the "natural" (a)biotic conditions that impacted the development of organisms, allowing a careful, protracted study of them (Przibram 1908; for discussion, see Wessely and Stobaugh 2019, 50–51). On the common project of researchers from the *Biologische Versuchsanstalt* to understand the organism-environment bond, historian Deborah Coen says, "Diverse as the Vivarium's scientists were in their experimental approaches and in their philosophical convictions, they converged in their search for a 'third way' between mechanical determinism and pure spontaneity, a framework that would do justice to the *complex interaction between organism and environment*" (Coen 2006, 496; emphasis added).

Furthermore, through their empirical work, organicist biologists recognized the central role of the environment during organismal development. Child, for instance, claimed, "As regards the nature of the relations between individual and environment, it has of course, long been known that . . . interchange of substances between organism and environment is necessary for continued active life. But investigations of recent years have shown us that environment is an essential factor, not only for continued existence, but also for the development of the form and structure and the physiological relations of parts characteristic of the individual" (Child 1927, 133). Importantly, as Esposito (2017) has reconstructed, for most organicists, the parts of organisms are shaped and constituted in a dynamic interaction involving the whole organism and its environment. Organisms, as "dynamic wholes," have to be conceived as active, self-constructing entities, capable of adapting and changing their morphology and behaviors according to external circumstances. In a representative example of organicist rationale, E. S. Russell asserted, "The life of an organism is essentially a unitary functional or dynamical process, in which *whole* and *parts* are inextricably connected. Both whole and parts are together the expression of the life of the individual" (Russell 1930, 149; emphasis in original). In that same vein, the holist Meyer-Abich put forward an additional facet of early twentieth-century authors' emphasis on the whole organism: it provided them with an integrated entity that could be juxtaposed with its environment:

> For 19th century's biology the organism as an active whole did not really exist. It appeared only as a state of cells with a very liberal constitution and with cells as its [sic] only and very real acting citizens. . . . Nowadays biology is always speaking about the *organism as a whole*. For many biologists still today it represents only a nice word, well to use in Sunday-sermons for the general public. But when the term has also some really important biological meaning, then it states only, that the whole of the organism represents an active and creative power, which controls actively the activities of the internal organs as well as the *relations of the organism with the environment*. (Meyer-Abich 1955, 67–68; emphases added)

However, what precisely constitutes this entity referred to as "the environment of an organism-as-whole?" This is the central philosophical question that has been at the background of this entire chapter. As we have explored, the "environment," when treated as a singular relatum, serves as a unifying ontological construct that reunites diverse external conditions—ranging from variables like temperature, light availability, and chemical gradients to factors such as the presence of conspecifics or potential predators—that collectively influence the attributes and operations of a specific organism (i.e., its counterpart relatum). Yet what does this mean? In stark contrast to the extensive body of philosophical and historical literature dedicated to the organism concept, relatively few scholars in the history and the philosophy of science have questioned what environments stand for or how one might characterize their ontological and epistemological dimensions within the context of biological research. In this sense, a couple of decades ago, the American ecologist Bernard C. Patten lamented the lack of reflection on what biological environments are:

> Something so universal and central to existence as environment must deserve better rendering than just external factors determining energy and matter movements across membranes, or unspecified selective forces vaguely shaping the course of evolution. When one poses the question "What is environment?" to audiences of some sophistication, the responses (or lack thereof) suggest this may be a question of similar difficulty to the philosopher's "What is reality?" As such, it is easy for [biologists] to concede the territory, perhaps too quickly, to the philosophical domain. (Patten 2001, 423)

Before proceeding at full steam with the integrated history and philosophy of science (&HPS) exploration outlined in chapter 1 (section 1.3), which concentrates on the demarcating lines and reciprocity between organism and environment, and potential asymmetries within the pairing, I consider it imperative to gain a more comprehensive understanding of the concept of the organismal environment. In particular, we need to explore further the particular metaphysical consequences that the environment concept brought to biology—consequences that were felt in both scientific theory and practice (see section 2.2.2).

The Organicist Movement unfolded in a productive epoch of theorizations with many conceptual alternatives and insights that will prove to be useful in debates on the organism-environment relationship. I will revisit these perspectives in due course, but in the next chapter, and with apologies to Patten, my primary focus will be to clarify—with a philosopher's hat—the ontological and epistemic underpinnings of the organismal environment itself. This is a necessary first step toward unraveling what kind of relationship is instantiated when organism and environment are considered relata.

3 What Environs an Organism? Environment as Relatum

> What is the nature of the agencies which determine the character of the environment?
>
> —American botanist and bryologist George Elwood Nichols (1924, 1)

> This is the pivotal issue . . . : what constitutes the environment?
>
> —American ecologist Thomas G. Balgooyen (1973, 1199)

What is the "environment" of an organism? What kind of elements and processes does it comprise, and how do scientists characterize them? How should we understand what an organismal environment is said to be, both ontologically and epistemically? Which frameworks are best suited to grasp how organismal environments are construed within biological science? These questions amount to the target space of this chapter.

A first clarification is in order: My only focus here is understanding how the environment functions as a relatum of token developing organisms. In that sense, throughout this discussion, "organism" will be shorthand for "token developing organism" and its type construal will not be invoked. Likewise, evolutionary considerations will be mostly bracketed for the time being (to reappear in chapters 5 and 6). Moreover, I must say that I have no pretense to extrapolate the philosophical analysis of these pages to the environment concept simpliciter or to more general qualified usages of this term (e.g., the manifold meanings it adopts outside the life sciences), or even to encompass other counter biological relata in particular (e.g., biomolecules, cells, organs, or ecological communities).[1]

Before moving forward, a second disclaimer should be put down on paper: This chapter is broadly concerned with the environment as the relatum of an organism qua scientific concept deployed in the theoretical frameworks, models, and practices of biologists, more often than not ambiguously specified and taken for granted. The issue of how organisms experience their surroundings, which could ground a different understanding of what environments *are*, addresses a distinct problem sphere, and thus a different battery of scientific and philosophical approaches and concepts (e.g., *Umwelt*, positionality, affordances, perception and sensory cue interfaces, embodiment, agency, experiential niche construction, cognition, situatedness) would be needed. This separate issue will only be partially broached in chapter 5 (experiential niche construction, section 5.4.2) and chapter 6 (affordances, section 6.5.5), with a preliminary historical discussion in chapter 4 (sections 4.2.2 and 4.3.3).

My aims here are, first, to paint some broad strokes on how organismal environments are construed within biological science and outline some desiderata that scientists have indicated should accompany their characterizations (section 3.1): (1) causal effectiveness (i.e., the partitioning or segregation of the elements and processes of the world that affect a particular organism); (2) relationality, understood as the mutual dependence of both relata (i.e., there are no organisms without environments and no environments without organisms), which is frequently expressed as a bijective function: to each organism corresponds an environment; and (3) dynamicity (i.e., the composition of heterogeneous biotic and abiotic factors of a given environment is variable and presents complex fluctuations as a function of time). Afterward, I will argue that an approach that does not shy away from tackling organismal environments from a metaphysical standpoint is needed to gain a deeper understanding of this scientific relatum, a project that hitherto has not received much attention in the philosophy of biology (but see Smith and Varzi 2001, 2002; Walsh 2015, 2022). In section 3.2., I adopt a "metaphysics *for* biology" stance (sensu philosopher Vanessa Triviño) and employ some tools from analytic metaphysics in the interest of clarifying what kind of entity an organismal environment is (Triviño 2022; see also section 3.2.1 for a discussion of this issue). I argue that the myriad biotic and abiotic elements that are circumscribed within a given environment do not possess any common ontological measure (i.e., a shared intrinsic property) but, rather, have individual extrinsic connections with a given organism. This opens the question of whether it is metaphysically consistent to treat the environing context of a developing organism as a unitary entity. My answer is that it is possible only in a restricted (abstract) sense: as an ontological construct that is best conceptualized as a *class* (in particular, a "class-as-many" in the Russellian sense) and not as a sortal or concretum. I will offer arguments for why an organismal environment is not an entity of the second kind by reviewing criteria of identification, identity, persistence, and countability. In particular, the indexicality of an organismal environment, whose composition is assumed to be sensitive to ontogenetic contexts and the activities an organism performs during its life cycle, complicates the picture. In contrast, I argue that my proposal to understand an organismal environment as a class-as-many—which considers at once all of its member factors causally acting on an organism and yet is not itself an entity whose extension is distinct from the individual, structured sum of these—is able to recover the desiderata laid down in section 3.1 and to accommodate the indexicality problem.

Finally, and as an important corollary that follows from the metaphysical appraisal, this chapter will end with an epistemological exploration of organismal environments that underscores a key feature involved in how scientists usually conceptualize them: the role of *abstraction* (section 3.3). I am interested in highlighting the noetic-epistemic processes that allow scientists to reach, in their theorizations and practices, the very idea of an "environment." In its mobilization in biological models, practices, and explanations, an organismal environment functions as an *abstractum*, a higher-order composite entity that is formed through a process of aggregation of diverse factors. I maintain that this view of the environment, of a plurality masked in a singular term, is indeed epistemically useful and heuristically fruitful (e.g., in terms of methodological affordability and explanatory convenience). In that sense, I will argue that the postulation of an organismal environment is, first and foremost, a *device of understanding*, a conceptual tool that scientists employ as knowledge makers in furtherance of grasping the complex settings of the world in which organismal ontogenies are situated and embedded. However, as we will see, problems ensue whenever scientists congeal their abstractions and treat them as ontic givens (sections 3.3.1 and 3.3.2).

3.1 The Environment of an Organism: What's in a Name?

To begin addressing the complicated roster of questions related to organismal environments I set to answer, allow me to start with a toy example. Think of what biologists assume to be components of the environments of two concrete specimens of two phylogenetically distant animals—say, a Pacific sea nettle (*Chrysaora fuscescens*) and a common leopard gecko (*Eublepharis macularius*). A gross description of abiotic factors within those environments would probably include things like light availability, temperature, pH and ion concentrations (for the cnidarian thriving in the ocean), oxygen levels, chemical compounds (e.g., endocrine disruptors), surface topography (affording locomotion to the squamate reptile), wind or water currents, and many more. Complementing this listing (see figure 3.1), a biologist might add an enumeration of biotic components, for example, appealing to the presence of, and the ecological interactions with, other organisms in their surroundings (e.g., heterospecific competitors, conspecifics, prey, predators, symbionts, commensals, parasites). This, of course, represents a nonexhaustive summary of the ontologically very heterogeneous components that (purportedly) partake in the environments of these organisms.

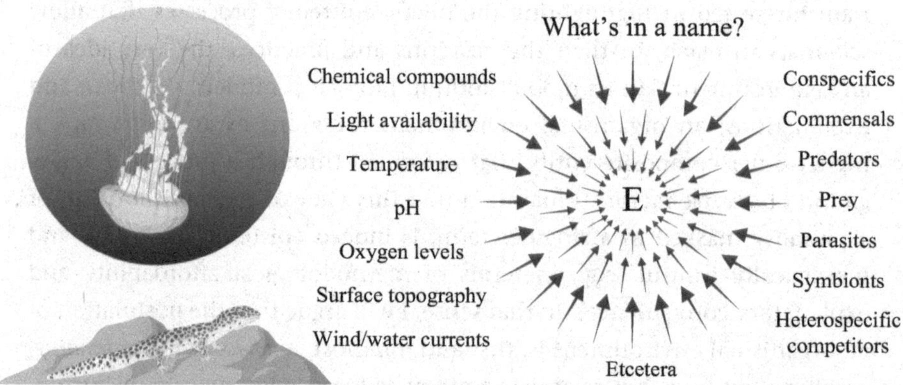

Figure 3.1
Abridged portrayal of environmental factors. A subset of the ontologically heterogeneous components comprising the environments of particular specimens of *Chrysaora fuscescens* (top) and *Eublepharis macularius* (bottom), which they could potentially interact with or be affected by at a time *t*. Source: Illustration by Fátima Sofía Ávila-Cascajares, created with BioRender.com (https://BioRender.com/e19l055).

Customarily, the definitions advanced to capture the environment of an organism are very broad and general, and many scholars resort to adding qualifiers with respect to the kind or type of environment they are referring to. In ecology, to pick a familiar example, one such distinction is between the "environment" sensu lato and the "effective environment" of an organism: "The *environment* of any organism consists, in final analysis, of everything in the universe external to that particular organism. Those parts of the total environment that are evidently of direct importance to the organism are regarded as constituting the *effective environment*" (Allee et al. 1949, 1; emphases in original; see also Allee 1941, 474–475).[2]

If the limits of the environment are left to encompass "everything in the universe external to that particular organism," it becomes unnecessarily wide-ranging and synonymous with the philosophical notion of "external world." When using the designatum of "environment," however, most ecologists—and most biologists from other disciplines that deal with token organisms, such as developmental biologists and physiologists—are (implicitly or explicitly) relying on a notion more akin to the effective environment (see, e.g., Caskey 1930, 52; Platt et al. 1964, 25; Niven and Liddle 1994, 317; Lewontin and Levins 1997, 98). In their conceptualizations, not everything from the outside world matters or affects a particular critter, and certainly not everything in the universe is factored in as a component of a singular environment.[3] Nevertheless, it should be made explicit that the biological concept of effective environment is not automatically synonymous with "immediate environment" (i.e., those heterogeneous factors that are present in close spatial vicinity to an organism in a particular time point), for, as American plant physiologist George J. Peirce fairly pointed out while embracing the overcomprehensive view of the environment as the external world,

> The environment of an organism is all, everything, that constitutes the world and the universe outside of itself. One may say that this definition is too comprehensive, that only *immediate* environment is meant when the word environment is generally used. Who can say that organisms . . . are affected only by their immediate environment? In fact we know that the contrary is true in certain particulars. We know, for instance, that we are daily affected by the sun—a remote body quite as much as by any part of our immediate environment. (Peirce 1904, 286; emphasis in original)

My quibble with Peirce is that, even if he was willing to grant that the sun is a component of an organismal environment, I think we ought to be more precise in characterizations of this sort. It is the sunlight, in the form of photon streams reaching circumscribed confines of our planet, entering into contact with the body of an organism (e.g., the aerial organs of any land plant or the skin of a member of *Homo sapiens* getting a tan on a crowded beach), and not the sun per se, that which is an abiotic component of a given effective environment. We do not need to subscribe the onerous vantage point of environment as *everything* that surrounds an organism across all possible spatiotemporal scales in order to make the following claim: A process or factor assumed to be part of an effective environment can still be causally relevant to an organism regardless of the actual physical distance separating them or their causal sources.

What is crucial in biological construals of the environments of organisms, then, is that their purported components could *potentially* or *actually* impinge upon said living entities—for instance, affecting (positively or negatively) their developmental trajectories or the functioning of their parts. In an early attempt by American botanist George Elwood Nichols to grapple with the environment concept, this facet came clearly to the surface: "The environment of any organism may be described as the sum total, or, perhaps better, the resultant of all the external conditions which act upon it. Representing as it does a complex of conditions, it is only by analyzing the environment into its various components that we can hope to understand its nature and to interpret the nature of its influence" (Nichols 1924, 1).

From this preparatory discussion, besides causal effectiveness per se, we can infer that there are two salient aspects of the organismal environment concept: (1) It is always *relational*; in other words, it must be juxtaposed to a counter relatum (i.e., that which is environed);[4] and (2) it performs a particular kind of metaphysical work partitioning the multifarious causal processes and components of the world in which an organism is embedded and that (could or do) causally affect it. Point 2 is actually a Spencerian legacy, as we saw in chapter 2 (sections 2.2.2 and 2.3.1). Let's now break apart these two theses in what follows (sections 3.1.1 and 3.1.2), and then I will introduce the third desideratum that biologists highlight for organismal environments—that is, their dynamicity.

3.1.1 One Environment, One Organism: The Relationality Desideratum

The restriction of the elements and processes that might act on an organism (i.e., the effective environment) is closely tied to the relational aspect highlighted in point 1 above. In fact, we can appreciate how some biologists frame this feature by their recourse to another qualifier of an environment—namely the "operational environment" (see also Brandon 1992, 81):[5]

> It must be continually emphasized that the operational environment is always the environment of a particular individual organism. There is no aspect of environment, in the sense of actively operating phenomena, that is not related to individual organisms. Even in cases where one organism is an environmental phenomenon of another organism, the viewpoint is always toward one of the organisms at a time. (Mason and Langenheim 1957, 331)

For many scientific thinkers, the relationality thesis (point 1) has been understood as a bijective function between an organism and its environment. To put it another way, although distinct organisms (e.g., conspecifics or heterospecific interactants) might be part of the environment of a particular token organism, every one of these has its own environment.[6] In the same physical medium (e.g., sixty cubic meters of ocean), different environments may take root. The German theoretical biologist Adolf Meyer-Abich, of whom we will learn more in chapter 4 (section 4.41), expressed that thought in the following way:

> Biotic environments are much more than pure physical surroundings. Many different organisms . . . are existing within the same physical surroundings but they live within their own biotic environments, which are biologically and qualitatively different from each other. A spider and her prey-animals exist in the same physical surroundings, but biologically they live in very different environments. For instance the prey-animals of the spider can not [*sic*] see and distinguish the gossamer of the cobweb; they are too fine for the eyes of the prey-animals. Of course these biologically different environments depend on the also different internal organisations of the organisms. (Meyer-Abich 1955, 66)

Beyond the possibility of having different environments in the same spatial, physicochemical settings, the relationality thesis has, as an important corollary that follows from it, the idea that *there are no*—there simply cannot be—*environments without organisms*.[7] Of course, there might be physical media in the world that are not inhabited by living entities, but a physical medium is not an environment on its own. The presence of and interaction with an organism—that is, that which would be environed—is a sine qua non criterion for a physical medium to count as an environment. Herbert Mason and Jean Langenheim push on this point:

> Water is only water until we introduce an organism. At that moment it becomes the environment of the introduced organisms. . . . If we now remove the organism, this water is no longer a part of the environment of this organism. If we now place several organisms in the body of the water and remove them one at a time, the water ceases to be the environment of each organism in turn as it is removed. . . . A similar demonstration can be set up to confirm the meaning of similar statements as they may pertain to any and every environmental phenomenon. (Mason and Langenheim 1957, 331)

This idea of the interdefinition of these relata has been framed as a "principle of mutuality" by some authors: "*An organism, by definition, requires an environment surrounding it and, conversely, an environment, by definition, requires organism(s) to surround*" (Reed and Jones 1977, 154; emphasis in original). A much more famous passage advancing the same proposition, but with a much stronger ontological bite, can be distilled from the works of evolutionary biologists Richard Lewontin and Richard Levins:

> *There is no organism without an environment, but there is no environment without an organism.* There is a physical world outside of organisms and that world undergoes certain transformations that are autonomous. Volcanoes erupt, the earth precesses on its axis of rotation. But *the physical world is not an environment, only the circumstances from which environments can be made.* . . . There is a non-countable infinity of ways in which the bits and pieces of the world might conceivably be put together to make environments, but only a small number of those actually have existed, one for each organism. (Lewontin and Levins 1997, 96; emphases added)

Lewontin and Levins's thoughts on what they christened "reciprocal codetermination"—or, in their words, "the role of the organism in the production of the environment" (Lewontin and Levins 1997, 96)—allow us to underscore an important complementary remark.[8] Organisms, by the very possibility of them being alive and maintaining themselves as such, cannot exist without being constitutively embedded in their surroundings. In this vein, we can emphasize that organisms unfold within, through, and due to environments, never in vacuo. The relation they hold with their environments is, first and foremost, an existential one (e.g., organisms need to harvest matter and energy from their surroundings in order to ensure their self-maintenance and persistence).[9] Although it might ring trivial to some ears, organisms *matter of factly* could not develop without their environing worlds. By the same token, environments come into being concomitantly as organisms do: The presence of developing systems is a *structuring precondition* of manifold circumstances that then can be said to constitute an environment. This should not be understood as organisms directly "causing" or "creating" fully formed environments. Sure enough, organisms bring about important causal consequences to the make-up of their surrounding circumstances (e.g., through their niche construction activities; see chapters 5 and 6), but this not to say that organisms are responsible for all the elements of the world that can affect them.

Environments, even though of utmost importance for organismal development, are considerably different from organisms, as we will see below (section 3.2). But we must first discuss the second thesis (point 2 above) that we reached through our preliminary analysis and then introduce the final desideratum that scientists champion when conceptualizing environments: their dynamicity.

3.1.2 Causal Partitioning and Dynamicity

As Balgooyen (1973) has suggested, it is perhaps wrong to ask what constitutes *the* environment, but rather, we should ask, "What is *an* environment?" (1199). Thus, with regards to the thesis (point 2) that I outlined above, environment as a single relatum performs unificatory metaphysical work by condensing disparate external conditions affecting the features and functioning of any given organism. This is the conceptual shift from the plural view of "circumstances" or "conditions of existence" of an organism to the singular term "environment" that I covered in chapter 2, which, as Pearce has convincingly argued, had metaphysical and methodological implications for the science of biology from the late nineteenth century until today:

> In terms of metaphysics, the successive transitions from individuated particular factors (e.g., climate), to a general plural term (e.g., "circumstances"), to a general singular term (e.g., "environment"), correspond to a progressive concealment of the different elements that make up the world outside the organism and the relations between these elements. This concealment, perhaps misleadingly, implies that the environment can be taken to be a single, unified cause. . . . However, the singular term "environment", like "organism", is an important heuristic for biologists, insofar as it gives them a way to talk about general causes without exploring the details of micro-level complexity. . . . Hence, the word "environment" does metaphysical work. (Pearce 2010, 249)

In terms of metaphysical partitioning, Brandon (1992) has claimed that the environment concept "serves to demarcate the external from the internal" (81). This is a common standpoint among scholars, but it is not the kind of partitioning I think an environment actually accomplishes when is considered the relatum of an organism. The problem of the separation between internal-external, or put better, between organism and environment, requires a different framing. This will be the subject of the next chapter. What I can say now is that the idea according to which the environment concept eo ipso separates the internal from the external begs the question of what's internal and what's external and prevents us from trying to understand what (if anything) enacts that distinction in the ontogenies of organisms.

Beyond the partitioning role hinted at by Pearce, biologists have long recognized that organismal environments are variable both synchronically (e.g., when comparing two conspecifics at a particular time point) and diachronically (i.e., throughout the life cycle of an individual organism). The dynamicity of environments is something important that they take into account. For example, for plant ecologists, it is commonplace to recognize that

> plants growing side by side . . . are not subject to exactly the same conditions. The immediate environment of the two may differ more or less as to water content, nutrients, light, temperature, humidity, wind, and other factors. Furthermore, if the environment of a single species is critically examined, it will usually be found to be far from uniform. During the early stages of development, seedlings, especially of trees and other tall-growing plants, and rosette forms, as those of evening primrose, occupy a very different habitat from that of the mature plants of these species. (Weaver and Clements 1929, 163)

Some authors have used the term "adaptivity" for the capacity that organisms have to remain viable in their environments while coping with changing conditions by regulating themselves (for discussion, see Di Paolo 2005; Bich et al. 2016; Menatti et al. 2022).[10] In ideas such as adaptivity, or the general subtending framework of phenotypic plasticity research (see section 1.1.1), the *dynamicity* of organismal environments, which do not stay fixed or unchangeable, is underscored. This idea is also present in many traditional sciences that study token organisms, such as certain subbranches of ecology or developmental biology. For example, despite the widespread assumption that marine ecosystems are less environmentally variable than their terrestrial counterparts, recent advancements in sensing technology have started to reveal ecologically significant variations in abiotic conditions across various temporal and spatial scales that affect the development of marine organisms. Contrast, for example, what is discussed in Steele et al. (2019) with Kroeker et al. (2020): The general dynamicity of environments does not rule out that some environmental conditions might be much more stable than others, especially with respect to the relevant timescale that is deemed relevant for a particular scientific study.

In this regard, the dynamicity of environments is certainly a desideratum that is highlighted much more often by scholars that work with organisms in "the field" (for discussion, see Grodwohl et al. 2018). In contrast, although laboratory settings enable researchers to control some environmental factors, they tend to isolate the individuals being studied and sometimes rely on unusual rearing conditions. Moreover, laboratory studies only capture a limited portion of the complex and dynamic environmental conditions that can causally affect organisms, which makes it challenging to determine whether the findings obtained there (e.g., in model organism-based developmental biology) are applicable to naturally occurring conditions (see, e.g., Poorter et al. 2016; Rivera-Yoshida et al. 2020). In sum, the elements and processes of the world that are set apart under the umbrella of an "organismal environment," are considered by scientists to be dynamic and nonstatic throughout ontogenetic time.[11]

As we saw in this section, organismal environments are not synonymous with the notions of "external world," "immediate surroundings," or "physical media" and require creatures to be environed in the first place to count as such. I maintained that scientists have advocated for three important desiderata that should be included in the characterizations of organismal environments: (1) causal effectiveness, which refers to the segregation or differentiation of the components and processes in the world that influence the development of a specific organism; (2) relationality, which refers to the interdependence of both the organism and its environment; and (3) dynamicity, which alludes to the fluctuating composition of various biotic and abiotic factors in a given environment over space and time.

There are still important pieces we must add to our thoroughgoing discussion if we want to understand what kind of relatum the environment of an organism is, especially in epistemic (and not only in ontological) terms. In section 3.3., I will argue that if we grasp the multiplicity of ontologically very heterogeneous factors that are included as components of a single entity, viz. the environment of an organism, we will face the realization that we are dealing with an *abstraction*, a higher-order composite entity formed through an epistemic process of aggregation. I will delve into the advantages that abstracting from the plurality of assorted conditions and processes that affect a developing organism yields to practicing biologists. Nevertheless, before dealing with the epistemology of scientific work, we need to take a small excursion—a jaunt for some, an exhausting trek for others, I reckon—into the metaphysics of environments. We have seen that organismal environments are dynamic and inherently relational and can be said to perform some sort of partitioning work of the causally relevant elements or processes of the world in which a developmental system is embedded. However, we have not answered the questions that opened this chapter: What is an "environment" for a particular organism? What kind of entity is an organismal environment? Through which kind of philosophical framework can we grasp the environments of token organisms? Those questions can be broached with the tools that metaphysics offers us.

In this chapter, I have been following very closely what scientists themselves have taken the environment to be, but staying at this level of discourse won't take us farther than what I already summarized in previous pages: Whatever an environment *is* still seems truly opaque, but we can extract that, at least, it is built from a relational basis and allows us to partition those dynamic components or processes of the complex world out there that have a causal toll on a developing organism. So far, so good—but we still need sharper philosophical tools to take a step ahead.[12]

3.2 A Metaphysical Primer of Organismal Environments

> Metaphysically, environments are much more problematic than organisms. . . . The word "environment" seems to refer to a mishmash of unrelated entities: sunlight, soil, climate, air, organisms, and so on.
>
> —American philosopher and historian Trevor Pearce (2010, 241).

3.2.1 Why a Metaphysical Analysis of Environments?

The metaphysics of science refers to the systematic study of the fundamental assumptions and concepts that underlie and inform the practices of different sciences. This includes questions about what is taken to exist according to scientific theories (and why such ontological posits are said to be warranted); the nature of reality; the structure, dynamics, and interactions of entities (e.g., through concepts such as law, causation, individuality, powers and dispositions, different instantiations of modality like necessity and possibility, reduction, and emergence); universal and existential generalizations in scientific discourse; the nature of particulars and properties; and the relationship between scientific theories and the world they seek to represent and explain. Within this purview, it is both possible to pursue metaphysical questions in the light of what the sciences beget and delineate scientific problems in metaphysical terms, all while the scope, aims and methods of the discipline remain open-ended and thus far under negotiation.

It must be said that, for most of the twentieth century, metaphysical analyses faced long-lasting disrepute and overt skepticism in the philosophy of science. However, in recent years, scholars have started to reverse that trend and reconsider their legitimacy and pertinence for understanding the activities and outputs of science (e.g., Mumford and Tugby 2013; Guay and Pradeu 2020 and references therein).[13] For philosophers of biology, the rising tide of metaphysical assessments of diverse problems and subjects, both traditional and novel, has been coming in slowly but steadily in the last fifteen years (e.g., Dupré 2021; Nicholson and Dupré 2018; Triviño and Cerezo 2015; Suárez and Triviño 2019; Alassia 2022). A significant number of philosophers have reached the realization that the sciences cannot be extricated from their metaphysical underpinnings. As Michael Ghiselin stated, "There is no way in which one can divorce science from metaphysics altogether: to deny metaphysics is itself a metaphysical thesis. One can no more have science without metaphysics than a drink without a beverage: the only choice is that between good metaphysics and bad metaphysics, good science and bad science" (Ghiselin 1997, 19). The British organicist Joseph Henry Woodger pointed out that those who believe themselves to be above metaphysics or consider its pursuit an utter waste of time are actually deeply immersed in it, stating that they are in fact "only a very little above it—being up to the neck in it" (Woodger 1929, 246). Indeed, scholars that refuse to interrogate or enquire about metaphysical presuppositions or address metaphysical problems have only inherited their metaphysical frameworks—perhaps tacitly and unbeknownst to them—from others. In scientific discourse, metaphysical statements become problematic, as Woodger contended, when they operate unnoticed because "they are entertained unconsciously, or their metaphysical character is not understood" (Woodger 1929, 27).

For the problem under examination, in the previous section I reached the conclusion that, according to scientists, we need to underscore the relationality aspect rooted in organismal environments, their dynamicity, and the partitioning of the (causally effective) surrounding conditions they perform. However, we usually speak of the "environments" of organisms as if these were completely transparent and unproblematic terms, but the idea of an environment, by itself, comes with hefty metaphysical baggage that we ought to unpack. My purpose here is not to disparage the metaphysical assumptions of philosophers of biology or scientists that employ the environment concept in their theoretical or experimental work but, rather, to try to gain a better understanding of what kind of relatum an organismal environment is by scrutinizing it from a metaphysical perspective. With this in mind, my investigation is not aimed at promoting a "revisionary metaphysics" of environments in the famous sense introduced by Strawson: "Descriptive metaphysics is content to describe the actual content of our thought about the world, revisionary metaphysics is concerned to produce a better structure" (Strawson 1959, 10). I take my approach to be more a "descriptive metaphysics" of the (organismal) environment concept as it is deployed, in the vast majority of times completely unspecified, in biological practice at large.

If we follow Vanessa Triviño's helpful classification of two complementary approaches within the "metaphysics of biology"—namely, "metaphysics *for* biology" and "metaphysics *in* biology"—the contributions of this chapter tilt more to the side of the former, although they try to grapple with both poles at the same time. Triviño explains this division of labor that by no means equals separation or precludes cross-fertilization:

> In Metaphysics *for* Biology, philosophers appeal to metaphysics in order to clarify a biological concept. In particular, they recur to the different metaphysical theories and concepts they might consider more accurate to explore and determine the ontological status of the entity to which the biological concept refers. . . .

> In Metaphysics *in* Biology, . . . philosophers explore and make precise the metaphysical commitments and implications that can be found in the different biological disciplines. In this case, therefore, philosophers of biology pay attention to biological theories, phenomena, and practices in order to disentangle the metaphysical commitments and assumptions that are given in them. (Triviño 2022, 4)[14]

In what follows, I will bring in some furnishings from analytic metaphysics to try to unravel what the environment concept (qua relatum of a token organism) entails and shed some light on its ontological status and the metaphysical assumptions that are baked in in its construal.[15] The battery of metaphysical tools I will employ range from the intrinsic-extrinsic properties distinction (section 3.2.2), the idea of "sortals" (section 3.2.3), and the notion of indexical classes (section 3.2.4).

3.2.2 Intrinsic and Extrinsic Properties of Environmental Components

> In order to understand the meaning of the term environment, it is first necessary to know just what comprises the complex series of conditions which surround both plants and animals.
>
> —American paleontologist Malcom Rutherford Thorpe (1924, xiii).

Throughout the recent history of biology, scientists have attempted to offer comprehensive or operational classifications of the components of organismal environments. For example, in the influential 1954 book *The Distribution and Abundance of Animals*, Andrewartha and Birch argued that, when the environment of an animal is restricted to whatever influences an individual's chance to survive and multiply, it can be analyzed in four major components: food, other animals (including predators and organisms causing diseases), weather, and the place in which the animal lives (Andrewartha and Birch 1954, 26–28; see also Andrewartha 1957). In direct response, Maelzer (1965a) suggested an alternative classification of what makes up an environment: (1) resources, (2) mechanical stressors, (3) weather, (4) members of the same species, and (5) predators, parasites, and pathogens. Later on, Andrewartha (1970) updated his taxonomy of environmental components and claimed that animal environments are composed of (1) resources; (2) mates; (3) predators, pathogens, and aggressors; (d) weather; and (e) "malentities."[16]

Regardless of the taxonomic efforts and frameworks of environmental elements one could erect and adopt, allow me to reflect on a related point that I also swiftly made when introducing the toy example of the abridged inventories of factors within the environments of the Pacific sea nettle and the leopard gecko (figure 3.1). When jointly peering into the character of these factors that are said to be components of an organismal environment, we can recognize, first, the subtending *ontological heterogeneity* of all of them. We could say virtually the same thing for other kinds of organisms—just look at the classificatory schemes above. For instance, biologists habitually discuss what is comprised in the environment of plants (e.g., moisture, temperature, light, chemical conditions, climatic conditions; see Nichols 1924, 4–31; Billings 1952, 253–255), but what is the common ontological measure that unites them?

When thinking of a token plant—say, a legume growing in the wild—what are the ontological criteria to lump together the sun rays hitting its leaves as they infiltrate through the canopy, the fleeting presence of bees visiting some of its flowers, gas concentrations in the circumambient air, moisture, acidity, water availability and the patchy nutrient distribution in the soil, the diazotrophic bacteria fixing nitrogen inside its root nodules, the manifold interactions it sustains with arbuscular mycorrhizal fungi underneath the silty loam?

There are no *intrinsic properties* these biotic and abiotic factors possess that could link them with one accord in a single entity: They are radically different in, among other things, material constitution, mereology, the spatiotemporal scales in which they unfold, and their sources and causation patterns. Hence, I contend that there is no intrinsic ontological measure that binds together the disparate elements that fall under the relatum "environment." This comes as no surprise if we take seriously the relational aspects of environments I hinted at in the previous section and the corollaries these bring. As a consequence of the relational nature of an environment, the elements that fall under its loose precincts share one *extrinsic* property: interacting, in one form of another, with an organism. When conceptualizing an environment, ontologically heterogeneous abiotic and biotic factors are extrinsically brought together by the relations they have (or could have) with a developmental system.

Bringing the intrinsic-extrinsic properties distinction (for an overview, see Hoffmann-Kolss 2010) to a discussion on environments might prima facie strike some philosophers as an odd tactic. After all, this distinction is not infrequently cast by some scholars by appealing to the environments of the entities under evaluation (e.g., Francescotti 2012, 91). Am I suggesting flipping the coin, and, if so, what would license this move? I should state that I am not claiming here that an *environment as a whole* has the extrinsic property of being related to an organism. Metaphysically speaking, the viewpoint from an environment-as-a-whole is a viewpoint from nowhere; it does not exist in the world, as environments are always framed and made intelligible through the lens of their counter relata—this is the relationality desideratum I discussed in section 3.1.2. My position with regard to the intrinsic-extrinsic distinction on this matter is that *all the myriad abiotic and biotic factors* that are considered elements of an environment *individually* share the extrinsic property of being related to a particular ontogenetic system. So, importantly, this extrinsic property should be predicated of the individual environmental factors themselves and not of the environment-as-a-whole. This overall modest claim, I think, adumbrates another important issue that will later allow us to move forward from the ontological to the epistemological terrain when dealing with the environment of organisms.

3.2.3 Environments Are Not Sortals

If there is no ontological common measure among its components (as these do share not intrinsic properties but an extrinsic, organism-directed one), can we metaphysically treat an environment as a single token entity? I think we can epistemically construe it as a *single abstract entity*, and there are good reasons for proceeding in that way (see section 3.3). This notwithstanding, I consider that, when taking metaphysics seriously, an organismal environment is better conceptualized as a class (in particular, a class-as-many; section 3.2.4) and not as a sortal or concretum; otherwise, we run the risk of pernicious reification (more on this last point in sections 3.3.1 and 3.3.2). Let me explain what all this means and why I bring this terminology to the fore.

Before submitting the idea of class-as-many, we first need to grapple with what the organismal environment is *not* in metaphysical terms. One possible avenue to capture the environment would be to frame it as a kind of "individual," as is customary in classic metaphysical parlance, or a sortal in particular. Drawing this contrast also serves as a fulcrum to reveal many important pieces we need to consider in a metaphysically informed characterization of effective environments: issues of identification, persistence, countability, and indexicality.

Within analytic metaphysics, a sortal term or predicate (sortal, for short) is usually one for which there is a counting criterion for tallying the entities to which its extension applies and a criterion for determining its diachronic persistence.[17] This crucially allows for the reidentification of the entities in a temporal series (for discussion, see Strawson 1959; Feldman 1973). In the room where you are probably reading this, you can identify some sortals: "book," "pen," "chair," "thrash can," "lamp," "laptop," or "backpack." We know familiar criteria for identifying and counting these objects, as well as for telling if these are the same objects we have been interacting with in past experiences. Sortals—namely, substantival general terms—differ from so-called adjectival terms, such as "blue," "hasty," "lame," and "unbending." Metaphysicians argue that adjectival terms only have application conditions, while sortals have these plus criteria of (persisting) identity (see, e.g., Lowe 1989). Importantly, then, to flesh out a sortal, one would need a criterion of identity (or an individuation principle) that should be able to discriminate whether a particular entity that corresponds to the sortal in question is numerically the same as another, either synchronically and/or diachronically. For instance, how many wine glasses do you have right now in your kitchen cabinets? Are these the same ones you used for your dinner parties last year?

How would this work for the environments of organisms? Let's discuss criteria of identity and issues of identification first and then switch to persistence criteria—which will take us to a problem that complicates easy attributions, namely, that of indexicality—and then cap off this subsection with countability conditions.

Identification and Identity Can we identify the environments of, say, two different organisms and find criteria of identity that would allow us to count and state, in a Moorean fashion, "Here is an environment, and here is another one"? Clearly, this framing is more than odd—if not plain mistaken. In this sense, the environment concept eludes an ostensive or deictic definition. We might ostensively define the term "green" if, while pointing toward a green object (e.g., a ripe Mexican *limón*), we say, "This is green."

Conversely, we cannot point with a suitable gesture at a single designatum of the term "environment" and make, at the same time, a statement of the type "This is an environment." If in a room full of people, you stood up and, while waving one of your hands, shouted, "Here is an environment!," we would intuitively understand what you mean, but, on second thought, the reference of *here*, in contrast to the color property of a *limón* that you can easily fit in your hand, is really hard to pin down. Does the *here* extend to the film of air surrounding your body, a portion of the room, the whole room, the building, the street, the city, farther? The biological equivalent of this would be to go out in a scientific field excursion and, in a particular place, assert, "Here is an environment." With this assertion, you would be pointing for sure to some environmental referents that simultaneously affect multispecies assemblages of the ecological communities that take root there, but no single organismal environment would be easily spotted.[18]

This is so because, at the outset, identifying environments is not a top-down endeavor for biologists. In a piecemeal fashion or through comparative reasoning with what is known for other phylogenetically closely related species, biologists locate certain biotic or abiotic factors that might affect token organisms (e.g., through field observations, measurements, interventions, and controlled laboratory settings) and then, by means of an extrinsic connection, postulate that these are environmental components alongside others previously identified.

For example, marine scientists characterize aquatic environments, first, by measuring specific properties such as water temperature, speed and direction of currents, precipitation, salinity, solar radiation, underwater sound, visibility, water-column height, water level and water quality, wave energy spectra, and wind direction, speed, and gust (see Nichols and Raghukumar 2022, 32). They garner their data, and through the inclusion of biotic factors as well, they stich up a partial set of components of a particular organismal environment for their research purposes (see, e.g., Goff et al. 2020).[19]

When characterizing environments, biologists employ a variable-driven strategy that highlights those environmental factors (more often than not in isolation) that are salient and relevant for the development of particular organismal phenotypes. This has evolved to become a methodological desideratum for biologists that study organismal environments in multiple fields of the life sciences, from developmental biology to ecology. In the words of embryologist Bernhard Dürken, "The environment of . . . the organism in general is, of course, very complex; it consists of many separate factors, each of which may be of importance in development. *Exact investigation of its action can only be carried out by testing the factors separately*; but that the environment is of importance in the growth of the organism is very clearly seen from its general effect" (Dürken [1932] 2016, chapter VI, paragraph 2).

In any case, the extension of the (effective) environment as relatum always consists of the sum of individual biotic and abiotic components that stand in a causal relation to an organism, not in an ontologically separate entity that exists on its own as a new whole and can begin to be identified at that level. Dissimilar individual factors can be extrinsically linked to an organism by the interactions and relations they hold with it (and thus be claimed to belong to the same environment); nonetheless, there is no criterion of identity or individuation principle that would allow one to carve out an environment-as-a-whole that does not resort first to this fragmentation and piecewise reassembly of components. That being said, I must stress that even figuring out how many and how dispersed these biotic and abiotic factors are in an environment is a very hard endeavor for scientists aiming for completeness. What are the limits or boundaries of an environment, if there are some to be drawn at all?

Persistence and Indexicality Even if, granting for the sake of the argument, we would be in the position to identify the complete list of environmental factors that affect (or could potentially causally impact) an organism at a time *t*, coming up with or pinpointing persistence criteria would still prove to be a thorny undertaking. Usually, criteria of persistence tell us when something continues to exist as the same entity and when it goes out of existence (e.g., turning into something else or perishing). In the philosophy of biology, it is at present popular to tackle problems of persistence from the angle of "genidentity." This refers to the identity through time of an entity that obtains given a well-identified series of continuous states of affairs (Pradeu 2018, 97; see also Lewin 1922). Scholars that embrace this view follow John Locke's idea that a "principle of individuation" is completely synonymous with "identity" and is to be found in an unbroken continuity of states (Locke [1694] 1975, section 6). Under these lights, the identity of a living being, for instance, is the continuity of one and the same "life" that, although it might undergo changes and alterations of different kinds, extends uninterruptedly over ontogeny. Applying this approach to organism environments would mean tracing the sequential continuity of environmental states, which would amount to capturing the identity of said environment.

Nonetheless, again, the relationality aspect of environments precludes a traditional treatment of diachronic persistence in terms of genidentity. Positively, there are certain environmental components that are semiautonomous given their particular causal sources, and it would be possible to trace an unbroken chain of continuity of their states in a time series (i.e., detecting factors affecting an organism at a time *t*, which causally change to a different configuration or persist with the same configuration at time t_{+1}, impacting the same token organism at that moment, and so on). I can mention some apparent examples, such as the wind or water currents that I brought up with respect to the Pacific sea nettle and the leopard gecko, or the gas concentrations in the surrounding air of a legume, or perhaps the temperature at a given hour of the day that accelerates or dampens the metabolism of another critter. But, besides this, the crucial point I want to underscore is that there is a strong organism-dependence on how the continuity of environmental states pans out. As I will argue extensively in the next chapter, what constitutes the boundaries of (developing multicellular) organisms and their environments changes throughout ontogenetic time, and what counts as environments (or, more precisely, the set of abiotic and biotic components that comprise an environment) differs as well during the development of organisms.

There are, of course, degrees to which the relativity of environments and their shifting reference applies to different organisms as members of particular lineages. Allow me to illustrate this with some examples. If we exclude its period of dispersal while it was still a seed (e.g., through anemophily, hydrophily, or zoophily strategies), the (abiotic) environment of, *mutatis mutandis*, a "sessile" plant might be broken down and analyzed as unbroken chains of states in a time series.[20] However, in sharper contrast, bring to your mind the image of a migratory bird: As it crosses the skies, what is its environment? If the bird traverses many kilometers through flight, experiencing drastically different surrounding conditions while airborne or perching, it makes little sense to claim that it has faced one and the same environment in time *t*, in t_{+1}, in t_{+2}, and so on and so forth (see, e.g.,

Kunz et al. 2008; Bauer et al. 2019, 866). Here, we see a case of an organism encountering distinct environments, although transiently, that are not continuous in an ordered time series. For instance, the winter setting from which a bird escapes, with the multitude of biotic and abiotic factors it harbors, is historically and causally semiautonomous from the factors in the sunny paradise the organism reaches when the migratory voyage comes to an end for the season.

The ontogenetic-relativity of environments is especially striking in the case of organisms with complex life cycles. A good example of drastic environmental shifts is the case of a typical digenean trematode (phylum Platyhelminthes; for an overview of their life cycle, see Poulin and Cribb 2002). Eggs released from adult worms in their primary vertebrate host become embryonated in fresh water and hatch into miracidia, which must find a suitable mollusk as a first intermediate host. Here, the first environmental shift occurs: from the environment afforded by living in water—with everything involved in that particular setting—to the gastropod's body—a different set of conditions of existence for the flatworm. In this scenario, genidentity fails to grasp the persistence and diachronic identity of environments: We cannot frame the water of a lake and the interior of a mollusk as continuous environmental states in causal terms (i.e., the body of the metazoan, which has an ontogenetic history of its own, is not genidentical to the surrounding water, whose components and embedded processes also have diverse causal histories), and yet it is perfectly rational—and scientifically coherent—to state that the intestine or the circulatory system of a molluskan host—with all the biotic and abiotic factors therein—form the circumstances out of which the new environment of the trematode is made. Biologists can legitimately reason in this way because, as we discussed in the previous section, any environment, as a multiplicity of biotic and abiotic conditions, is *extrinsically defined* with respect to an organism. The fact that water is the environment of a token trematode at a concrete moment in its life cycle and then later bodily interiors are its new milieu is a good example of Lewontin and Levins's (1997) "reciprocal codetermination" principle in action—environments require environed organisms to count as such, and these, in turn, cannot develop without suitable surroundings. Inside the host tissue, organisms experience developmental transformations from sporocysts to radiae to cercariae. Afterward, the second environmental shift materializes for the next stage of the life cycle:

from the mollusk's interior back to water. Free-living, but short-lived, cercariae emerge from the first intermediate host and must locate a suitable second intermediate host—where another environmental shift ensues. For example, in the case of *Fasciola* spp., cercariae encyst (forming metacercariae) on aquatic vegetation. At that moment, metacercariae are ingested, along with the second intermediate host, by an appropriate definitive host. Immature flukes from *Fasciola* spp. excyst in the duodenum of ruminants, penetrate their intestinal wall, and migrate through liver parenchyma to biliary ducts. The adult flukes, in their new environment, then reproduce, passing the eggs along the feces, bringing a loop of the life cycle into completion. Again, an analysis in terms of unbroken chains of environmental genidentity would be problematic for all these stages in the life cycle of a digenean trematode (e.g., embryos, miracidia, sporocysts, radiae, cercariae, and metacercariae). Throughout the developmental histories of these organisms, contrasting environments—from bodies of water to the squashy interiors of vertebrates and invertebrates—whose biotic and abiotic components have semi-independent causal histories and trajectories, arise as they are populated by active developmental systems that move and change their constitution. So there is a strong dependence of what an organism does or what developmental changes it is undergoing—in addition to what phenotypes it has—and what sort of surroundings environ it and can causally affect it.[21]

Coming back to Balgooyen's idea (1973) that the question of what constitutes *the* environment is misguided and ill-defined and that, rather, we should ask, "What is *an* environment?," we are now in a position to add an important nuance. In order to better understand the environing worlds of organisms, we should ask, "What is *an* environment *at a particular stage or time slice* of its life cycle?" The reference of an organismal environment is inescapably context-dependent, a changing bundle of causal relevant factors without clear proximal limits. Another way to express this characterization is to claim that an environment exhibits *indexicality* (figure 3.2).

The notion of indexical, grosso modo, alludes to a term or a linguistic expression whose reference (or its extension) can shift from context to context.[22] Since organismal environments are indexicals, the claim "the environment of organism X is Y or is composed of Zs and Ts" is never absolute. To know what is being indicated or implied, we certainly need the context of an organism's ontogeny.

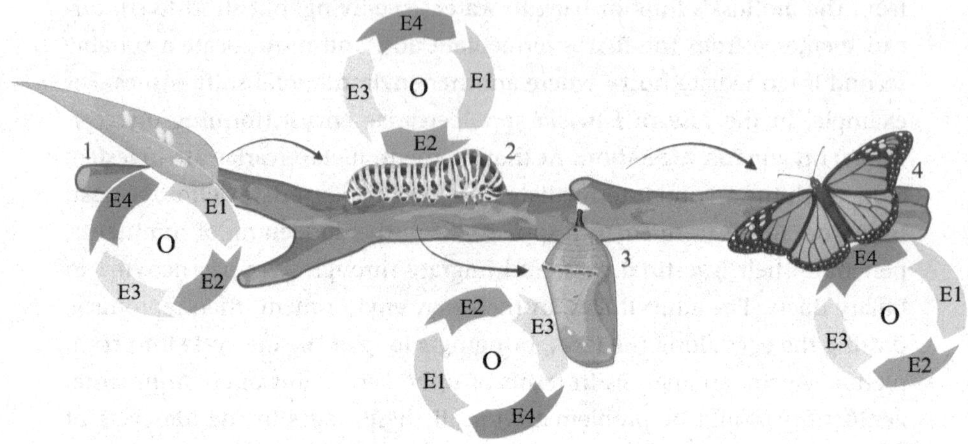

Figure 3.2
Illustration of the indexicality of organismal environments in a life cycle. A developing monarch butterfly (*Danaus plexippus*)—transversing from egg (1) to caterpillar (2) to chrysalis (3) to adult butterfly (4)—faces distinct environmental conditions. An answer to the question "What is the environment of a token butterfly?" depends on the ontogenetic context of the organism. Illustration by Fátima Sofía Ávila-Cascajares, created with BioRender.com (https://BioRender.com/i79r335).

In our trematode example, what amounts to the causally relevant environmental factors for an organism depends on what moment of its life cycle it finds itself in: different environmental compositions affect the same organism as it develops, all while it is implementing different ecological strategies and instantiating distinct morphological and physiological traits either in its diverse hosts or in freshwater.

In sum, tracking the diachronic persistence of environments is a tricky endeavor and precludes traditional reasoning precisely because of the problem of indexicality. Now it is time to go over the final criterion that upends sortals to keep showing that scientific construals of organismal environments dodge a metaphysical characterization of this kind.

Countability Even though, through the relationality desideratum, we assert that there is a corresponding environment for every organism, can we count environments by means of unequivocal criteria? We have granted that every organism is constitutively embedded in its surroundings, and some scientists interpret this claim as being synonymous with having an effective environment that subsumes multitudinous causal components. Accordingly, reckoning environments requires recognizing and counting how many organisms are developing and facing their conditions of existence in a particular setting. A number of environments could only be approximated indirectly by counting organisms—assuming, of course, that we have clear criteria to count the creatures in question.[23] So the counting criteria do not belong to environments per se but depend entirely on the fact that organisms, by themselves, exhibit the properties that sortal concepts demand (see Okasha forthcoming). However, as we just mentioned, given the indexicality of environmental components, one can also make the plausible claim that there might be *several environments* for a single token organism if our frame of reference is a life cycle. After all, not every ontogenetic stage faces the same abiotic and biotic components, and, as we saw, there are causal gaps between certain components if one takes continuity from one time point to the next. How many environments can we count for a migratory bird or a monarch butterfly moving across North America? How many for the trematode of our example? Counting environments, assuming someone would want to do it seriously and thinks this practice even makes sense, is also not an uncomplicated task.

In a nutshell, identification, counting, and persistence conditions are not fulfilled when considering environments on their own. To get broad, rough approximations of these facets, one would need to introduce elements that affect the counter relata—that is, the organisms being environed. Organisms act, then, as counting balances and as developmental sifters for the identification of effective sets of abiotic and biotic factors and are important for tracing which environmental components are semiautonomous in their causal trajectories and which depend entirely on organismal activities, whether behavioral, physiological, or morphogenetic, in particular moments of the unfolding life cycles. Moreover, the indexicality of environments seems to be an inescapable element that needs to be reckoned with. It is clear, then, that organismal environments are not aptly described by sortal concepts, and I hope to have marshalled sufficient evidence and reasons for showing that questions about identification, counting, and persistence cannot be given a conventional metaphysical treatment when dealing with the environing contexts of organisms. So, if environments do not fall under the category of sortals, what could organismal environments be, metaphysically speaking?

3.2.4 Organismal Environment as Indexical Class-as-Many

Philosophers have argued that, because sortals are countable common names, employing this framing reveals something about the very nature of their referents (i.e., its intension, or, in other words, what kind of entity they are or what necessary and sufficient properties they instantiate to qualify as such). Equally, I think that also not being able to make sense of something as complying under "sortal" is enlightening about its underlying ontology. After having discussed some options for how to metaphysically frame organismal environments and the difficulties they face, in what follows I advance a positive proposal for how to construe them that does justice to their context-dependency to particular stages of life cycles (i.e., indexicality) and fulfils the three desiderata laid down in section 3.1.

As we have seen, although phrased in the singular, the environment is, rather, a collective term. On this point, plant physiologist Peirce, whom we mentioned in section 3.1, averred,

> The word *environment* is used ignorantly by everyone, for *no one has ever succeeded in making a complete analysis of what is meant by this collective term*. Furthermore, although we speak of an organism as reacting to its environment, it does not react to its environment as a whole but to each of one of the separate influences which are the factors of its environment. It is, therefore, very important to know what these factors are, and what are their effects. (Peirce 1903, 285; first emphasis in original, second emphasis added)

In opposition to Peirce, I do not think the environment concept is "used ignorantly" by scientists and philosophers. The fault line, in my opinion, is that the history of how we came to incorporate that term as one of our chief thinking tools (covered in sections 2.2.2 and 2.3.1) has definitely slipped through the cracks of the life sciences, and in present-day scholarship, many people simply assume that we already have full clarity about what environments are or what they refer to. The environment concept is treated as if it were translucent. It is healthy for our knowledge-producing ventures to shake the hornet's nest of certainty with philosophical scrutiny once in a while, though. In particular, going behind appearances, what at first glance something *feels like* to us, is one of metaphysics's long-lasting legacies as a discipline that we ought not to disregard.

Multiple factors of our complex world—which are ontologically very heterogeneous, as I have stressed repeatedly throughout this chapter—indeed concurrently affect an organism at a particular point in time, and as we will see in chapters 5 and 6, organisms are also able to respond to the affordances of their surroundings and actively shape worldly components. Yet how can we make sense of them acting collectively? Scientists investigate individual environmental factors in their experiments and field studies as a methodological desideratum (see section 3.2.3 on identification and identity); however, when they employ the environment concept in their theories, models, and practices they want to postulate it as a useful designatum or entity that somehow includes all of these factors at once. For a biologist, an organismal environment hardly could be reduced or accounted for by one or few of its individual chunks and parcels. In other words, it is injudicious to declare that an environment can be reduced to discrete components, whether biotic or abiotic, or that one component takes ontological precedence over the rest. Ecologists, for instance, employ the notion of "holocoenotic environment" to capture this intuition and argue that environmental complexes should be treated as wholes (see Allee and Park 1944, 167; Cain 1944, chapter 1; see also Agarwal 2008, 24; Matthews 2014).

So how do we make sense of that additional theoretical desideratum, paired on top of the methodological one, in a way that is also metaphysically sound for biological science? Moreover, what metaphysical framework can attend to the indexicality of environments? I consider that one promising avenue is to be found by appealing to the idea of "class." In this section I do not wish to take a deep dive into the metaphysics of classes, but simply to propose a possible understanding of what an organismal environment is that could be debated and refined in future works. As I said, I think that, metaphysically speaking, an organismal environment can be fruitfully thought of as a class; in particular, using a notion pioneered by philosopher Bertrand Russell, it could be construed as a *class-as-many*. Here, I am following and expanding the insights that two philosophically savvy American plant ecologists, Herbert L. Mason and Jean H. Langenheim, had in the late 1950s and that unfortunately have gone almost completely unnoticed by subsequent scholarship.[24]

In order to get the gist of the distinction I introduced above, we need to take into account that, for Russell, a class is not a sort of iterative (mathematical) structure but, rather, refers to a conjunction of entities denoted by concepts (e.g., think of the class of humans denoted by the concept of "professional baker"). Dealing with classes thus requires paying attention to the extension of said concepts (e.g., the class of professional bakers would be the catalog of all the professional bakers that exist) while pairing them with some definition or rule (i.e., their intension) that tells us in a principled way whether an element belongs to the class or not (e.g., someone is a professional baker if and only if they consistently bake bread while being employed in an authorized bakery; see Russell 1903, chapter VI). According to Russell, it is possible to distinguish between classes-as-one and classes-as-many:

> A class . . . is distinct from the whole composed of its terms, for the latter is only and essentially one, while the former, where it has many terms, is . . . the very kind of object of which many is to be asserted. The distinction of a class as many from a class as a whole is often made by language: space and points, time and instants, the army and the soldiers, the navy and the sailors, the Cabinet and the Cabinet Ministers, all illustrate the distinction. (Russell 1903, 68)

Concerning the class-as-one, the class (as a whole) itself is assumed to be an entity, beyond its individual members. In contrast, a class-as-many considers all the separate members of a class at once and nonetheless is not itself an (actual, concrete) entity. The collection denotes a concept but not an entity whose extension is something different than the sum of its individual components. For example, in an organism's surroundings—say, of any plant of your liking living in the wild—there are sun rays, gases in different concentrations, close-by conspecifics, threatening herbivores, and so on, but they do not form a new supra-entity (i.e., an environment-as-a-whole) that transcends the individual existence of the individual component phenomena. Russell says,

> The notion of a whole, in the sense of a pure aggregate which is here relevant is . . . not always applicable where the notion of the class as many applies. . . . In a class as many, the component terms, though they have some kind of unity, have less than is required for a whole. They have, in fact, just so much unity as is required to make them many, and not enough to prevent them from remaining many. A further reason for distinguishing wholes from classes as many is that a class as one may be one of the terms of itself as many, as in "classes are one among

> classes" . . . , whereas a complex whole can never be one of its own constituents. (Russell 1903, 69)

Discussing Russell's distinction,[25] Kevin Klement provides a more palatable rephrasing:

> When the members of a class do form a complete whole, and can be treated as a single thing, the single thing involved is the class as one. The class as one is an individual, and, as the label implies, counts as one. A class as many, however, is not a thing at all, or at least not one thing; it is as many things as there are members of the corresponding class as one. The most direct way to speak of a class as many is extensionally, by directly naming each of the members, and adjoining the names with "and." (Klement 2014, 2)

Through the arguments of section 3.2.3, we have seen that an organismal environment does not count as an individual qua sortal. Likewise, it would be hard to spell it out as an individual whole by relying on other metaphysical frameworks because it lacks a common intrinsic ontological measure (section 3.2.2). An organismal environment is not integrated as a larger whole because it depends, for its emergence and relative stability, on the extrinsic connections that the individual components hold with a token organism, the counter relatum that makes it intelligible. Returning to Russell's characterization, the ontologically heterogeneous component phenomena making up an organismal environment have "just so much unity as is required to make them many, and not enough to prevent them from remaining many": There are indeed important reasons for treating them together, as they share the unity of action on an organism and are not completely isolated factors.

Of course, I do not intend to deny that, at the margin of what a particular token organism does or is subjected to (or what sort of developmental traits it exhibits), there are (somewhat or fairly autonomous) dynamic causal processes that tightly interlink certain suites of abiotic factors, as well as some biotic factors, that would be considered ingredients of a singular effective environment. Certainly, the complexity of the varied, interacting factors contributing to the effective environment of plants, animals, and other developing organisms have long been recognized by biologists, and it is one of the cornerstones of their thinking. For instance, early on, Clements (1928), one of the founding fathers of modern ecology, claimed that the factors that make up a habitat are "almost inextricably interwoven" (238), and biologists have been trained with that mindset. This is true for many abiotic factors that are causally dependent on each other, but it is not for every one of them and undoubtedly not for all biotic factors as well, which include other organisms that have their own individuality, are embedded in their own contexts, and might move flexibly across their surroundings.

I consider that there are sufficient reasons to contend that an organismal environment might be metaphysically viewed as a class-as-many that partitions the elements of the world in which an organism finds itself embedded and developing, with the criterion of membership for the changing set of biotic and abiotic environmental components (dynamicity) having an actual or potential causal relation (extrinsic connection, effectiveness) with the token organism (relationality) at a moment of its life cycle (indexicality). Mason and Langenheim (1957) argued something similar with respect to the effectiveness and the relationality theses (see figure 3.3), without, however, arriving at the conclusion of metaphysical partitioning and

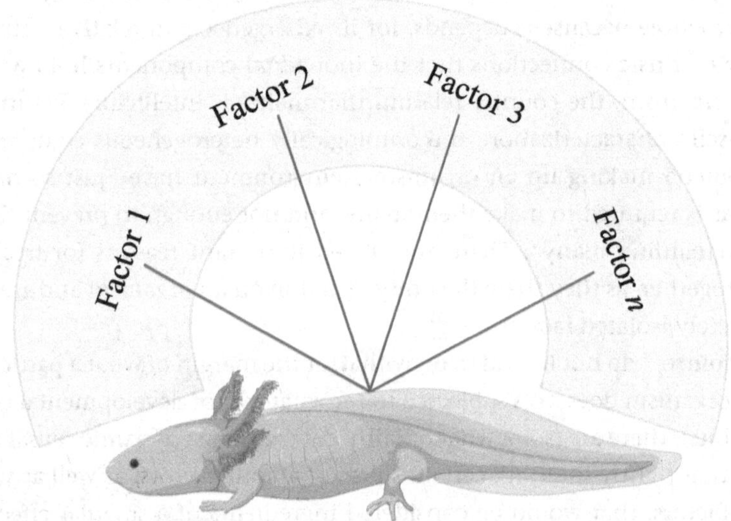

Figure 3.3
Diagram representing Mason and Langenheim's idea to conceptualize the (effective, operational) organismal environment as a class-as-many. The collection of environmental phenomena (1, 2, 3, etc.), each holding relations with a token organism (in this case, a specimen of *Ambystoma mexicanum*), represents the class "environment." Illustration by Fátima Sofía Ávila-Cascajares, created with BioRender.com (https://BioRender.com/l32f056).

indexicality that, I have claimed, also comes with the idea of an organismal environment:

> When we employ the word "environment" we often introduce confusion into our language because it refers both to the class as defined intensionally and to the class as defined extensionally. . . . About all that the environmental phenomena have in common is that they impinge in some significant way upon the organism. This, we pointed out, is the defining property of the class environment; it will be noted that it is based upon extrinsic relational properties of the constituents of the class. More significant to "environment" as we employ it, however, is the *way* in which each phenomenon impinges upon the organism, *and the fact that each phenomenon impinges differently. This is a signal that we should be viewing environment as a class-as-many.* (Mason and Langenheim 1957, 333; first emphasis in original, second added)

The last point mentioned by Mason and Langenheim is also noteworthy: The fact that every environmental component impinges differently on a token organism is further evidence that we might not be dealing with a class-as-one. In particular, how would we be able to sum up the effects of a purported environment-as-a-whole in way that does not ontologically collapse and flatten the individual effects of manifold component phenomena and processes? This is a conceptual tension that lies deep in biological thinking—discursively stressing the wholeness of the environment while recognizing its causal fragmentation: "To the working ecologist *the environment is holocoenotic*, that is, it is a unit composed of many parts, *as a rope is many of many strands*. Even though holocoenotic, the different parts may at times assume control as the concentration of one of them approaches the minimum or maximum which the animal can tolerate and hence acts as a limiting factor" (Allee and Park 1939, 167; emphasis added). If we abide by the metaphor suggested by Allee and Park, we can confidently say that the (abstract) holocoenotic environment-as-a-whole does not pull an organism in a single direction because there is no single rope; there are multiple entangled ropes pulling it from different directions at a particular moment of its ontogeny. (On how organisms actively pull the ropes back, see chapters 5 and 6.)

Besides the foregoing propositions, and apropos indexicality, I want to evince that it is still possible to keep the contextual association and the organism-directedness of environments by means of the notion of class-as-many. A way to spell this out is to claim that the elements that are included as members of the class-as-many shift throughout ontogenetic time as different causal relations and settings unfold—either kick-started by organismal actions or developmental transformations, through dynamic changes in the manifold processes of the world that we construe as "environmental" that bring about effects on a particular organism, or through a reciprocal interanimation of both. In that respect, as we have seen, what counts as environmental components depends on the context of concrete organismal ontogenies. Within this metaphysical vantage point, diachronic identity could also be thought about in terms of classes-as-many: a token organism could be said to be juxtaposed to an environing class-as-many, and the number and identity of its individual members (i.e., abiotic and biotic phenomena) might persist or change throughout time as development ensues. Along the same lines, environments in toto should not be treated as metaphysically separate entities whose extension is something different from the collective, structured aggregation of its members (i.e., all the ontologically heterogenous environmental factors that are impinging upon an organism in a concrete instant).

For all of these reasons, I maintain that the environment qua scientific relatum is a plural abstract entity—or, better put, a plurality of conditions that co-occur in an extrinsically generated, context-dependent class—linguistically masked in singular yet not conforming to this on ontic grounds. This metaphysical conclusion is equivalent to the historical thread that we uncovered in chapter 2 related to the shift that ensued in biological science since the second half of the nineteenth century, moving away from views of plurality of contexts (e.g., conditions of existence) and toward speaking about "environment" in singular.

Importantly, however, this should not be taken to mean that I want to outlaw the use of "environment" altogether because it is ontologically unbefitting—that would certainly be foolish, as it would take away a valuable conceptual remit—or that we ought only to speak of the world *in se* or how our best metaphysical assessments suggest it might be. This is because, especially in the natural sciences, it is epistemically very useful to reduce pluralities into singularities for the sake of methodological tractability and explanatory expediency.

In this chapter I have argued that a good inroad to understand how an organismal environment functions as *relatum* comes via an approach of metaphysics *for* biology, but for the remainder of the chapter I will move away from that viewing platform and venture into epistemic pastures. Please note that I have been constantly referring to the environment as an *abstract* entity when it functions as a relatum of an individual organism. Now it is time to put that qualification into the forefront. In what follows, to conclude this chapter, I will turn to a brief epistemological exploration of organismal environments that underscores a key feature involved in how scientists usually conceptualize them: *abstraction*. This can also be interpreted as an epistemic corollary of the metaphysical analysis of organismal environments that was advanced in these pages. The reason why scientists fruitfully talk about "environment" in singular is due to their common recourse to abstraction. I am thus interested in highlighting the noetic-epistemic processes that allow scholars to derive the very idea of an environment and posit it in their theories, models, and practices. I will argue that environments, as epistemic constructs that are conjectured as relata of token entities, are *devices of understanding*, conceptual tools to deal with the plural settings of the world, extending through multiple spatiotemporal scales, so vast and complex that they defy our expectations, in which organismal ontogenies are situated. The next section, then, will try to offer some pieces that I consider important to start developing a thorough epistemology of organismal environments.[26]

3.3 Toward an Epistemology of Organismal Environments: Environment as Abstraction, Environment as Device of Understanding

> The meaning of "environment" is so complex that it must be broken down into simpler components which can be considered separately.
> —Australian population ecologists Herbert George Andrewartha and Louis Charles Birch (1954, 26).

> A collection is [a] hypostatic abstraction, an *ens rationis*.
> —American philosopher Charles Sanders Peirce (1931, 5:334).

> The enormous success of . . . scientific abstractions . . . has foisted onto philosophy the task of accepting them as the most concrete rendering of fact.
> —British philosopher Alfred North Whitehead (1929, 70).

As I argued in the previous section, conceptualizing the environment as a single entity (e.g., as a sortal or class-as-one) is ontologically misleading but it comes with some epistemic purchase because biologists are able to grasp the plurality of surrounding conditions in which organismal ontogenies are situated and embedded and simply work with two singular terms in juxtaposition: organism *and* environment. As Trevor Pearce has claimed, from the second half of the nineteenth century onward, "the concept of organism–environment interaction . . . became popular because it was an *abstract, portable dichotomy* produced by the conjunction of two singular terms" (Pearce 2010, 242; emphasis added).

The idea of "environment" qua a singular relatum is tailored to meet the expectations of how empirical studies are usually conducted within the bounds of biological science. An organismal environment in an experimental setting, for instance, does not have to be decomposed in all of its variables at once; nonetheless, those elements or processes that are not intervened in or controlled for are assumed to be acting, in a regular way, on the organisms under study.[27] Likewise, when an environment features as an explanans or as part of the explanantia in an explanation for certain developmental or ecological phenomena, biologists can treat it as a surrogate for comprehensive and heterogeneous causal sources: they do not need to make explicit what exactly was altered in the heterogeneous processes of the world in order to make significant claims of the sort "environmental conditions affect the development of phenotype X" or "the environment is responsible for limiting the growth of organisms in a certain period of the year." Equally, the "environment" writ large suggests to biologists more focused components and factors they could study further to come up with more precise causal explanations or predictions; that is, its use in singular linguistic form is a heuristic guide for future research. Talking about the "environment" is also a useful shorthand for communication processes across different epistemic communities of scientists.

Accordingly, I want to suggest that the organismal environment concept is a very important epistemic tool to deal with the multifarious conditions in and through which development happens. A visual representation of the expansive complexity that is fruitfully subsumed under the idea of "environment" can be seen in figure 3.4. Instead of bringing in the plurality of processes and factors at once, there are important epistemic reasons why biologists condense them in the singular relatum "environment."

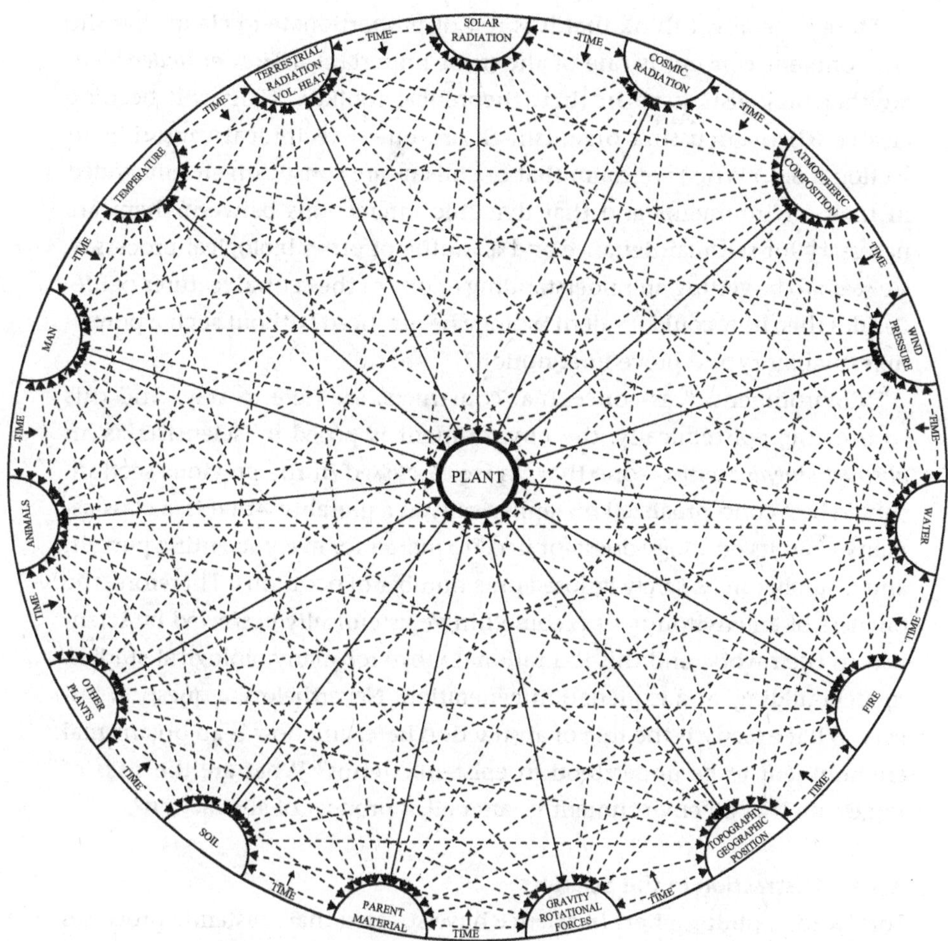

Figure 3.4
Visual representation of an interconnected "environmental complex." The figure only portrays a condensed catalog of (biotic and abiotic) environmental causal factors interacting with a token plant throughout developmental time. Source: Diagram reproduced with permission from Billings (1952, 256), *The Quarterly Review of Biology*. Copyright © University of Chicago Press—Journals.

More generally, I think that it is not disproportionate to claim that the environment concept is one of the most important *devices of understanding* that biologists have in their theoretical toolkits. (This will become clearer when discussing organism-environment reciprocal causation in section 5.5.2.) When we claim that organisms are constitutively embedded in their environments and that dynamic interactions between them are necessary for self-maintenance and a swathe of other biological processes, we seem to be getting an understanding of one of the central features of life. Besides this, how could modern biology be practiced without such a powerful explanatory and procedural notion?

"Environments," as conceptual constructs, are here to stay. And this so because, epistemically, the environment is posed by scientists as an *abstract singular entity*—one that, as we discussed in the previous section, should best conceptualized metaphysically as a plurality—that is, a class-as-many—but its ontology does not need to matter for every scientific pursuit, and it surely can be safely ignored on a number of occasions. Therefore, the oneness of an environment is something epistemically bestowed by scientists in their work, and this is a rational move for many biological studies, model building, and explanation generation. Nevertheless, a question lingers: What exactly is the role of abstraction here, and how is an organismal environment to be understood in epistemic terms? What are the risks of forgetting that an environment is, after all, a product of abstraction?

3.3.1 Abstraction in the Sciences

For decades, philosophers of science have stressed that epistemic processes and operations of abstraction are widespread in the sciences, especially in the construction of scientific theories and models (for overviews, see Jones 2005; Levy 2021; Portides 2021; on the cognitive aspects of abstraction processes, see Carrillo and Martínez 2023). Traditional philosophical accounts of abstraction emphasize the addition, subtraction, or omission of properties of the target system in the ensuing representations of it that are mobilized in scientific practices.[28] In abstractions, it is argued, subtractions or omissions are justified by virtue of excluding irrelevant particularities or idiosyncrasies of the object under study that might muddle the analysis or add unnecessary complications in the design of experiments or in the assembly of models, or simply because those features are not directly related (e.g., causally) to the explanandum at hand.

The multifaceted epistemic roles that abstraction plays in scientific endeavors have been a matter of debate among scholars.[29] Some of them highlight their positive aspects for science, and others concentrate on their potential shortcomings. In the first camp, for instance, through his notion of *hypostatic abstraction*—sometimes also referred to as subjectal abstraction—Charles Sanders Peirce argued that the main purpose of this epistemic process is to open up paths of inquiry by means of creating novel research objects or new subjects for logical analysis. According to Peirce, hypostatic abstraction is achieved when a quality or property shared by a collection of particulars is abstracted from that set and taken into account *in* and *for* itself as a property worth studying further. This construction, now at the foreground, henceforth becomes the subject of so-far-unexplored scientific work. On this, he claims that "the transitory elements of thought . . . are made substantive elements. . . . It thus becomes possible to study their relations and to apply to these relations discoveries already made respecting analogous relations. In this way, for example, operations become themselves the subjects of operation" (Peirce 1931, 3:64; for discussion, see Zeman 1982).

In contrast, not all deliberations surrounding abstraction underscore its generative epistemic outcomes in the sciences (e.g., how abstraction can launch or adumbrate difficult subjects as a consequence of giving them a manageable form for future analysis). A central problem with some abstractions, heuristically and explanatory powerful as these surely are for particular scientific fields, hinges on the danger of ontologically overinterpreting them and thus turning abstracta—otherwise useful in how scientists grasp, communicate, and generate knowledge—into concrete entities isolated from their contexts, faithfully representing reality as it is structured "out there." The *fallacy of misplaced concreteness*, as Whitehead memorably called it, happens when "we have mistaken our abstractions for concrete realities" (Whitehead 1929, 78). In the pragmatist tradition, this is sometimes called "pernicious reification."

For example, in evolutionary biology, a potential peril scientists face is one of pernicious reification of the explanatory and mathematical models used in population and quantitative genetics to convey evolutionary dynamics, which unavoidably derive from operations of abstraction to render the objects of study tractable: making use of an analogy, the maps employed by biologists—the simplified models tailored for particular epistemic aims—sometimes are taken to be the (evolutionary) world (Morrison 2002, 64; see also Andrews 2021). Biologists can mistake the abstract for the concrete, thus reifying it (see Winther 2020). For traditional evolutionary models, populations are cast as ensembles of abstract (gene) types, an epistemic move that definitely does not capture the ontic correlate of a population in the world in its entirety: that is, assemblages of material, developing conspecifics interacting with themselves, organisms from other species, and their complex and fluctuating surroundings. Construing populations as ensembles of abstract gene types for certain explanatory purposes (e.g., tracking allele frequency changes) is nice and fine as long as gene dynamics are not treated as ontological equivalents of *populations of living organisms*. What sounds prima facie obvious is not always acknowledged explicitly by biologists: Any population is more than its gene pool, and, as such, the population level is not exhausted by alleles and their evolutionary kinematics (see Walsh 2019; see also section 5.3.3).[30]

Now, how are discussions on abstraction, including its generative upshots and its latent perils, related to organismal environments?

3.3.2 Organismal Environments as Products of Aggregative Abstractions

Keeping in mind the cautionary metaphysical arguments of section 3.2. while being conscious of how biologists employ the concept of "environment" as an explanatory useful "whole" in their models, theories, and practices, we come to the realization that organismal environments are abstracta and not concreta. Multiple ontologically heterogeneous factors of the world are indeed affecting an organism at a particular point in time, but when scientists want to understand these interactions, they unavoidably resort to abstraction. This is an epistemic process that grants intelligibility to scientific research. Abstracting the plurality of surrounding conditions yields two singular, workable terms in juxtaposition: organism and environment as relata. Mason and Langenheim (1957) have something to say on this point: "Actually we employ the concept [environmental] *factor* as an abstraction for unifying *in language* a one-many relation [i.e., between organism and surroundings] by reducing it to a one-one relation [i.e., between organism-environment] in order that we may grasp the significance of correlations between environmental phenomena and biological processes" (334; emphases in original). But what kind of abstraction product is an organism's environment?

Jones (2018) has distinguished three forms of abstraction outcomes: generalizations, which simplify by generalizing details, structurations, which simplify by contextualizing details, and aggregations, which simplify by condensing details. A generalizing abstraction takes an element of a representational target and poses it as a class of elements, and potential instances of the class might include elements absent in the target. A structuration gives to some element or feature of a representational target a position in a structure, but potential occupants of this position might include elements not present in the target. In turn, aggregative abstractions, according to Mexican philosopher Sergio Gallegos Ordorica, link multiple elements of the representational target as a single, higher-level composite object: "An aggregation process may have as a goal the creation or development of a high-level aggregate object, and this may take place either by relying on structural or functional similarities or by relying on relations of temporal or spatial proximity" (Gallegos Ordorica 2016, 164). I think that organismal environments conform to the third category of abstraction products—namely, aggregations.

In the case that interests us here, multiple elements of the representational target (i.e., an organism's complex, fluctuating surroundings) are construed by biologists as a single composite object (i.e., an organismal environment in a specific moment of its life cycle). The "environment" becomes hypostasized and hence transforms into a subject of scientific research on its own: The elements linked together in an organismal environment share (extrinsic) causal connections with the token organism in question. As I mentioned, the environment concept has been a very powerful epistemic tool for biologists in modeling approaches, interventions, and explanatory endeavors. As Pearce says, "The singular term 'environment' . . . is an important heuristic for biologists, insofar as it gives them a way to talk about general causes without exploring the details of micro-level complexity" (Pearce 2010, 249).[31] By construing it as a composite, higher-order object, the environment is a receptacle of causal variables that can be included in different models and approaches without spelling out everything that is comprised within it. It works as a sieve for experimental interventions (e.g., "The environment was kept constant and we measured or modified Z factor."), and as I have argued, it is a common ingredient of causal explanations (e.g., as explanans: "The environment caused the appearance of developmental trait X." or "Changes in the environment allowed the evolution of Y.").

Its many useful epistemic advantages notwithstanding, there are many reasons why we should avoid the pernicious reification of an organismal environment. For one, the ontological heterogeneity of its components precludes simplistic causal reasoning; in fact, it would be misleading to treat an organismal environment as "a single, unified cause" (Pearce 2010, 249). Second, as we saw in section 3.2.3, an organismal environment is not free of indexicality; it depends on particular contexts of ontogenetic stages to be assembled (epistemically, this is accomplished through aggregation to form higher-order composite objects). The "environment of organism X" is an expression whose reference shifts across contexts and time points. For biologists, it is epistemically fruitful to reason by appealing to singular environments, but there are no singular environments of a token organism (neither synchronically nor diachronically): There are many shifting surroundings with myriad ontologically heterogeneous processes nested within them across different spatio-temporal scales. Just look again at figure 3.4. Which parameters of which abiotic and biotic factors are included in an environment is never a context-free determination. Therefore, treating "the environment" as a concrete thing is an ontic mistake given its dynamic, relational character. Moreover, the fact that we can conceptualize the coexistence of multiple organismal environments in the same physical medium through the relationality desideratum might also be interpreted as another reason in support of the idea that environments are abstracta. The metaphysical partitioning that the environment performs with respect to the causal elements that might potentially or actually affect an organism leaves out from its provinces many elements of the world that exist in close (spatiotemporal and causal) vicinity with the selected elements, and this is yet again another sign of abstraction.

In this chapter, my objective was to provide a general overview of how biological science construes organismal environments as relata, as well as to outline basic desiderata that scientists had expressed in relation to their characterizations (section 3.1). These include causal effectiveness, which entails the partitioning or segregation of the elements and processes of the world that influence the development of a specific organism, relationality, which refers to the mutual dependence of the pairing of organisms and environments and is frequently expressed as a bijective function, and dynamicity, which captures the variability and complexity of the heterogeneous biotic and abiotic factors that compose a given environment as a function of time. Subsequently, I argued that it is necessary to adopt a metaphysical approach to gain a deeper understanding of the organismal environment as scientific relatum. In section 3.2, I employed tools from analytic metaphysics to clarify what kind of entity an organismal environment is. I contended that the biotic and abiotic elements that are circumscribed within a particular environment do not possess any common ontological measure and instead share individual extrinsic connections with a given organism. This raised the question of whether it is metaphysically consistent to treat the environing context of a developing organism as a unitary entity. My response is that an organismal environment is an ontological construct that is best conceptualized as a class-as-many rather than a sortal or concretum. I provided arguments for why an organismal environment is not an entity of the second kind by reviewing criteria of identification, identity, persistence, and countability. I also addressed the indexicality of organismal environments, which complicates simple attributions given that its composition is assumed to be sensitive to ontogenetic contexts and the activities an organism performs during its life cycle.

Finally, I explored the epistemological implications of my metaphysical appraisal, highlighting the role of abstraction in how scientists typically conceptualize organismal environments. I fleshed out the noetic-epistemic processes that allow scientists to reach the idea of an "environment" in their theorizations and practices. I argued that an organismal environment, as an outcome of abstraction, is a higher-order composite entity that is formed through the aggregation of diverse factors. This view of the environment as a plurality masked in a singular term is epistemically useful and heuristically fruitful; thus, I emphasized that we should think about an "environment" primarily as a device of understanding, a conceptual tool that scientists employ to grasp the complex settings of the world in which organismal ontogenies are situated and embedded. An organismal environment is an important conceptual tool for biologists, an "ens rationis" as Peirce (1931, 5:334) would say, that allows them to make sense of the complex settings in which ontogenies unfold and open new avenues of research.

However, we should always remember that it is, at the end of the day, an abstraction, a limited conceptual device of our making.

With all of this in mind, in the next chapter I will broach the problem of organism-environment separation: What instantiates, if anything, the segregation of these two relata in development? I will offer an overview of distinct positions that have been advanced in the history of biology to deal with this issue and articulate my own answer.

4 Drawn, Erased, Renegotiated: Organism-Environment Separation in Past and Present Biology

> Enough hard cases come readily to mind to make it obvious that it is not a simple thing to draw an outline around an individual organism.
>
> —American psychologist and philosopher Susan Oyama (2006, 279)

> What is an organism? A sheaf of times. What is a living system? A bouquet of times.
>
> —French philosopher Michel Serres (1982, 75)

4.1 The Unsettled Interface(s) Between Organism and Environment

In *Biology for the Modern World*, the British embryologist and geneticist Conrad Hal Waddington stated, "One of the major distinctions which biologists are in the habit of making is between the animal or plant, and its environment—that is to say, all the things, outside the organism's body, which affect it" (Waddington 1962, 88).[1] The decades behind it notwithstanding, this assertion could well be made in any contemporary textbook or specialized volume of many biological subdisciplines. Undeniably, the distinction between organism and environment matters for manifold biological practices, theory building, and developmental, physiological, ecological, and evolutionary explanations. As philosopher Marjorie Grene put it, if we could not "separate the individual from its environment, we could neither begin nor continue to practice biology at all" (Grene 1968, 56). But what actually constitutes such an important biological distinction in the first place or enacts it during the ontogenies of organisms?

A common framing to tackle the separation of these relata within the life sciences and the philosophy of biology has been to appeal to the notion of a segregating, mediating, or separating *boundary*. Therefore, that is the conceptual space from which I will begin broaching the issue of organism-environment separation.

An intuitive answer has been to construe this presumed boundary as a separating "surface," and point to the skin or to other covering, insulating tissues or layers at the body's edge (e.g., exoskeletons in arthropods or the cuticle of plants; for discussion, see Stroll 1979). Nonetheless, one might contend that skin-like coverings constitute an arbitrary or not fully cogent criterion to embody, in point of fact, the ontogenetic correlate of organism-environment separation. First of all, layers like these have a limited phylogenetic scope—that is, not all organisms have something (functionally or structurally) equivalent. In addition, these tissues or layers are not present throughout the life cycles of all organisms (e.g., we encounter phenomena like ecdysis or metamorphosis); and, most importantly, a body surface criterion can be deemed an aprioristic trap that matches our gross intuitions about organismal individuality (i.e., it already presupposes that we know what an organism is and where it ends).[2] Furthermore, skin-like tissues (usually) do not partake in any active, mediating role with many environmental factors.

If not the skin or body surface, does it make sense to speak of an organism-environment boundary? In the mid-twentieth century, some scholars thought the issue at stake was a mere matter of semantics, simply revolving around how the environment is operationally defined. For example, in an exchange with Theodosius Dobzhansky on the environment concept in *The American Naturalist* (see also section 5.3.3), Thomas Morrill (1957, 269) wrote, "On the one hand, 'environment' may be thought of as ending at the organism's skin, as distinct from the organism, autonomous in itself. On the other hand, 'environment' may be regarded as the organism's 'environs', to some extent organized around and conditioned by the organism, inseparable from it."[3] Besides the terminological and semantic quibbles, many authors were willing to give up the separation between organism and environment altogether on the grounds that body surface was a lousy indicator of living processes (e.g., Angyal 1941, 97).

Even though strong, contrasting standpoints were advanced and defended, the organism-environment boundary problem was not settled during those decades, as we will see below. Looking at our current juncture, the unresolved issue of organism-environment separation has certainly become again a contentious topic in contemporary life sciences and the philosophy of biology.[4] Empirical data and conceptual insights coming from manifold fields and research areas such as epigenetics, niche construction theory, ecophysiology, ecological evolutionary developmental biology (eco-evo-devo), immunology, and microbiome research, among others, are inviting us to rethink the traditional segregation of organisms and environments qua relata. Organisms are increasingly conceived as developmentally plastic and open to environmental influences (see, e.g., West-Eberhard 2003, 2005; Sultan 2015; Pfennig 2021). Additionally, scientists contend that the environment should no longer be understood as a one-way, external (selective) force acting on organisms. Instead, they accentuate that the properties of environments are also actively modified by organismal action (Moczek 2015; Uller and Helanterä 2019). Through their agency as niche constructors, organisms causally affect the properties of their environments, and these, in turn, have a considerable influence on them. Hence, organisms and environments are increasingly understood as entangled in reciprocal causal loops that shape developmental and evolutionary outcomes (for discussion, see Fábregas-Tejeda and Vergara-Silva 2018a, 2018b; Schwab et al. 2019; Baedke et al. 2021; see also chapter 5). Moreover, many authors have contended that a range of different biological phenomena complicate a clear-cut distinction between these relata, such as epigenetic determinants, widespread developmental symbioses and phenotypic plasticity, biotic entrenchment, extended physiologies, developmental niche construction, and other forms of organism-environment reciprocal causation (see, e.g., Baedke 2018; Gilbert et al. 2012; Nicoglou 2011; Hernández and Vecchi 2019; Turner 2002; Stotz 2017; Sultan 2015). But if we grant that organisms are inextricably linked with and fully embedded in their (partially constructed) environments, how should we deal with the "organism-environment boundary" question? Where does one start and the other end?

When tackling the organism-environment relationship, most contemporary philosophers have concentrated on the side of biological individuality, with "organismality" being a subproblem embedded within this larger debate, or more recently, by exploring the epistemic underpinnings and practice-based variability of the environment concept.[5] However, the interface between token organism and surroundings—and what enacts the distinction between them—has not been sufficiently investigated and has broader implications beyond these aforementioned debates.

In this chapter, my main aims are, first, to clarify some of the positions at stake in this problem, which have been articulated and defended in different points in the history of biology (i.e., in early twentieth-century organicist biology and contemporary debates), and second, to offer a novel perspective that synthesizes the advantages of former positions and points to directions to move the debate on organism-environment separation forward.[6] Understanding this issue is of paramount importance for adding another piece to the jigsaw puzzle of what kind of relationship is instantiated when organism and environment are considered relata. As my initial historiographic-philosophical contribution to this chapter, I propose a taxonomy of four different broad stances apropos organism-environment separation that hitherto have not been identified and systematized in the literature: (1) single boundary seekers, i.e., those committed to finding a singular, definite boundary between organism and environment and thus specific, comprehensive, and fixed criteria to individuate organisms from their surroundings; (2) boundary skeptics—that is, those that disavow a meaningful distinction between organism and environment due to the recurrence of inextricable codetermination couplings and instead embrace a view of inseparable organism-environment systems; (3) boundary renegotiators—that is, those that argue for reframing the issue of "internal-external" segregation and the construal of a mediating "boundary"; and (4) boundary pragmatists—that is, those arguing for a practice-first view of "individuation," wherein how one draws the line between an organism and its environment is more a function of which scientific practices are mobilized and which explanatory projects are pursued, rather than a reflection of self-established divisions in organic systems.

In sections 4.2–4.4, I recount some historical examples of biologists from the Organicist Movement that vouched for positions 1–3. In section 4.5, I expound the contemporary debate on organism-environment separation and present defenders of positions 1, 2, and 4 (as no boundary renegotiators have emerged in recent times). Throughout this chapter, I will raise some criticisms and point to some limitations of these positions. In section 4.6, I will argue that a fifth position that expands and integrates valuable insights of former perspectives is tenable for scientists and philosophers: namely, that one can be a *shifting boundaries defender.*

According to that vantage point, as an alternative to espousing a single, all-encompassing or an overtly blurred organism-environment boundary (i.e., what positions 1 and 2 defend, respectively), *different boundaries* and relations between an organism and its surroundings could be instantiated, both synchronically and diachronically, and these might undergo *dynamic shifts throughout life cycles*. Hence, my proposal is an ontogenetically and phylogenetically informed perspective on shifting organism-environment limits that considers life cycle variation. Building from that theme, this account will be put forward with the help of case studies of diverse organisms that exhibit complex life cycles within and across generations. Finally, I discuss which advantages are afforded by the shifting boundaries perspective in contrast to previous positions.

As we saw in chapter 2 (section 2.3), the organism-environment relationship was a central framing device in many discussions at the onset of twentieth-century biological science. How past biologists discussed the organism-environment boundary is so far an uncharted component in the historiography of the Organicist Movement and in the history of debates on biological individuality (see, e.g., Lidgard and Nyhart 2017; Lyons 2020). In the following sections, I explore how this alleged "boundary" was framed, expunged, and renegotiated by organicist and holistic biologists from different disciplinary orientations and theoretical standpoints.

4.2 Introducing Early Attempts at Single Boundary Seeking

In the history of biology, general theories of biological boundaries have been sought by means of different perspectives (e.g., theorematic approaches that draw on hierarchical frameworks and insights from the study of complex systems; see, e.g., Whyte 1949, 108; Platt 1969; Bunge 1992). For the case of organisms, some authors have argued that it is possible to find criteria to satisfactorily (or even wholly) characterize the organism-environment boundary. I call them *single boundary seekers*. For them, the construal of a single, definite organism-environment boundary makes sense, and thus it is worth pursuing a robust characterization of its scope, causal regimes, confines, and functions. In particular, according to their broad position, the purported boundary can be carved out by single (albeit very complex) criteria. To count as such, each criterion chosen must meet three requirements:

(1) this criterion should be able to refer to causal processes that fulfill the functions of a definite, single boundary with the environment; (2) it should be *generalizable* across organisms; and (3) once this boundary is set in place, it must remain *fixed* by means of constant adjustments, in the sense that the boundary itself does not change into something else that should then be described using other criteria; in other words, it is presupposed that the boundary picked up by the selected criterion remains *one* throughout a life cycle.

Some authors have sought general and abstract definitions of the organism-environment boundary, aimed at comprehensive domains of application. Typically, this boundary is framed as having a regulatory role of matter-energy exchanges and buffering disturbances, as in the work of the theoretical physicist-turned-organicist biologist Lancelot Law Whyte (1896–1972; see Whyte 1949, 104–105). In general, for scholars that have embraced systemic viewpoints, the outer boundary is the external limit of regulated matter-energy transfer processes through which the environment and the system mutually influence one another and through which the organism ensures its persistence. In what follows, I offer two examples of concrete approaches to draw singular organism-environment boundaries in early twentieth-century biological science.

4.2.1 The Metabolic Answer of Charles Manning Child

A good example of single boundary seeking within organicist biology can be uncovered by exploring the ideas of American developmental physiologist Charles Manning Child (1869–1954). After completing a master's degree at Wesleyan University, he went to the University of Leipzig, where he obtained his PhD in 1894 under the guidance of Rudolf Leuckart with work on the morphology and histology of the sense organs of dipterans (Hyman 1955, 717; for biographical sketches, see Hyman 1957; Maienschein and MacCord 2022, chapter 3). When he returned to the United States, Child became an active member of an influential group of embryologists (for context, see Maienschein 1991). From his appointment as an associate in zoology at the University of Chicago in 1896 until his retirement as full professor in 1934, Child was a pioneer in the study of metazoan regeneration and physiological and morphological regulation (see, e.g., Child 1906a, 1906b, 1908).

In his 1915 book, *Individuality in Organisms*, Child aimed to tackle "the problem of the nature of the unity and order in the organism, . . . [and] the maintenance of individuality in a changing environment" (Child 1915, x). Indeed, Child granted that the individual organism was not "independent and self-determining in its origin" but instead arose from the "relations between living protoplasm and the external world" (Child 1915, 41; for historical background on the protoplasm concept, see Liu 2019). Organism and environment tightly interact to bring about living processes, and this was something he emphasized throughout his published works: "Every organism lives in an external world, an environment which acts upon it in various ways, and to this action the organism reacts with changes of one kind or another. . . . As a matter of fact, a living organism is inconceivable except in relation to environment" (Child 1927, 126, 132). The starting point of Child's theorizing was that, in contrast to a machine, an organism constructs itself by functioning right from the beginning of its ontogeny (Child 1915, 16). For Child, this is an outcome of the continuous processes of regulation and equilibration between a developing organism and its surroundings, which he deemed a key ingredient of what makes life what it is and likened it to the flow of a river (Child 1911a, 186).[7]

In this constant flow of activity, how does an organism attain a certain degree of independence from the surroundings that are shaping it? Interpreted as such, the problem at hand that Child was concerned with relates to the question of *self-individuation*—that is, how an organism develops to be or become, to a certain extent, different from its environing context. For Child, one of the puzzles at the forefront of development was that of individuality; as "the organic individual appears to be a unity of some sort, its individuality consists primarily in this unity, and the process of individuation is the process of integration of a mere aggregation into such a unity, for this unity is not simply the unity of a chance aggregation, but one of a very particular kind and highly constant character for each kind of individual" (Child 1915, 2). From Child's vantage point, individuality and development went hand in hand, as the former was an outcome of the latter (for discussion, see Maienschein 2011, 76–77). For views such as this and the explicit usage of the term, Prieto (2024, chapter 5) has contended that Child is one of the first authors to talk about the "physiological individuality" of living entities in its modern sense (see Child 1916).

Child considered the definitiveness and constancy of the morphological and physiological relations that take place through the spatial conjunction of the organism's parts and the sequence of changes unfolding throughout its ontogeny to be criteria of some degree of unity ascribable to organisms. One of the main theses that he wanted to flesh out in his 1915 book was that the foundation of organismal individuality—understood in a contrastive sense to an environment—must be found in metabolism—in particular, metabolic relations interlocked in time and space. In that sense, as per Child's rationale, "the problem of individuality resolves itself into the problem of the nature, origin, and maintenance of these [metabolic] relations" (Child 1915, 5).

Child argued that a token organism is a dynamic system of relations unfolding between a physical substratum and chemical reactions. The subtending thesis that articulates his book is that "metabolism is the formative agent in the organism" (Child 1915, 25). He is known for advancing the idea that, in early developmental stages, axial gradients get established. According to his experimental results, the origin of gradients involves an increase in metabolic activity in a particular region of the embryo, usually anteriorly positioned (see, e.g., Child 1911b, 1913; for discussion, see Child 1914). From the point of origin with the highest rate of reaction, called the apical end, an upsurge of increased chemical activity extends much like a wave spreads in a pond (Child 1915, 31). In Child's purview, metabolic gradients are the key factors in determining the direction of growth and differentiation of embryogenesis and are "the simplest and primary form of the organic axes of so-called polarity and symmetry and the starting-point of the mysterious 'organization'" (Child 1915, 35). In a nutshell, the origin of gradients involves an increase in metabolic activity in particular regions of the embryo, and Child conceived axial gradients as the simplest instantiation of physiological unity and order. Building from this idea, Child asserted that the extension of dominance during development and "differences in metabolic rate, constitute . . . the foundation of unity and order in the organic individual, the starting point of physiological individuation" (Child 1915, 40).

In that sense, Child theorized that the means through which a developing organism separates itself from the flux of changing environmental conditions is the establishment and maintenance of a metabolic gradient, and "the degree of individuation depends upon the permanency of the gradient, the metabolic rate in the dominant region" (Child 1915, 40–41). Along those lines, for Child, the organism-environment boundary is constituted by self-maintaining metabolic gradients that individuate the developing organism from its encircling medium. A single criterion (i.e., persisting metabolic gradients) was enough for him to account for the separation between developing organism and environment.

Child's gradient theory grew out from his studies of planarian regeneration (see Mitman and Fausto-Sterling 1992; Esposito 2016, 145–178). In these animals, regeneration occurs in a graded process along a main axis, "wherein each physiological process seemed determined by its location along the axis." Child extrapolated this phenomenon to embryonic development in general given that, for him, "the reconstitution of pieces into new individuals is fundamentally the same process as embryonic development, and the same relation of dominance and subordination exists in both" (Child 1915, 125). We will now see that an appeal to metabolic gradients in development was not the only criterion through which a single organism-environment boundary was drawn by authors of organicist affiliation that operated primarily within the space of distinct biological subdisciplines.

4.2.2 Organisms Carve Out a Space in the World through Perception, or the Position of E. S. Russell

A contrasting case of boundary drawing can be found in the works of the Scottish zoologist and philosopher Edward Stuart Russell (1887–1954). As Russell is an author that will reappear constantly throughout this book (especially in section 6.3.1), I will devote some space to contextualize his bio-philosophical and theoretical contributions before recounting his views on organism-environment separation.

In the historiography of biology, Russell is mostly remembered as the author of the pioneering book *Form and Function: A Contribution to the History of Animal Morphology* (Russell 1916). Russell's excursus into the history of biological thought is frequently cited as an authoritative source to understand protracted debates on the study of morphology, and, as claimed by some historians, laid the foundations of how the form-function dichotomy has been construed in the historiography of the life sciences ever since (on the latter, see, e.g., Appel 1987, 3; Ochoa and Barahona 2014, 29–30). Nonetheless, besides being an influential historian of biological science,

Russell was a quite well-rounded scholar: he was a scientist and philosopher that contributed to multiple areas (see Esposito 2013, 2014).

Russell went to Glasgow University, where he studied zoology under John Graham Kerr's supervision. From a very young age, Russell began publishing observations and anatomical descriptions of marine invertebrates on the shores of Scotland (e.g., of Stauromedusae and Hydrozoa; see, e.g., Russell 1904, 1907). Entrusted by D'Arcy Thompson, Russell gave notice of and described the cephalopods collected by the fishery cruiser *Goldseeker* (Russell 1909a). Around that time, Russell started his scholarly career as a theoretician publishing short articles in the *Revista di Scienza, organo internazionale di sintesi scientifica,* founded by the Italian philosopher and writer Eugenio Rignano in 1907. The journal, later called *Scientia,* was chiefly devoted to publishing avant-garde and theoretically audacious pieces related to all the natural sciences (Turbil 2018, 25). Rignano was a vocal defender of neo-Lamarckism and the constitutive purposiveness of organismal activities (e.g., Rignano 1930), so it is not a surprise that Russell found a safe space for developing his ideas there. (I will come back to this facet of Russell's work in section 6.3.1.) Russell published dozens of book reviews and short articles in the journal, which helped to establish an international reputation for him.

After his graduation, Russell became assistant of the famous Scottish naturalist John Arthur Thomson in Aberdeen, and his interest in the philosophical aspects of biology grew from his close interactions with the Scottish polymath Patrick Geddes (Esposito 2013). Shortly thereafter, the Board of Agriculture and Fisheries, which had taken over the North Sea fishery investigations, advertised a position for a zoologist to assist in dealing with the large amount of material collected. Russell joined the Board of Agriculture and Fisheries in the spring of 1910. From 1911 onward, he wrote *Form and Function* in his spare time. Later, the Board of Agriculture and Fisheries was elevated to a ministry, and in 1921 Russell was appointed director of fisheries investigations—and, later, chief scientific officer—of this ministry. The Fisheries Laboratory at Lowestoft was under his aegis for the rest of his career. In parallel to his daily work in fisheries, he started to advocate for a methodological and ontological stance of "functional biology" in his articles and books ("Fisheries Scientist" 1954; Esposito 2013; for Russell's biographical details, see Graham 1954).

In a debate Russell held in 1923 with the Scottish physiologist John Scott Haldane (1860–1936) at the Aristotelian Society, he argued that organisms appear to show a persistent and enduring individuality of action that has no parallel in the inorganic world. The activities of organisms (e.g., a salmon ascending a stream or the growth of a seedling plant) are to a high degree adjustable to varying circumstances. This meant, for Russell, that "the organism must be regarded then not as an arbitrary or artificial unit, existing as such only 'for us,' as does a machine, but as *a real unity existing in its own right and showing a measure of independence of its surroundings*, an ability to go its own way" (Russell 1923, 144; emphasis added). In opposition to authors such as Child, Russell argued that physicochemical criteria, such as metabolic gradients, "cannot ascribe to the living thing any but a conventional unity or individuality, cannot separate it from the general material flux save provisionally and for convenience of preliminary study" (Russell 1924, 13). Biologists had to look elsewhere for answers, given that the separation from the environment is a precondition for an organism to have agency to act on its surroundings.

Russell contended that "*it is mainly through perception that life becomes individualized and separates itself out from the environing flux*. Through perception the organism clears, as it were, a space around it in which to live" (Russell 1924, 59; emphasis added). Following this notion, the ontological nature of organism-environment separation must be found, according to Russell, in the distinct perceptual experience of each organism, and that is what really sets them apart from their environing conditions.

Interestingly, Russell claimed that individualized activity does not have to involve consciousness or higher forms of cognition (Russell 1923, 146; 1924, xiv). He claimed, "One may think then of all living creatures . . . perceiving its own objective world and reacting to this perceived world in such a way as to satisfy its needs. . . . Each organism would have its own individual view-point upon the universe . . . mirroring in its own way that aspect of reality which is accessible to it" (Russell 1924, 13). In this regard, Russell maintained that it was important to avoid anthropomorphism when conceptualizing the perceptual capacities and activities of nonhuman organisms. He defended that we must regard the activities of a given individual organism from its own point of view, carried out in response to its own *sensed environment* (Russell 1923, 147).

For Russell, the fact that a unit (i.e., an organism) can perceive and sense different facets of the world (i.e., an environment) constitutes an ontological disjunction. Hence, in the scope of Russell's bio-theoretical work, what separates an organism from its environment is experiencing the external world *as an environment*. This, it must be said, is a nonspatialist understanding of a single boundary that does not accord to the contours of a body, but at the same time, it is a conceptualization that still relies on a single criterion (i.e., perception) that purportedly tracks that which enacts the separation between these two relata. This is a very different answer compared with the one provided by Child; nevertheless, in broad strokes, this positioning places Russell as another exponent of single boundary seeking.

Let's now switch to some early-twentieth-century authors that argued that there is no principled way to draw a boundary between organism and environment and that, instead, both should be regarded as inseparable and commingled.

4.3 The Outlooks of Boundary Skeptics

> It is impossible to separate living creatures from their surroundings. To do so in fact is to kill them; to do so in theory is to turn biology into necrology, a vice which has always too largely infested our science, paralyzed its thinkers, sometimes even skeletonized or mummified them.
>
> —Scottish biologist and sociologist Patrick Geddes and biologist J. Arthur Thomson (1911, 192–193)

The stance that organisms cannot be disentangled from their environments, and consequently that a clear line of separation between them cannot be drawn, has been proposed repeatedly throughout the history of biology (see Rattasepp 2010; Baedke 2019; Baedke et al. 2021). For instance, at the turn of the twentieth century, in his *Problems of Biology*, George Sandeman wrote,

> [The] environment is just as much a part of the creature as are its morphological proportions. . . . At what point, or on what principle, does the oxygen and the moisture of the air cease to be mine and become that of the environment? Or, in speaking of the skin, can we find any essential difference between the kind of its relations to that which is beneath it, and the kind of its relations to what is above? . . . We come to the old difficulty, that *biology is drawing hard divisions where research knows of none*. (Sandeman 1896, 150–151; emphasis added)

A few decades later, the American ichthyologist Francis Sumner, as many more had done before him, called for a disavowal of the separation between organism and environment:

> [The] organism and the environment interpenetrate one another through and through. The distinction between them . . . is only a matter of practical convenience. Should not such considerations affect our attitude toward the propriety of neglecting the environment as an object of biological research? I make no pretense, of course, that such neglect is universal. An active and important group of present-day biologists is giving its chief attention to the organism-environment complex. (Sumner 1922, 233)

But who were some of the biologists that Sumner referred to?

4.3.1 John Scott Haldane and Holistic Viewpoints on Organisms and Environments

> If . . . we attempt to take the organism to pieces, or separate it from its environment, either thought or in deed, it simply disappears from our mental vision.
> —Scottish physiologist John Scott Haldane (1919, 72–73)

John Scott Haldane is a prime example of someone paying attention to what Sumner called "organism-environment complexes." Haldane received his education at the Edinburgh Academy, Edinburgh University, and the University of Jena. After graduating as a physician from Edinburgh University in 1884, he was appointed as demonstrator in physiology at University College, Dundee. While there, he conducted research into the air quality of dwellings and schools. In 1887, Haldane joined his uncle, John Burdon Sanderson Haldane, at Oxford University. His uncle was the Waynflete professor of physiology, and John Scott worked with him as a demonstrator. (John Scott Haldane later chose his uncle's name, John Burdon Sanderson, for his own son, who would become one of the famous founders of theoretical population genetics.) The measurement of partial pressures of oxygen and carbon dioxide in smaller quantities of blood was previously unfeasible, until Haldane devised a method to measure the gases contained in one cubic centimeter of blood. At Oxford, Haldane focused on the suffocative gases present in coal mines and wells and discovered their dependence on spontaneous oxidation processes occurring in the coal and soil. He conducted further studies that led to an understanding of constant exposure to foul air and hazardous gases—in particular, the physiological effects of carbon monoxide. In 1896, for instance, Haldane submitted a report to the home secretary that detailed the causes of death in colliery explosions and underground fires, which formed the basis for the development of measures to prevent such dangers and others related to mining (e.g., the use of canaries as living detectors of the presence of carbon monoxide). Haldane continued his research into mining problems and maintained his association with the mining profession for the rest of his life (for detailed biographical accounts of Haldane, see Douglas 1936; Goodman 2008; Sprenger 2019; see also Holmes 1969, 99; Butler 1988, 481; Edsall 1972, 229–231).

Haldane, well acquainted with Russell's work, as we just saw, criticized his organicist outlook for not seriously integrating the important roles played by the environment in sustaining life. He averred,

> At first sight it might perhaps seem that there is no difference between Dr. Russell's position and my own, so I should like to point out the difference. Dr. Russell's standpoint is rightly described as "organismal," and he does not apply his reasoning to the external environment of organisms. When I was a young man I became familiar with the same organismal point of view . . . ; but . . . I also saw that *in considering life it is impossible to separate an organism from the external environment with which its life is wrapped up at every turn*, and I therefore broke away definitely from the organismal standpoint in favour of the conclusion that . . . we are interpreting the *whole* of the phenomena . . . , and not merely the bodies of living organisms. . . . Environment is not something outside the unity of a life, but within it. (Haldane 1931, 142; first emphasis added, second in original)

As we saw in chapter 2 (section 2.4.1), Haldane had been one of the main popularizers of the notion of "organicism," but as his career progressed, he became dissatisfied with how the organism-environment relationship was treated by some of its main proponents, such as Russell. Drawing on insights gained from his empirical research on respiratory physiology, often with himself as a test subject, Haldane looked askance at the possibility of finding absolute demarcation criteria between organism and environment (see Haldane 1917). A central difficulty was pinpointing exactly where one entity starts and the other ends if both, at all times, are bringing forth living processes. The immediate environment is actively maintained just as is the organic structure itself: "In fact we cannot distinguish between normal living structure and its normal environment, and consequently cannot say where living structure ends and environment begins. We can thus discover no spatial demarcation between what is living and not living" (Haldane 1929, 11).

Haldane vouched for a view of *ontological coconstitution*, wherein organism and environment are entangled and form a single interacting, living whole that cannot be unknotted (for discussion, see Baedke et al. 2021). In that regard, he claimed, "The life of an organism, including its relations to internal and external environment, . . . persists actively and as a whole. . . . What persists is neither a mere definitely bounded physical structure nor the activity of such a structure. *There is no sharp line of demarcation between a living organism and its environment*" (Haldane 1917, 98; emphasis added). Or, in the form of a dictum: "An organism and its environment are one" (Haldane 1917, 99). Haldane rejected the existence of a single, definitive organism-environment boundary and espoused similar theoretical viewpoints throughout his career. He argued that our intuitions concerning mutual action or reciprocal causation go astray when dealing with the organism-environment pairing: "How its environment acts is determined by the organism, and how the organism acts is determined by its environment; but the conception of mutual action fails to express the integration which is evident in the relationship, since we are unable to distinguish separately the factors concerned" (Haldane 1929, 12). Accordingly, a different type of relationship must be theorized. Haldane generically referred to it as "reciprocity" (Haldane 1884, 32–33), although it can be qualified and understood in a strong *ontological coconstitution* sense. As we will see in the next chapter (section 5.2.3), Haldane expanded and applied Immanuel Kant's view of reciprocity as the organizational pattern of organisms to the relation between organism and environment. In Haldane's holistic viewpoint it is impossible "to distinguish separately the factors concerned," and life is regarded as an "integrated unity covering both organism and its environment" (Haldane 1935, 12). Haldane was not a lone voice in early-twentieth-century biology advancing a position of boundary skepticism regarding organism and environment.

4.3.2 Marcel Prenant and Dialectical Biology

Besides Haldane, early dialectical biologists like the French parasitologist Marcel Prenant, professor of zoology at the Sorbonne, argued against the staunch separation between organism and environment, albeit with quite distinct philosophical sources. In 1935, Prenant wrote the book *Biologie et Marxisme* (translated into English as *Biology and Marxism* in 1938). There, he reasoned, "The chief duty of dialectical thought is that of never losing sight of the complex interrelations between the living being and its environment. From the common-sense point of view the living thing very often seems almost independent of what surrounds it, or at least possessed of a high degree of autonomy" (Prenant 1938, 60–61). Prenant was, like Haldane, a critic of the shortcomings of mechanistic standpoints. He diagnosed that the concept of an "organism-environment complex" was usually lost or disregarded in most of them. He retorted, "What [mechanists] forgot was that if the environment reacted on the organism, the organism in its turn reacted on the environment" (Prenant 1938, 61). In his dialectical outlook, in which any boundary gets blurry throughout protracted interactions, Prenant agreed with Haldane in granting unity to the phenomenon of life, manifested in organism-environment interpenetration: "Without losing sight of the unity of all matter, Marxist biology still insists that the phenomena of life have their own characteristics. . . . Each living thing has a material structure which is peculiar to it and which operates alongside the environmental factors to produce the concrete results of life" (Prenant 1938, 62). Prenant defended a strong ontological view of dialectical coupling between organism and environment, characterized as "reciprocal action of phenomena in which it is not possible to state with certainty which is cause and which is effect" (Prenant 1938, 213). In other words, Prenant garrisoned the primacy of ontological coconstitution between organism and environment.[8]

Haldane, Prenant, and some of their organicist and holistic colleagues were, in my terminology, *boundary skeptics*—that is, authors that disavow a meaningful distinction between organism and environment due to the recurrence of inextricable codetermination links and in its place embrace a view of inseparable organism-environment couplings.

It must be said that the forswearing of an organism-environment boundary was also questioned by several scholars of the time. E. S. Russell, for instance, commenting on Haldane's outlook, criticized the physiological merging of an organism with its environment: "While the morphologist studies the form and structure of the organism in almost complete isolation from its environment, the physiologist merges it completely into its surroundings and robs it of all independence . . . it cannot consider the organism as a living and independent whole. Its laws have no reference to individuality" (Russell 1924, 12–13). For Russell, a middle path between morphological fixity and absolute physiological merging should be possible.

Furthermore, the famous Joseph Needham wielded an explicit attack on Haldane's holistic views on methodological grounds. Needham claimed that "we easily see that his demand for the unification of the organism and its surroundings is a methodologically impossible aim, for if no line can be drawn between organism and immediate surroundings, no better line can be drawn between immediate surroundings and far-off surroundings" (Needham 1936, 10–11). That means that the problem of individuation, which Child had talked about at length (see section 4.2.1), is not resolved but merely pushed aside. I will come back to this issue in section 4.5.1.

4.3.3 Other Intellectual Traditions Defending Organism-Environment Inseparability

Before proceeding to the next arc of the story, I simply want to point out that many other authors from different theoretical backgrounds within the Organicist Movement vouched for entangled organism-environment complexes in the first decades of the twentieth century. One of the most important traditions is represented by the physiologist and philosopher Jakob von Uexküll (1864–1944) and his followers. In that sense, a comprehensive survey of positions that erased the separation between these relata in the early twentieth century should include a treatment of this influential school of thought (for discussion, see, e.g., Brentari 2015; Michelini and Köchy 2019). I will only make some brief remarks on Uexküll's vantage point as presented in his *Umweltlehre* (see also section 5.2.5).

In his first writings, centered on empirical studies of animal physiology, Uexküll did not use the term *Umwelt* but captured the relationship between organism and surroundings through the French term *Milieu* (with Germanized capitalization). Later on, Uexküll postulated the notion of "surrounding world" (*Umwelt*) to apprehend the deep and particular intertwining of each organism with its environing conditions, which goes beyond the idea of an *Aussenwelt* (external world) that gives an organism material support and a space in which to thrive.[9] Uexküll was heavily influenced by the work of Immanuel Kant—especially by the "Copernican turn" unwrapped by the *Critique of Pure Reason*, which, instead of proposing an objective world that exists independently of a subject, reverses the framing to point out that it is the subject who establishes the conditions of possibility of knowledge of the world. Uexküll expanded Kant's views by asserting that not only are human beings determined by the structure of a priori elements of pure intuition and the categories of the understanding that shape the experience of their sensible surrounding worlds, but that the same is true of other (nonhuman) animals. For Uexküll, as for Russell (see also section 6.5.5), organisms are not stimuli-responsive machines but active subjects that experience their worlds (see Jaroš and Brentari 2022).

With respect to the extension of the Kantian project, Uexküll claimed, "The task of biology consists in expanding in two directions the results of Kant's investigations:—(i) by considering the part played by our body, and especially by our sense-organs and central nervous system, and (2) by studying the relations of other subjects (animals) to objects" (Uexküll 1926, xv). Accordingly, in connection to (i), he conceptualized the *Umwelt* as constructed through the sensorial explorations of organisms. He argued that each metazoan is reciprocally connected with its own surroundings, which it subjectively perceives and acts on. Thus, an *Umwelt* contains a "perception world," which is experienced through sense receptors and processed neurally, and an "effect world," in which the organism causally acts on its surrounding medium through various traits and behaviors. The organism's perceptions and actions together create a feedback loop, or "function-circle" (*Funktionskreis*), between the perception world and the effect world. The organism (subject) perceives a "carrier of a feature" of the "environment" (object) in its perception world and reciprocally acts on and shapes the same environment (as a carrier of this effect) in its effect world (Uexküll 1926, 155–161). For example, when organisms explore new environments, they attempt to establish new functional circles (e.g., by changing nutritional habits). If this action is not rewarded (i.e., a functional circle cannot be closed, for instance, when the new diet is not nutritious), the organism gives up this environmental setting (switching back to old nutritional habits or exploring new ones). During this process, the environment causes the organism to rewire neural connections and physiological receptor-reaction associations.

Uexküll's view of reciprocity conceptualizes the organism-environment relationship as highly individualized and mediated through experience (see Brentari 2015; see also Baedke et al. 2021, 4–5). As Claire Schiller (1957, xiii) recounts, "[An *Umwelt*] is the world around an animal as the animal sees it, the subjective world as contrasted with the environment."

For Uexküll, any organism is always perfectly adapted to the notifiers appearing in its *Umwelt* through the structure of its sensory organs and its effectors, and this constitutes the basis of its existence. Each animal forms an inextricable unit with its surrounding world. This view was in tension with the Darwinian view of evolution because, according to Uexküll, Darwin's theory compared each living being with the whole external world that is accessible to us humans, even though each animal is related to only a small portion of the things and processes that are salient in the world of *Homo sapiens* (Uexküll 1945, 76–77).

In the next chapter, I will cover some of the reasons that explain why the study of experienced environments fell out of favor in biology after the mid-twentieth century and why it is making a comeback in contemporary debates through concepts such as "experiential niche construction" (see sections 5.3.3 and 5.4.2).

So far, I have illustrated diverse views on how the presumed boundary between organism and environment was drawn, and I covered some stances that disavowed the existence of organism-environment separation. To cap off the historiographic examination of this chapter, and before moving forward with the philosophical analysis of contemporary controversies on organism-environment separation, I will recount how certain positions did not abide with this dichotomous split between organism-environment that a boundary could, in principle, achieve and, instead, sought to reframe the organic limit between the external and the internal through novel lenses.

4.4 Renegotiating the Internal-External Distinction

As we have seen, one way that biologists have tried to spell out and frame organism-environment separation is by resorting to the idea of a mediating or separating boundary between developing systems and their surroundings. However, not all authors in the history of biology have been satisfied with this framing. Why should this separation be enacted by a single boundary, either construed as a physical surface of some sort or a functional edge or margin that individuates an organism from its conditions of existence? People that reject the framing of a "boundary" and instead advance different construals of organism-environment separation are christened, in my terminology, *boundary renegotiators*. In the next subsection, I revisit the arguments of one scholar that tried to renegotiate the issue of the internal-external distinction in development: the German theoretical biologist of holistic inclinations Adolf Meyer-Abich.

4.4.1 Adolf Meyer-Abich and the Complementary Principle in Biology

Adolf Meyer-Abich (1893–1971), who used the shorter name Adolf Meyer before 1938, was an influential German theoretical biologist and philosopher of science (for biographical details, see "In memoriam Prof. Dr. A. Meyer-Abich" 1971; Amidon 2008).[10] He studied philosophy and natural sciences in Göttingen and Jena from 1913 to 1917 and worked as a librarian in Göttingen, Kiel, and Hamburg. From 1930 to 1958, he held an academic position in Hamburg and periodically taught in other places such as Santiago de Chile, the Dominican Republic, El Salvador, and the United States of America (Ehrle 1965). When Uexküll was the head of the Institut für Umweltforschung, Meyer-Abich was also working at the University of Hamburg. In fact, in the 1930s they led together some sporadic seminars on natural philosophy and epistemology. During the interwar period, Meyer-Abich played a significant role in the early stages of German theoretical biology, where he strived to replace mechanistic-reductionist and vitalist biological theories with holistic approaches that integrate evolution and development (Meyer 1936; Meyer-Abich 1934; see also Meyer-Abich 1936).

In the 1940s and onward, Meyer-Abich proposed an evolutionary theory of holobiosis (*Holobiose*). This theory contended that organisms, in the grand sweep of evolutionary history, develop through biontic processes (i.e., parabioses, antibioses, symbioses, and finally, holobioses), which involved the interaction of simpler and lower life forms to produce more "complex" and "higher" organisms (Meyer-Abich 1948, 237; for an analysis, see Baedke et al. 2020). Meyer-Abich placed his theory of holobiosis in opposition to traditional evolutionary views, which gravitated around what he dubbed "eogenesis," a general view of descent that starts "from finished adult forms" (Meyer-Abich 1943, 61). In contrast, he suggested that to explain evolutionary change, biologists must flesh out the interdependent assemblage processes of organisms, which first exist as symbiotic partners and then progress into holobionts (*Holobionten*) and culminate in organs or systems of organs that contribute to a larger integrated whole (a

Holobiome, in his parlance; Meyer-Abich 1950; see also Meyer-Abich 1942, 201–202). During the evolutionary phase of holobiosis, the interdependent organismal units cannot develop, grow, or reproduce autonomously. For Meyer-Abich, holobionts are able to realize unique traits together (e.g., novel metabolites and reproductive bodies) that no individual organism could produce on its own (Baedke et al. 2020).

According to Meyer-Abich's theory, symbiotic organisms create *new environments*, which they both act on and experience in a novel manner. For him, organisms do not only construct their own individual environment but also construct integrated and shared environments for one another (see also Meyer-Abich 1964). Meyer-Abich expanded his thoughts on the organism-environment relationship later and, inspired by Uexküll's work, argued that biotic environments are much more than pure physical surroundings (as we saw in section 3.1.1): "Organisms . . . are existing within the same physical surroundings but they live within their own biotic environments. . . . Different environment . . . of an organism means a different internal world of this subject too. In this sense *organisms are always subjects, not objects*" (Meyer-Abich 1955, 66; emphasis added).

Meyer-Abich framed the "environment/surrounding world-innerworld" (*Umwelt-Innenwelt*) distinction as part of what he, inspired by the Danish physicist Niels Bohr's work, called "the complementarity principle in biology" (for background, see Beller 1992; Roll-Hansen 2000). In direct analogy to what happened with some physical phenomena—such as the wave-particle duality—he argued that to attain "complete knowledge of an object we need at least two different and autonomous branches" of science looking into it (Meyer-Abich 1955, 58). He exemplified this with the everlasting form-function dichotomy in biological thought. Meyer-Abich claimed that form and function, just like corpuscle and wave, are "complementary biological categories and therefore morphology and physiology [are] complementary . . . sciences" (Meyer-Abich 1955, 65). This reading departed radically from Russell's (1916) confrontationalist account of structuralism and functionalism.

In addition, Meyer-Abich claimed that another complementarity is to be found in the *internal-external relations* that living organisms establish with their worlds (Meyer-Abich 1955, 67). Meyer-Abich used the example of a spider to illustrate the external-internal world distinction. He said that, for a token spider, "its environment embraces all other living beings, with whom it is in living contact, and all inorganic factors important for its life (water, air, etc.), whilst its internal world is represented by the totality of all its organs, tissues and cells as a living whole" (Meyer-Abich 1955, 66). Meyer-Abich claimed that the external and internal world of an organismal subject cannot be studied simultaneously in biological science because, as soon as you focus on one, you lose sight of the other, so that is why they are complementary, nonsubsumable categories in scientific inquiry:

> If we study its environment, then the spider represents to us an individual whole subject which only as such appears as the center of its environmental relations. Doing so we are never occupied with the investigation of the particular environmental relations between each of the inner organs of our spider and its particular environment within the spider. [All] mutual relations of internal organs with environmental factors are controlled by the *whole* of the organism. Then all external factors affect first and directly only the whole organism, and the consequence for different internal organs is never the result of direct interrelationships between any inner organ and the environment but is only the result of the relations of the organism as a *whole* . . . with its own internal organs. (Meyer-Abich 1955, 67–68; emphases in original)

For Meyer-Abich, there was no organism-environment boundary, in the sense that this construal proves to be ill-defined if we pay attention to the *inward* and *outward worlds* of organisms. At the same time, he did not subscribe the idea that organism and environment form an indissoluble system where the components cannot be extricated, as boundary skeptics such as Haldane stressed. According to Meyer-Abich, the organism as a whole is, *in itself*, the mediating interface between the external world and the internal world of an experiencing subject. He asserted, "The whole of the organism . . . controls actively the activities of the internal organs as well as the relations of the organism with the environment" (Meyer-Abich 1955, 68). He elaborated further on this thesis:

> We can investigate the environment of any organismic subject only with reference to our subject as a whole; there are no direct relations between environmental factors and inner organs; this innerworld of our subject disappears immediately, when we take into account its environment. And the same is true with the environment, when we are investigating the innerworld relations of our organismic subject. *Environment and innerworld of the same subject never meet each other directly, but only through the intervention of the holistic entirety of the organism*. . . . So we have to conclude: *Environment and innerworld are complementary biological categories within physiology*. (Meyer-Abich 1955, 69; first emphasis added, second in original)

Meyer-Abich's position on organism-environment separation might count as a rara avis in the landscape of early-twentieth-century biology, but since he attempted to re-form the terms in which this issue was usually framed, he counts in my taxonomy as a *boundary renegotiator*. Of course, he was not the only scholar in German-speaking contexts that renegotiated the traditional casting of organism-environment separation. In a more encompassing historical narrative that transcends what I presented here, the ideas of the German zoologist and philosopher Helmuth Plessner (1892–1985) and the Swiss zoologist Adolf Portmann (1897–1982) should be reconstructed as well (see, e.g., Grene 1968; Kugler 1967; Michelini et al. 2018; Jaroš and Klouda 2021).

We have seen that the issue of a purported organism-environment boundary was a matter of contention in early-twentieth-century biology: It was drawn, erased, and renegotiated. Positions differed (1) on the nature of the boundary itself and by what criteria it can be drawn (from metabolism to perception), (2) on the purported inexistence of the boundary due to ontological coconstitution (highlighting the notion of inseparable organism-environment complexes instead), and (3) on the question of the adequacy of the construal of a boundary to capture the internal and external worlds of organismic subjects. Importantly, by charting these positions, we enrich the historiography of the organism-environment relationship in early-twentieth-century biology, and this historical understanding can inform contemporary philosophical work, as we will see in the next sections.

4.5 Current Controversies on Organism-Environment Separation

In what follows, I will show that the taxonomy of positions dealing with organism-environment separation that I derived from my analysis on organicist biology, despite historical differences in concept usage and differing techno-material settings in which science is carried out, is projectable in an abstract fashion to our current juncture.[11] In particular, we will see that two of the stances I identified in previous sections (i.e., boundary skepticism and single boundary seeking) are thriving in the contemporary debate (sections 4.5.1 and 4.5.3) and that a new position—which had no correlate in early-twentieth-century biology—has been advanced by philosophers of biology (i.e., boundary pragmatism; section 4.5.2). Likewise, it is worth noting that boundary renegotiators shine for their absence. We turn our attention next to the contemporary debate on organism-environment separation.

4.5.1 Organism-Environment Entanglements or the Rationale of New Boundary Skeptics

In recent decades, we have seen a resurgence of the claim that organisms are not sharply distinguished from the environing conditions they inhabit. A famous example comes from developmental systems theory (DST). This standpoint impugns the assumptions of the genome harboring a complete *representation* of development (i.e., a full-fledged genetic program with causal privileges) and there being a one-to-one correspondence between genotype and phenotype (Oyama 2000c; for an analysis, see García 2005). Instead, developmental systems are viewed as responding flexibly and contingently to manifold interdependent ontogenetic resources (e.g., genetic, cellular, environmental) that are equally necessary for the completion of a life cycle and the transgenerational stability of form (Griffiths and Gray 1994, 2001). Hence, in the purview of DST, developmental information is distributed throughout the entire developmental system and reconstructed anew in each life cycle. DST rejects the thesis according to which an organism and its environment can be characterized as wholly independent of one another (Oyama 2000a). In contraposition, Oyama's stance furthers the claim that nothing *form-giving* (e.g., the environment or genes) is separate and autonomous from that which is formed (i.e., the developmental system). In this sense, DST denounces the artificialness of the dichotomy between "internal" and "external" causal factors in development. Elements of the organismal environment—or, better said, what hitherto were considered "environmental"—are reassembled under DST as genuine parts of the processes by which development comes about. In this view, organisms and environments are treated as "interdefining" and "interdependent" (Oyama 2000b). DST proponents go further and assert that "there is no distinction between organism and environment" (Griffiths and Gray 2001, 207). When DST defenders embrace the idea of commingled, coconstituting organism-environment systems, they become boundary skeptics.

Along these lines, Smith (2017) contends that it is misleading to address development independently of its environmental contexts, and hence, even though organisms are *whole* systems, they are not sharply distinguished from the worlds that they inhabit. For Smith, organisms are not separate from their environments, and the traditional conception of organisms gives an ill-guided metaphysical picture: "Typically, organisms are understood as products of the relation between internal parts and processes and the environment, which is taken to be something that is separate and distinct from them. But *this bifurcation of the world into organism and environment is misleading*. . . . The constitutive embeddedness of organisms in their worlds suggests that the line drawn between 'organism' and 'environment' is . . . metaphysically misleading" (Smith 2017, 7–8; emphasis added). When discussing past boundary skeptics, I mentioned that Joseph Needham launched an attack on Haldane's holistic views on methodological grounds: If we cannot draw a line between organism and immediate surroundings, we are not in a better position to draw a line with far-off surroundings (section 4.3.1). Even though Oyama's and Smith's viewpoints are indeed very important and valuable, as we should not disregard the key roles played by environmental factors and contexts during development—something biologists have been studying at least since the beginning of the twentieth century (see section 2.3.3)—I consider that Needham's objection has an ontological upshot that can be waged against their accounts and to boundary skeptics in general. Within their standpoints, the problem of individuation is not resolved, and central concerns are not attended to: The boundaries of an entangled organism-environment system must still be drawn elsewhere.

An alternative to that complication is to endorse an ontological view of a continuous organic world wherein there are no possible distinctions. This resolution would be problematic for several reasons, one being that when seen from a mereological standpoint, a continuous organic world would yield a potentially infinite array of odd gerrymandered objects (e.g., biological creatures separated by thousands of kilometers and existing in different spatiotemporal scales being brought together in one and the same entity). This is an issue that, when discussing Haldane's views on reciprocity, some scholars directly objected to in the past:

> The point is that if the whole is the organism plus the environment, then because there is no boundary in nature marking off an organism's environment from the rest of the universe, the planet Mars and Mount Everest and the teeth of a mouse are all part of the environment of any and every biological organism. And if the theory holds that the parts can only be understood if considered in relation to the whole, then the attainment of knowledge becomes impossible. (Phillips 1970, 428)

If, on the contrary, the boundary of an intermingled organism-environment system is established by resorting to a particular criterion, then this approach becomes identical to that of a single boundary seeker: There would still be a boundary separating an inside and an outside of the system under consideration. Along those lines, a boundary seeker could simply incorporate multiple elements, once thought of as "environmental," within her conceptualization of the proper limits of an organism and reach the same conclusions as a (former) boundary skeptic.

Moreover, the postulation of an organism-environment complex with no bounds whatsoever is not experimentally tractable or translatable into biological interventions. Of course, a conciliatory solution is to remain a boundary skeptic in the ontological domain but accept a form of boundary pragmatism in epistemic and methodological affairs in accordance with the needs of particular biological practices. Scholars could grant that no definite organism-environment boundary will ever be characterized—because there is none—although some compromises may be taken to resume biological inquiry from some standing, tenuous as it might be. For instance, Smith (2017) spells out her ontological commitments that line her up with the camp of boundary skeptics, but she is, nonetheless, mindful of the contextual practicality and convenience of sometimes drawing a distinction between organisms and environments in scientific practices. For that reason, Smith is, on the epistemological terrain, a boundary pragmatist.

However, insofar as being a boundary skeptic leads you to either (1) embrace a wholly continuous world, (2) become a boundary seeker of the limits of organism-environment systems, or (3) accept some form of boundary pragmatism, there are sufficient reasons to hold that this position is not a satisfactory comprehensive solution to the problem posed in the first place.

Now let's cover the practice-oriented positioning of present-day boundary pragmatists.

4.5.2 Meeting Boundary Pragmatists and Their Practice-First Stance

Boundary pragmatists place a special emphasis on the praxis of individuation and the epistemic standing of scientific practices through which the construal of individuals is rendered possible, paying special attention to the aims that ultimately guide these endeavors. Boundary pragmatists are those scholars arguing for a practice-first view of "individuation," wherein how one draws the line between an organism and its environment is more a function of distinct explanatory projects and scientific agendas than a reflection of univocal divisions in nature (for discussion, see Smith 2017).

Philosophers of science who subscribe to the so-called "practice turn" have recently proposed that, instead of being on the lookout for overarching theories of biological individuality, scholars should focus on *individuation practices* to better understand the myriad purposes that lead scientists to individuate biological systems in the first place and how these practices subserve their epistemic aims under structured problem agendas (e.g., Love 2018). Adapted to the problem of organism-environment separation, that stance translates into the inspection of *organism individuation practices* in different biological disciplines that deal with token developing organisms (e.g., developmental biology, ecology). A boundary pragmatist is concerned with how scientists de facto single out organisms from their surroundings in experimental settings and field studies. A key question for them concerns not what the boundary per se is (i.e., ontologically) but, rather, how scientists segregate variables in their studies (e.g., which variables count as "environmental" and which as "organismal," and what epistemic or non-epistemic contexts guide those choices) and through which practices individuation is made possible.

In line with what other scholars have claimed for the wider puzzle of biological individuality, organismal boundary pragmatists can maintain that there are different concepts as to what constitutes the organism-environment boundary, and none of them will be equally appropriate for all investigative contexts; biological reality is complex and can be cut up in different ways. In that regard, if the individuation of organisms in their environments serves manifold epistemic roles in the sciences, why commit ourselves to a monist framework regarding the threshold or edge of difference between an organism and that which environs it? Won't it be better to allow for multiple structural, functional, temporal, relational, or material standards to appoint the organism-environment boundary, each tailored to its own scientific needs? A boundary pragmatist need not subscribe to a particular theory of organismality—but some might—and, in general, she can remain ontologically agnostic regarding organism-environment separation (or the lack thereof) and other related matters (e.g., the ontological status of the organismal environment concept).

Having presented the general position of boundary pragmatism, let me offer two reasons to not fully embrace this vantage point as a complete solution to the problem of organism-environment separation. I agree with Kovaka (2015) that not having a general theory of biological individuality—and, one must add, of organismality—is no barrier to good empirical work. However, for the case of organisms, some scientific questions can only be formulated if our focus does not elude grasping the properties and conditions of possibility of bona fide, human-independent organism-environment separation. For example, the establishment of a boundary that grants a degree of autonomy to living systems is a very important subject in origin of life studies (see, e.g., Ruiz-Mirazo et al. 2004; Bich 2019). How did entities, which we would qualify as "living," evolve to be (to a degree) distinct from their environing flux of physicochemical conditions? In other words, how did organisms evolve in the first place? How did cells arise, and how were the limits with their environments established? These are some questions that cannot be answered by restricting oneself to peruse individuation practices in present-day life sciences.

More examples of empirically open questions include "How have organism-environment limits changed throughout the major transitions in evolution and across different clades?" As a case in point of that shift, organismal interfaces with surrounding conditions certainly changed with the onset of multicellularity, and these evolutionary phenomena are worth studying in themselves. Studying the developmental processes that enact and maintain some forms of organism-environment separation is an important avenue of research that should not be curtailed by the presumption that current scientific individuation practices capture all that is relevant about organism-environment interactions and the renegotiations of the limits of both relata that unfold in ontogenetic time. Focusing on how scientists individuate their research systems for different epistemic aims does not solve all that there is to say about organisms.

Of course, boundary pragmatists might argue that they could simply specify the relevant boundaries relative to the purposes of research questions such as "How have organism-environment limits changed throughout the major transitions in evolution and across different clades?" A more sophisticated version of boundary pragmatism can contend that boundaries are domain-relative to scientific purposes (in the broadest sense) without succumbing to an anything-goes relativism, akin to what some authors have maintained could be the case for "natural kinds" (see Slater 2015). This would be a reasonable position, but I think the real issue at stake in organism-environment separation is the *self-individuation* carried out by a (nonhuman) living system embedded in certain surrounding conditions.

The second reason to resist a tout court boundary pragmatism is related to this point: Individuation practices might not always coincide with how organisms, qua living entities and agents (see chapter 6 for discussion), individuate themselves from their surroundings to become, to a certain extent, different from them. The aims of scientific individuation are not always aligned with patterns of organism-environment interactions, and even less so with the actual causal regimes that keep out of an organism certain elements of its environment that are not contemplated in the setup of particular experiments—which have precise targets and variables to measure—but that still affect it in particular moments of its life cycle. Importantly, the problem of what enacts the separation between an organism and the manifold biotic and abiotic factors that environ it is not simply a matter of epistemic expediency. In its extreme form—which fortunately is not advanced in extant scholarship—a practice-only account of individuation could deny to organisms any *real individuality* beyond human strictures. If taken to this extreme, boundary pragmatism could be detrimentally anthropocentric because the self-individuating processes and intrinsic normativity of organisms (e.g., pursing goals of their own, such as surviving, maintaining their organization, or reproducing) could get erased. We might forget that organisms, as evolved entities, are not toys of our own making. Assuming that admittedly heterogenous individuation practices performed by scientists amount to or capture all boundaries a developing organism could have with its environment is a rather restraining viewpoint. In that regard, although boundary pragmatism follows biological science very closely, it still has some shortcomings that scientists and philosophers would want to avoid if this stance is taken to be the definitive answer to the problem at hand: It is not able to assert something projectible about what the ontological referents of organism-environment separation during development are, or the general nature of organisms and their evolution.

Boundary pragmatism is a valid and sophisticated position for many problem spheres within biology and the philosophy of biology, but I do not consider that it qualifies eo ipso as the ultimate solution to understand how organisms develop to become different (to a certain extent) from the environing flux of conditions in which their ontogenies are situated.

As we have briefly seen, boundary skeptics do not resolve the issue of organism-environment separation because they push the individuation problem one step away on the ladder, and boundary pragmatists also suffer from limitations, and their standpoint might prevent us from tackling this issue head-on. Could contemporary single boundary seekers give comprehensive accounts of what enacts the separation between a developing organism and its environment?

4.5.3 Abstract and Concrete Accounts of Single Boundary Seekers

In contrast to boundary pragmatists, and as we have seen through the historical examples of section 4.2, single boundary seekers understand self-individuation as an intrinsic feature of organismality, something organisms constitutively *do* to define their identity and diachronic persistence, and not something that purely stems from polyvalent choices in scientific endeavors. In recent years, for instance, Montévil and Mossio (2020) argued that "the identity of an organism refers to its individuation, *i.e., the fact of possessing those properties that allow drawing its boundaries and discriminating it from the surroundings*" (1–2; emphasis added).

When dealing with this problem, some authors have sought general and abstract definitions of the organism-environment boundary, aimed at comprehensive domains of application. Besides the case of Whyte that I briefly covered at the beginning of section 4.2, another famous abstract formulation of what counts as the organism-environment boundary comes from the theory of autopoiesis and its later developments in the hands of the so-called autonomy school in the philosophy of biology (for an overview, see Moreno and Mossio 2015 and references therein; see also section 6.5.4). Autopoiesis refers to the distinctive self-recursive character of the organization of living systems, which maintains constant certain relations between components through a network of metabolic processes of production that yields a concrete unity in space (Maturana and Varela 1972; Varela et al. 1974). For Maturana and Varela, the dynamic, productive transformations of metabolism lead to the persistence of a living system while concomitantly forming its limits (e.g., in the case of cells, their membranes). Meincke (2018) highlights this facet of the theory: "The network of processes of production of components produces the components which in turn produce the network of processes of production of components" (5). The self-recursion of autopoietic organization is what establishes a bounded "inside" and thus an "outside" in all living systems (for discussion, see Meincke 2020). According to this view, one of the defining characteristics of organisms is that they are able to self-generate an internally defined identity that is not fully governed by the external processes of their environment (Varela 1979). In turn, this property is generally explained as a consequence of the organizational closure of the system. In this sense, Mossio and Moreno (2010) assert that organisms possess the capacity to sustain themselves by creating an intricate network of structures that take part in reciprocal constraining activities with regard to their boundary conditions. This interdependence among a set of constraints is what they term "organizational closure" (Mossio and Moreno 2010, 276). From this perspective, then, organizational closure is the criterion that tracks organism-environment separation (Ruiz-Mirazo et al. 2000, 217). As Isaac Hernández and Davide Vecchi claim, from the organizational approach, "the *limits of the individual are thus the limits of this closure,* which becomes the fundamental criterion for tracing the boundary between individual and environment" (Hernández and Vecchi 2019, 2; emphasis added). This view was endorsed by Varela, who believed that an organism's boundary with its environment "is indissolubly linked to the operation of the system. If the organization of closure is disrupted, the unity disappears" (Varela 1979, 55).

In contrast to these highly abstract approaches, which aim at theoretical generality, other authors that have tried pinpoint what enacts the distinction between organism and environment in concrete instances of ontogeny.[12] An interesting example of single boundary seeking comes from the work of Pradeu (2010). He has argued that the immune system qualifies as a bona fide boundary-maker for the organism. Through its diachronic surveillance activity, the immune system exerts a rejection of those (environmental) entities that are not part of the organism. In this sense, immunogenicity—namely, if the entity encountered by the immune system elicits a response or is tolerated as a part of the organism—constitutes an individuation criterion and, incidentally, a good measure for it. For Pradeu (2011), the immune system delineates the organism, which must be conceptualized as a heterogeneous entity, made up of both endogenous and exogenous constituents such as symbiotic bacteria that are immunotolerated. Following this viewpoint, certain microorganisms that have crucial roles in the development of a host organism, customarily considered elements of the environment, must be regarded as genuine organismal constituents: "In other words, the immune system defines a boundary between the 'inside' and the 'outside' of the organism, but this boundary is not equivalent to the boundary between the endogenous . . . and the exogenous" (Pradeu 2011, 84). To sum up, in Pradeu's view, the immune system establishes the partially open, highly regulated, constantly negotiated boundary of the (genetically mixed) organism.

Let me now present some objections to the accounts of (abstract and concrete) single boundary seekers as full answers to the quandary of organism-environment separation. First, one could argue that, when taking a very abstract framework, the criteria chosen by single boundary seekers (e.g., that which enacts organizational closure) are too far removed from actual scientific practice (e.g., the purported boundary is hard to trace and intervene on). Although many biological examples can be interpreted a posteriori under the conceptual frameworks of the autonomy school, the opposite direction (i.e., undoing abstraction and kickstarting empirical research) has been hitherto relatively rare. How could we trace the organism-environment boundary of a developing multicellular organism if we construe it as that which enacts organizational closure? As a consequence of the trade-off, one has to be willing to pay for gaining generality regarding all living systems, and the abstractness of the accounts of autonomy defenders collapses several limits a multicellular organism could possess with its environment into a single one. This, in turn, might mask interesting organism-environment dynamics and active, qualitative changes in the bounds different organisms set with their environments (all subsumed under the generic heading of "organizational closure") that could be studied by scientists. For example, alluding to closure in every instance of organism-environment couplings in the life cycle of a token monarch butterfly might not say something meaningful about how the individual organism experiences radically different environments, or changes physiology and ecological interactions, and so on, from one stage to the next: It might allow us to tackle important problems such as diachronic identity, but organizational continuity is not—and should not be posed as—the only thing that matters in these phenomena.[13]

In a different register, single concrete criteria also exhibit important limitations. For one thing, they have limited phylogenetic scope. The immune system, for example, cannot be the organism-environment boundary for all organisms. Immune systems work differently depending on the metazoan clade, and things get even more complicated outside the animal kingdom. One might grant that immune systems still count as organism-environment boundaries in some animals and plants, although functioning very differently, but perhaps the same cannot be claimed for unicellular eukaryotes. I am not suggesting that these organisms do not have immune responses; I am simply maintaining that their immune systems might not qualify as conclusive, policing boundaries with their surroundings.

Of course, one can choose to embrace a variety of single boundary seeking approaches that dispense with requirement 2 introduced at the beginning of section 4.2—that is, the generalizability of the boundary. This position might be called *single-criteria boundary pluralism*: organisms will have different boundaries with their surroundings depending on the clade they belong to, but they all still rely, in some form or another, on enacting a singular criterion. However, and more importantly, single criteria, even in a pluralist, clade-specific reading, only carve out a limited subset of abiotic and biotic factors (or blurry in-between factors) encompassed under the organismal environment; bring to mind the metaphysical disquisitions from sections 3.1 and 3.2.2. All of these elements are not usually parsed out by, say, immune systems or by membrane-like tissues.

This leads us to a point concerning the metaphysics of environments that is usually neglected. As we attested in chapter 3, the "environment" as a single relatum performs unificatory metaphysical work by segregating and condensing disparate external (biotic and abiotic) conditions affecting the features and functioning of any given organism, the counter relatum to which it is juxtaposed. Anything that is posited as *the* organism-environment boundary should be, in principle, capable of accounting for the separation of an organism from all of these heterogeneous factors—in particular, those that are causally relevant for the particular organism considered, as we saw with the desiderata of the previous chapter.

On top of the inability to account for the ontologically eclectic components encompassed by any organismal environment, any single concrete boundary criterion is not fulfilled throughout the ontogeny of multicellular organisms (e.g., the development of skin or the differentiation of the immune system are subsequent events to early cell divisions in a vertebrate embryo). It does not seem plausible to claim, for instance, that before the differentiation of the immune system or epithelia, a vertebrate organism did not have a limit (or a set of limits) with its environment in place. Evelyn Fox Keller, for instance, has argued that the single-cell stage in multicellular organisms (i.e., when the boundaries of cell and organism coincide) poses an interesting challenge for approaches like DST; she worries that, as ontogeny progresses, the boundary gets subsequently erased by DST frameworks (Fox Keller 2001, 302). With the efforts of single-criterion boundary seekers, an inverse scenario would be implicitly assumed: a diffuse or nonexistent organismal boundary with the environment (e.g., in early developmental stages) until the singled-out factor develops (e.g., an immune system or the skin) to ensure that the only bona fide "boundary" is functioning properly.

Taking these points into account, along with the other criticisms made to boundary skeptics and boundary pragmatists, in the next section I develop a novel position in this long-standing debate in the history of biology, one that defends shifting organism-environment boundaries and integrates life cycle variation.

4.6 Introducing the Standpoint of Shifting Boundaries Defenders

The projects of past and contemporary boundary seekers are indeed very valuable—in both their abstract and concrete formulations—but I consider that it is necessary to renegotiate our understanding of organism-environment separation, echoing the spirit, but not the direction, of the philosophical project of Meyer-Abich (section 4.4). I consider that there are two complementary roads to do so, hinted at by the objections posed in the previous pages and the view of the organismal environment that emerged from chapter 3: (1) Its epistemic advantages notwithstanding (see section 3.3), the dichotomous framing of the organism-environment pairing misdirects us into thinking that *there must be a single boundary* between these relata or no separation at all. However, given the ontological heterogeneity subtending the organismal environment concept, we must grant that, for most multicellular organisms, there must be *several boundaries* at particular ontogenetic stages, and not a singular boundary that can be carved out through a single criterion. (2) We must pay attention to the dynamical character of boundaries throughout a life cycle. Moreover, as we learn by recounting the arguments of boundary skeptics, we should be mindful of the manifold roles played by environmental factors and contexts in development. At the same time, as boundary pragmatists have stressed, it is also important that our proposals are not extricated from scientific practices and the actual explanatory projects of biologists. Ideally, they could even inspire scientific work or open new questions for inquiry. So any proposal that could advance the discussion on organism-environment separation should aim at integrating roads 1 and 2. This endeavor should retrieve some of the heuristic benefits intrinsic to the practice-informed, contextually grounded viewpoint advocated within boundary pragmatism. Simultaneously, it should attempt to articulate a stance regarding the ontology of the manifold kinds of organism-environment interactions that have evolved in diverse lineages and how organisms self-individuate in development to become distinct from that which environs them, as authors from the Organicist Movement such as Child or Montévil and Mossio (2020) in recent times have stressed. In a nutshell, we need a position that integrates the valuable insights that many scientists and philosophers that have been engaged with the problem of organism-environment separation have called attention to.

In the past, other scholars have emphasized the dynamic character of organismal boundaries. For example, philosopher Julio Tuma maintained, "It is a remarkable and sobering regularity of nature that organisms remain whole because their *parts are constantly being built up and broken down, and this is no less true of their boundaries*" (Tuma 2009, 289; emphasis added). I want to expand and refine some of these previous insights in what follows.

According to the vantage point of *shifting boundaries defenders*, as an alternative to espousing a unitary or an overtly blurred organism-environment boundary, different boundaries and relations with the environment could be instantiated in distinct moments of the life cycles of organisms. By positing a single organism-environment boundary, the burden of proof for the adequacy of any single criterion becomes very heavy. This is another reason why boundary skeptics correctly indicate that it is not possible to identify a sole, definitive boundary for multicellular organisms in their environmental settings. Nevertheless, once we recognize that *multiple* boundaries with the environment could be instantiated in a developing organism, both synchronically and diachronically throughout its ontogeny, the challenge of boundary skeptics becomes weaker as well: from the fact that no all-encompassing, single organism-environment boundary exists, it does not follow that organism and environment form an inextricable whole. Organism-environment separation, then, is not an ontological problem that can be solved by an appeal to a single boundary or the complete absence of one. The key starting thesis of my position is the following: *In the ontogeny of multicellular organisms, separation from environing conditions is achieved through multiple synchronically instantiated boundaries to different biotic and abiotic factors that also might change diachronically throughout life cycles* (e.g., through organogenesis, developmental processes such as metamorphosis, or by what counts as an environment in a particular ontogenetic moment due to its indexical moving reference).

To illustrate the first point of the multiboundary nature of organism-environment separation, think of a very familiar case: an adult *Homo sapiens*. In our species, the immune system is a boundary for nonimmunotolerated entities (e.g., bacteria and allergens); the respiratory mucosa is a boundary that filters and traps pollutants of the air; the gastrointestinal tract does its fair share of filtration of ingested biotic and abiotic components; the skin counteracts sun rays and water exposure and minimizes the ingress of microorganisms and toxins; the kidneys take part in the control of fluid osmolality, various electrolyte concentrations, and removal of toxins; the excretory system at large ensures the removal of manifold materials from body fluids; and a lot more. This simple example also allows us to introduce the importance of road 2—namely, the dynamic character of organismal boundaries throughout life cycles. A human organism does not have the same boundaries at all stages of its life: a zygote has a cell membrane but certainly not an epidermis or excretory systems; a late blastocyst has a trophoblast but no single delimiting membrane; a gastrulating human embryo has, for instance, an allantois that exchanges gases and handles liquid waste, a yolk sac that mediates blood supply, and an amnion in contact with amniotic fluid; an embryo undergoing organogenesis begins to exhibit some of the aforementioned "textbook boundaries" in different degrees of functioning (e.g., immune, renal and excretory systems), and so on in subsequent developmental stages. What constitutes the boundaries of a human organism and its surroundings—which also vary drastically, beginning with the womb as a temporary environment—changes throughout ontogenetic time.

Outside the recognizable human case, what counts as a particular boundary for particular environmental factors at a particular moment of a life cycle will depend on the organism under consideration and what the environment *is* for said organism; bring to mind the discussion on organismal environment as an indexical class-as-many from chapter 3 (section 3.2.4). The stance of shifting boundaries defenders grants that organismal boundaries themselves have evolved throughout evolutionary history and have diverged into different clades of critters that have faced changing surrounding conditions. Thus, we need a life cycle perspective, devoid of "adultocentrism" in the sense criticized by Minelli (2003), to grasp the dynamicity of organism-environment boundaries in developmental trajectories. Against this background, the proposal of multiple shifting boundaries is an ontogenetically and phylogenetically informed perspective that considers life cycle variation (see Fusco and Minelli 2019). Besides conceding the multiplicity of synchronic boundaries, a central tenet for my account is that an organism's boundaries with its environment shift and are constantly renegotiated: They are adjusted, change their permeability, break down, and might be formed anew throughout the organism's ontogeny.

Let's keep adding more flesh to the bones of the position by means of additional examples. Take, for instance, some cases of metamorphosis. After a fairly standard development that begins with equal spiral cleavage, the embryo of some ribbon worm species (phylum Nemertea) turns into a pilidium, one of the most peculiar planktotrophic larval types among marine invertebrates. The juvenile forms inside the body of the larva from a series of isolated rudiments located close to the stomach, called imaginal discs. The next stage in the life cycle is what scientists have dubbed a "catastrophic metamorphosis": the juvenile breaks through the larval tissue, puncturing it completely, and then proceeds to devour the larval remains (Maslakova 2010). What happens with organism-environment limits during the coexistence of the budding juvenile and the larva? What happens with them after the juvenile breaks through the amniotic membrane and the larval epidermis and moves freely? Pondering on the nemertean case, the famous zoologist Julian Huxley suggested that "the whole *raison d' être* of a metamorphosis is the restriction of the animal to one environment in one period of its life, to another and a wholly different environment in another period. *Different environments require different structures;* and the metamorphosis is the time when the old structures are destroyed" (Huxley 1912, 77; emphasis added). Do some boundaries also get destroyed in the larvae and new ones arise with juveniles?

It might be plausible to suggest that, in nemertean metamorphosis, a shift in organism-environment boundaries takes place in this transition. For example, the outer tissues that interact with water are very different and are related to distinct modes of locomotion: the larval epidermis is composed of multiciliate epithelial cells—including the prototroch, a distinct preoral equatorial ciliary band involved in larval swimming that is lost after metamorphosis—while in the juvenile and the adult the outermost layer of the body consists of a ciliated and glandular epithelium containing rhabdites, which deposit the mucus in which the organisms glide on the benthos. Additionally, after metamorphosis, the feeding habits of nemerteans, their environments, and the affordances of what can be acted on as prey are drastically altered: a planktonic microalgal grazer (the pilidium) becomes a benthic predatory juvenile that stays that way until adulthood. Juveniles come into contact with nutritious large animal prey (e.g., annelids, clams, crustaceans, and other invertebrates) in their sediment-ridden benthic environments through an extendable proboscis enclosed in a rhynchocoel (a long fluid-filled cavity that extends longitudinally for most of the length of the gut); in contrast, the pilidium larva employs very different structures—which are dispensed with during catastrophic metamorphosis—to feed on much smaller prey (e.g., planktonic unicellular organisms) that thrive in the water column through an external ciliation mechanism generated by a current that carries suspended prey past the primary ciliated band. Subsequently, the larva conducts coordinated deformations of its body to redirect captured cells into a vestibular space between the lappets, resulting in these being transported by currents within a buccal funnel toward the stomach entrance (see von Dassow et al. 2013).

These are just some examples, although the exact nature of all the boundaries of nemerteans, as well as the patterns of organism-environment interactions sustained throughout their ontogenies, remains an exciting puzzle that could be taken up by scientists. Moreover, this scenario of shifting organism-environment boundaries could also be extended to all holometabolous insects that undergo metamorphosis from imaginal discs.

In general, it seems reasonable to suggest, as a hypothesis that should be put to the test, that drastic changes in environment (such as those afforded by metamorphosis and other developmental transformations in complex life cycles) might require the establishment of new organism-environment boundaries. For example, the indexicality of the environments of digenean trematodes I discussed in chapter 3 (section 3.2.3) is a good example of a complex life cycle with drastic environmental shifts. As with the case of the pilidium larvae and the nemertean juveniles, different stages in the life cycle of digenean trematodes (e.g., embryos, miracidia, sporocysts, radiae, cercariae, and metacercaria), which take place in contrasting environing conditions—from bodies of water to the squashy interiors of vertebrates—exhibit variable organism-environment boundaries.

The existence of shifting boundaries might also extend across generations, as in the case of multigenerational life cycles in animals and the alternation of generations in plants. Land plants exhibit a haplodiplontic life cycle, which means they alternate between haploid and diploid generations (Willson 1981). Gametes develop into a multicellular haploid gametophyte and, subsequently, fertilization paves the way for a multicellular diploid sporophyte, which produces haploid spores via meiosis. Across generations, the environments encountered by plants from different groups might also be drastically different, so it is not implausible to suggest that different organism-environment boundaries might be required and instantiated in the concrete ontogenies of token plants. Significantly for the stance of shifting boundaries defenders is that complex life cycles with multiple environmental shifts are far from being the exception in the tree of life. For example, the vast majority of known animal species have complex life cycles whereby metamorphosis mediates a rapid, drastic change in physiology, morphology, and ecology (for discussion, see Wilbur 1980; Fusco and Minelli 2019).

Another line of evidence to suggest that organism-environment boundaries might shift throughout ontogeny is that, according to mounting evidence across animal taxa, there is differential sensitivity to environmental stressors during life stage transitions such as a metamorphosis (Lowe et al. 2021). This differential vulnerability to environmental fluctuations extends to whole life stages as well (for discussion, see Bateson 1979).

A different swathe of biological phenomena might also suggest that shifts in the "permeability" of organism-environment boundaries are also possible, but this time, by reducing the impact of environmental influences. "Dormancy" is an umbrella term for pervasive manifestations in the tree of life that capture periods in an organism's life cycle when growth, development, and metabolic activity temporarily come to a halt or are considerably reduced. The dormant condition is usually triggered by (actual or anticipated) harsh environmental conditions, such as nutrient scarcity, predator presence, or extreme temperatures. This is a very common phenomenon in many types of seeds that interrupt their germination until the conditions are ripe for resuming development. This also happens with embryos from many species of metazoans, including insects, rotifers, brine shrimps, copepods, and many fish species (see García-Roger et al. 2019 and references therein). In general, dormancy is manifested in postembryonic life stages as diapause, estivation, hibernation, or brumation in the animal kingdom. With dormancy, it is as if an organism, at whatever stage of its life cycle it finds itself in, becomes impervious to environmental factors, narrowing, reshaping or shutting off its boundaries. Achieving dormancy suggests that the organism has successfully managed to limit its interactions with the environment, sometimes by creating insulating layers (e.g., in mammalian hibernation), by altering its general structure (e.g., in sclerotial metamorphosis in filamentous fungi), or by constructing physical buffers (e.g., in lungfish estivation), among other strategies to reduce the permeability and causal influence of environmental factors (for recent overviews, see Tarrant 2019; Wilsterman et al. 2021).

Combining these examples, one might suggest that the dormant state is also another line of evidence for the general claim that organism-environment boundaries might experience shifts during ontogeny, at least during three instances: when entering the dormant state, for the duration of dormancy, and when exiting this phase. These shifts in organism-environment boundaries are quite specific to the environmental conditions encountered, and, on occasions, signal a differential loss of permeability to single stimuli. A spectacular case of this selective loss of permeability is cryptobiosis as it happens in tardigrades, which comes in different varieties depending on the adverse environmental condition faced (e.g., anhydrobiosis, anoxybiosis, chemobiosis, cryobiosis, and osmobiosis).

In sum, evidence from research on metamorphosis, complex life cycles in multiple clades (including multigenerational life cycles), differential sensitivity to environmental stressors during life stage transitions and particular life cycle stages, and dormancy (broadly construed) suggest that organism-environment boundaries might be altered and reshaped, and some might be built anew in the course of development.

4.6.1 Advantages of Being a Shifting Boundaries Defender

Now we could ask: Could the shifting boundaries view just be a minor reformulation of former approaches that, after scrutiny, appears to be like another position at its core? In the foregoing discussion, I argued that the rationale for adopting a multiple boundaries view for multicellular developing organisms is twofold: (1) Given the ontological heterogeneity of the biotic and abiotic elements that comprise the environment of a token organism at a particular point of its life cycle (cf. discussions in chapter 3), the construal of a single boundary is inadequate, as it poses a too heavy ontological load for any entity or process to be able to qualify as the singular boundary that sets an organism apart from its surroundings; rather, I think, there is enough evidence to suggest that different environmental components are parsed out by different boundaries. And (2) if we take seriously the ontogeny of organisms (especially those with complex life cycles), we can notice that, throughout development, new structures arise with different mediating capacities and differential, nonstatic permeability with particular environmental components (and what counts as environment in certain stages of the life cycle changes as well due to the phenomenon of indexicality we have been fleshing out in this book). To track the changes in boundaries, and how these are formed, dissolved, and altered throughout developmental time, biologists would certainly rely on different criteria that are specific to the structures and developmental processes under study. In that sense, my position does not collapse into "single boundary seeking" with one univocal criterion or metacriterion that would be able to fully adjudicate the separation between organism and environment. Moreover, I am suggesting that there is no deeper time-invariant criterion to track organism-environment separation, but ontogenetically and phylogenetically variable criteria that nevertheless can be fruitfully investigated by biologists. This is, in a sense, closer to boundary pragmatism than to single boundary seeking. If biologists want to tackle the developmental processes through which organism-environment separation is instantiated in the ontogeny of particular critters, they could resort to some features of the developmental systems under study, thereby extracting several metacriteria to guide their inquiries that are sensitive to periods in life cycles, changes in physiology, the action of fluctuating environmental factors, and so on. Even though being a shifting boundaries defender might be experimentally challenging, this stance cannot be translated into boundary skepticism, as organism-environment separation is not disavowed both synchronically and diachronically in the ontogeny of organisms. After all, I consider that organisms self-individuate during their development, and they are distinctive entities compared to the plurality of conditions we call their "environments."

Likewise, my position does not collapse into boundary pragmatism because my argument is that those variable time-sensitive criteria to track organism-environment separation—which are specific to the lineages of the organisms under consideration—are not necessarily recovered by the heterogeneous explanatory aims of scientists that work with token organisms—although there could indeed be many overlaps. The problem of organism-environment separation in development thus cannot be solved only by looking at individuation practices in the biological sciences and assuming epistemic pluralism about boundaries, as boundary pragmatists assert. In a way, my proposal attempts to renegotiate our understanding of the separation between token developing organisms and their surrounding conditions, so that places it well within the general spirit of boundary renegotiators like Meyer-Abich. However, I reject his radical reframing of the problem (i.e., the whole organism acting as the mediating interface or boundary between internal and external worlds) because I think it falls into a trap of an unsuitably ontological construal of an environment-as-a-whole (which I criticized in section 3.2). We can still show how different developing structures act as boundaries for particular environmental factors, even when this happens throughout the dynamic unfolding of ontogenetic contexts. My proposal is a renegotiation of the terms in which organism-environment separation is cast, pluralizing them and allowing for their diachronic dynamicity to come into the picture, and an outright rejection of the idea that a single structure or process—be it a concrete boundary like the immune system or the skin or a holistic one such as an organism-as-a-whole—is able to instantiate and enact said separation during ontogeny. This also sets shifting boundaries defenders apart from boundary renegotiators.

What advantages are afforded by the shifting boundaries perspective? This stance is not completely detached from former positions, as it builds from and reconciles important elements of them: ranging from the sensitivity and embeddedness of developing organisms in their environments to the phylogenetic diversity of life cycle variation and different organism-environment interactions to retaining the actual practices of biologists (e.g., identification of life cycle stages and openness to experimental interventions), suggesting new lines of inquiry and pushing forward for a more adequate and renegotiated understanding of the subtending ontology of the organism-environment relationship during ontogeny within a breadth of evolutionary scenarios. An issue of special importance is that this position does not recant the project, which was very much alive during the Organicist Movement as we saw in previous sections, of finding out how organisms individuate themselves to become ontologically distinct from their surroundings. Due to its importance for scientific practices in a wide range of biological disciplines that study token developing organisms, the shifting boundaries account keeps organisms and environments separate, as advocated by single boundary seekers; however, it remains sensitive to the deep bonds of organisms with their surroundings and to the significant reciprocal interactions between them that span entire life cycles, as authors that uphold boundary skepticism would emphasize.

The account of shifting organism-environment boundaries, while being mindful of the advances in the biological sciences and being tethered in them, is still abstract enough—but without falling into extremes—to pursue the generalizing goals of an operational conception of organism (*qua individual* distinct from its environment) that metaphysicians and philosophers of science have recently argued is needed (see Guay and Pradeu 2020, 2).

As I have stressed, the specific developmental and physiological causal processes through which organism-environment boundaries become instantiated in particular slices of the life cycle of any particular organism might be, of course, variable among taxa (and this remains a stimulating topic that biologists could investigate), but the dynamicity and multiplicity of organismal limits with environing surroundings stills holds as a projectible quality that I want to highlight. In that sense, even though the proposal advanced here lacks the concrete specificities of what are acting as shifting organism-environment boundaries in most of the creatures alluded to, it still bears weight as a good starting point for theorizing and collecting more empirical evidence. The shifting boundaries view is both an ontological position and a theoretical hypothesis that might be supported or undermined by further empirical investigations and testing.

What I could add here is that, even if there would be strong grounds to contend that a particular species or broader taxon has a constant, single organism-environment boundary, that scenario would, nonetheless, still count as a special evolved case of the general phenomenon of shifting boundaries in organisms. A similar thing happens with phenotypic plasticity as a developmental phenomenon: just as the presence of flat, unresponsive norms of reaction to changes in environmental conditions—as canalized, evolved scenarios—does not invalidate the widespread reach of phenotypic plasticity in organisms, a case of constant boundaries or even a single static boundary throughout a monogenerational life cycle won't disprove the varying character of organism-environment boundaries in the majority of multicellular critters. I argue, on the contrary, that a case of this phenomenon precisely might be worth investigating further. For instance, under which conditions can a constant, single organism-environment boundary evolve, and through which cellular and eco-physiological causal regimes might it be sustained and enacted?

In contrast, if scientists and philosophers alike presuppose that a (presumed) single organism-environment boundary will remain in place throughout the life cycle of an organism (or that there is indeed a single, definite boundary), this might prevent them from investigating some salient phenomena pertaining to organism-environment interactions and how the separation between these relata is achieved in development (e.g., how a living system continuously negotiates its existence and limits with the world that environs it, or how it overcomes adverse conditions by dynamically reshaping its individuality, or what counts as a self-individuation process in the first place).

Finally, the shifting boundaries stance also invites us to be more precise about what is circumscribed under the term "environment," as a singular metaphysical category that is juxtaposed to that of "organism," in particular ontogenies. This forges a strong connection with the metaphysical examination of environments that I presented in the previous chapter.

4.6.2 Recapitulation

Distinguishing what enacts the separation between organism and environment throughout development is no easy task. Tuma argued that "there is already a great deal of interactional complexity . . . involved in boundary formation prior to taking into consideration the layers of complexity involved in any epistemic or theoretical understanding of these boundaries" (Tuma 2009, 288). Boundary drawing exercises in biology, as Haber and Odenbaugh (2009) have contended, demand "revisions as new evidence and hypotheses are discovered or adopted" (219). With this chapter, I hope to have marshalled the necessary critical reflexivity with respect to the general discussion on organism-environment separation in the history and philosophy of biology, inviting appropriate revisions and renegotiations in our theoretical understanding of this problem.

I offered a historically informed taxonomy of four distinct positions that can be identified with respect to the issue of organism-environment separation in ontogeny: boundary skeptics, boundary pragmatists, single boundary seekers, and boundary renegotiators. Most of these positions were advanced and defended by holistic and organicist biologists in the first decades of the twentieth century. By appealing to developmental and physiological evidence, boundary skeptics reject the distinction between organism and environment due to tight codetermination and embrace a view of inseparable organism-environment systems; boundary pragmatists argue for a heuristic, contextual role of "individuation," which closely follows scientific practices and the explanatory projects that are ultimately pursued; single boundary seekers postulate specific and comprehensive single criteria to individuate organisms from their surroundings; and boundary renegotiators advance proposals to reframe what is "internal" and what is "external" for an organism. All of these positions recover important elements from the issue at stake, but all suffer from shortcomings as comprehensive answers to the problem of organism-environment separation; due to this, I argued, it is necessary to advance a new view that integrates their valuable insights but expands them in important ways. I introduced the stance of *shifting boundaries defenders* by appealing to phylogenetically and ontogenetically diverse case studies. According to this stance, there is no single, definite organism-environment boundary but *multiple boundaries* to ontologically heterogeneous environmental factors, and these shift and are renegotiated throughout life cycles: they are adjusted, break down, and change their permeability, and new boundaries might be formed in the ontogeny of organisms. This approach is empirically grounded in biological knowledge of development, and it is potentially testable by scientific work and open to being refined and expanded by theoretically minded scholars. Being a shifting boundaries defender affords accommodating several advantages of former positions while opening new lines of inquiry into the breadth of phylogenetically distinct ways of how organisms self-individuate from their environing surroundings and renegotiate their limits.

After having covered the ontological and epistemological underpinnings of organismal environments (chapter 3) and plunged into the thorny subject of organism-environment separation, in the next chapter I will focus on another important dimension that needs to be explicated and historically contextualized in order to advance our understanding of the organism-environment relationship: reciprocity.

5 Of Reciprocity or the Contested Symmetry of the Organism-Environment Relationship

> The effect of taking the dialectic between organism and environment seriously is, if anything, even more profound for evolutionary biology than for studies of development.
>
> —American evolutionary biologist Richard Lewontin (2001, 56)

5.1 Organism and Environment Are to Be Found on a Two-Way Street

Typically, biologists assume that the organism-environment nexus is bi-directional, a result of mutual influences—in short, reciprocal. In their famous textbook on animal ecology, for instance, W. C. Allee, Orlando Park, Alfred E. Emerson, Thomas Park, and Karl P. Schmidt highlighted the widespread reach of organism-environment reciprocity and the importance of studying it as part of the overarching agenda of biological science:

> *The reciprocal relations require especial attention. The interaction of the environment and the organism is obvious in almost every field of biology.* Physiological processes are correlated primarily or secondarily with environmental fluctuations: energy for life is derived from the environment; growth and development show relationship to environmental factors; environmental forces and substances impinge upon the sense organs of animals and the reactive systems of plants; behavior patterns in large part are responses to environmental patterns; distribution of plants and animals is determined by variations in the environmental complex; . . . and the environment has acted as a selective agent in determining the survival of organisms and populations. (Allee et al. 1949, 1; emphasis added)

These reciprocal interactions are not exclusive of animals and plants: organisms from all taxonomic groups are said to interact in a reciprocal manner with their environments, and these two-sided processes have important outcomes in physiology, development, and evolution. However, on top of this widespread recognition, key questions linger: If organism and environment are said to stand in a relation of reciprocity, what is entailed by this claim? What is the most suitable construal of organism-environment reciprocity for biological science? Is the relationship between organism and environment wholly symmetrical, or are there some particular characteristics of these relata that break the symmetry of the pairing? In this chapter, I broach these questions and assess different understandings of organism-environment reciprocity in past and present debates across the life sciences.

My first objective is to systematize divergent positions on organism-environment reciprocity and to develop a taxonomy of different formulations of this notion (section 5.2). As part of my historical analysis, I will show that organism-environment reciprocity, and the perspectives it opens for approaching biological problems, once popular in early-twentieth-century research, fell out of favor in postwar biology (section 5.3). Only in recent years has this notion resurfaced as a salient issue for philosophical and scientific inquiries (section 5.4). Unlike the preceding chapters, the upcoming pages will place greater emphasis on evolutionary considerations in combination with the overarching developmental viewpoint that I have thus far been articulating in this book. The reason for this shift of emphasis, as we will come to see, is that biologists have long held that the reciprocal interplay between organisms and their environment plays a crucial role in shaping the course of evolutionary processes.

At the outset of this discussion, it must be said that the subject of "reciprocity" has been theorized in many ways throughout the history of biology. First, a diversity of relata have been conceptualized as engaged in reciprocal associations: for instance, organism-environment, gene-environment, gene-population, and population-population. From that perspective, it would be a mistake to think that the notion of reciprocity is limited to the dynamic braid between an organism and its environment: reciprocal connections and dependencies in the biological world extend beyond the organism-environment relationship (see, e.g., Reznick 2013; Ware et al. 2019; Pausas and Bond 2022). Second, different kinds of reciprocal relationships have been posited for these varied relata. As we will see in detail in this chapter, for the organism-environment relationship, biologists have postulated (1) ontological coconstitution, (2) mutual structural fitting, and (3) reciprocal causation, among others, as different hallmarks of reciprocity.[1] In pursuance of the first objective of this chapter, I will expound how these three ideas were interpreted and advanced by organicist and holistic biologists in the first decades of the twentieth century (section 5.2). Historian Georg Toepfer maintains that reciprocity is "the relationship of relatedness of parts of a system to each other; it may consist of mutual influence (interaction) or mutual dependence (interdependence) (through mutual production or maintenance)" (Toepfer 2011b, 738; my translation). If we follow Toepfer's general framing, both (1) ontological coconstitution and (2) mutual structural fitting correspond to two different views of reciprocity based on the postulation of *interdependence* of the relata in question. In contrast, (3) reciprocal causation posits *mutual interaction* between organism and environment.

I focus on these three different construals of reciprocity, which are by no means the only theoretical possibilities (e.g., "concomitant reaction"), because they were actively endorsed and debated by organicist and holistic biologists; other views of organism-environment reciprocity that were sanctioned by philosophers and psychologists (e.g., the pragmatist view of "transaction") will not be covered in what follows.[2]

My second general objective for this chapter is to evaluate the underlying differences in how this key notion is understood and to argue that *reciprocal causation* is the most appropriate way to construe organism-environment reciprocity, both ontologically and epistemically (section 5.4).

In what follows, I start by contextualizing the biological debate on the reciprocity between organisms and their environments with a historical overview of the three different accounts of organism-environment reciprocity (1–3). My analysis then turns to the gradual eclipse of these accounts during the latter half of the twentieth century (section 5.3), only to witness their resurgence within the intellectual currents of recent decades (section 5.4). Finally, I will present my original contribution on how to come to grips with organism-environment reciprocity (section 5.5).

5.2 Different Understandings of Organism-Environment Reciprocity: A Historical Survey

5.2.1 Organism-Environment Reciprocity in Historical Focus

There is a long historical pedigree of ideas underscoring organism-environment reciprocity (see, among others, Müller 1994; Toepfer 2011b, 738–763; Baedke 2019; Baedke et al. 2021; Prieto and Fábregas-Tejeda

2022). Svensson (2018) has attributed the idea of causal reciprocity and the view of organisms as active modifiers of their environmental conditions to Levins and Lewontin's (1985) influential book. However, as we saw in section 4.3.2, the tradition of dialectical biology, with self-identified members such as Marcel Prenant, precedes Levins and Lewontin by several decades. And even more importantly, ideas on organism-environment reciprocity have a longer theoretical history in biological science.

The German zoologist Ernst Haeckel serves as an excellent illustration of this long-standing tradition. In his overview of zoology, Haeckel introduced his popular concepts of "ontogeny," "phylogeny" and "ecology" and considered both the "static" and the "dynamic" aspects of animal life (Haeckel 1866a, 11, 238; for discussion, see Levit and Hossfeld 2019). Haeckel grasped a key message from Darwin's *Origin of Species*, which emphasized the importance of considering the link between organisms and their (biotic) surroundings in determining the positive contributions of specific traits to survival and reproduction (see also section 2.2.2). By recognizing the interplay between organism and environment, Haeckel argued, naturalists can be in a good position to understand the advantages and disadvantages conferred by particular biological features. Along these lines, as early as 1866, Haeckel asserted that "reciprocity of every single individual with its entire surrounding leads to the adaptation of its individual characters" (Haeckel 1866a, 154). Haeckel referred to "ecology (*Oekologie*)" as the system embracing development, evolution, and environment (see Watts et al. 2019). In his *Generelle Morphologie der Organismen*, he described it as "the entire science of the *relations of the organism to the surrounding outside world*, to which we can count all 'conditions of existence' in a broader sense. These are partly organic, partly inorganic in nature" (Haeckel 1866b, 286; emphasis added; my translation). The organism-environment pairing thus took the center stage in the Haeckelian understanding of ecology, and he even stressed that, first and foremost, ecology is the study of the reciprocal relationship between organisms and their environment (see also Haeckel 1868, 539; Haeckel 1870, 365). Therefore, reciprocity was posed as a foundational benchmark of ecological science, and many authors have accentuated this theme ever since (e.g., Drude 1906, 186; Tucker 1953, 51). For instance, the influential textbook *Principles of Animal Ecology*, which I mentioned at the beginning of this chapter, has as its incipit the following assertion: "Ecology may be defined broadly as the science of the *interrelation between living organisms and their environment*, including both the physical and the biotic environments, and emphasizing interspecies as well as intraspecies relations" (Allee et al. 1949, 1; emphasis added).

While Haeckel did explicitly touch on the notion of organism-environment reciprocity, he did not do so in detail, for instance, developing a comprehensive theory on how this relation should be spelled out. However, other scholars during the late nineteenth and early twentieth centuries further explored and expounded on this concept. As we will see below, organism-centered biological perspectives, such as organicism, (German) holistic biology, and dialectical materialism, developed sophisticated, although divergent, views of organism-environment reciprocity (see also the discussion in section 2.4.2).

Esposito (2017) argues that, as a general baseline, organicists shared the theoretical tenet that organismal parts are "shaped and constituted in a *dynamic interaction with the whole organism and its environment*. These dynamic wholes have to be conceived as teleological and self-sustaining entities, able to adapt and change their form and behavior according to the external circumstances" (226; emphasis added). As one example of these commitments, in an anonymous *Nature* review of Przibram's *Experimental Zoologie*, the leitmotif of reciprocity clearly came to light: "*Living implies action and reaction between the organism and its environment*; the continuance of the creature's activity depends on its structural organisation and on its specific series of chemical processes" ["(1) Experimental Zoologie (2) Mechanism, Life and Personality" 1914, 193; emphasis added].

In that same article, another book was also reviewed: *Mechanism, Life and Personality*—a book penned by the holist J. S. Haldane. As mentioned in chapter 4 (section 4.3.1), Haldane, even in the early stages of his career, emphasized that the organism is "no more determined by the surroundings than it at the same time determines them. *The two stand to one another, not in the relation of cause and effect, but in that of reciprocity*" (Haldane 1884, 32–33; emphasis added). How should organism-environment reciprocity be understood? In the subsequent sections, I will survey how reciprocity was construed in the work of different organicist and holistic thinkers.

5.2.2 Ontological Coconstitution: A Holistic Conception

In section 4.3.1, I introduced the research background and holistic ideas of John Scott Haldane. I recounted that, informed by his empirical investigations into respiratory physiology, Haldane cast a discerning gaze on the quest of demarcating an organism from its environment. Central to this challenge was delineating where one entity starts and the other ends, given that both continuously contribute to the manifestation of life. According to Haldane, the immediate surroundings of an organism, much like organic structures themselves, are actively sustained, blurring a clear-cut distinction between these relata.

For this idea, Haldane found himself influenced by Claude Bernard's conceptual framework of regulation, particularly by his emphasis on the constancy of the "internal environment" (*milieu intérieur*; see Allen 1967). At that time, this notion served as a theoretical organizing principle for general physiology (Holmes 1963, 1986) and, in Haldane's view, held the potential to be extended to embrace the "immediate surroundings" of any organism. Inspired by Bernard's ideas, Haldane recognized the significance of maintaining a stable internal environment and saw parallels in the context of an organism's immediate external *milieu*. Haldane asserted, "If we look, not merely at the internal, but also at the external activities of an organism Claude Bernard's generalization seems still to hold. . . . Regulation of the external environment is in fact only the outward extension of regulation of the internal environment" (Haldane 1917, 81, 93). According to Douglas (1936), "Claude Bernard's idea of the constancy of the internal environment of the body became in Haldane's hands a far wider conception, embracing the ever-varying relations of cell or organism with environment and throwing into relief those subtle adaptive changes which are so characteristic of life" (137; see also Haldane 1929, 41, 57).

In Haldane's view, organismal environments (both internal and external) and organic structures are actively maintained as components of a larger, unitary living whole: "The life of an organism is ultimately just as much bound up with its external as with its internal environment" (Haldane 1931, 74). While Bernard's renowned assertion hypothesized that "the fixity of the internal environment is the condition of free, independent life" (Bernard 1878, 113; my translation), Haldane, on the other hand, posited the viewpoint that "the life of an organism must be regarded as an objective active unity which embraces its environment" (Haldane 1936, 149). In Haldane's perspective, the organismal environment was not an external entity separate from life; rather, it formed a constitutive component of it.

For Haldane, to reason through the lens of "reciprocity" meant to overhaul the ontological status of organismal environments. In his words, by "bringing the phenomena of life under the category of reciprocity . . . the surroundings of an organism are to be looked on, not merely as objects in ambient space which may act on, or else be acted on by, the organism, but as these objects in so far as they participate with other objects in forming with the organism a system whose parts reciprocally determine one another" (Haldane 1884, 34). Haldane regarded the environment as something inherently relational to particular token organisms and akin to the "effective environment," a distinction that we contrasted in chapter 3 (section 3.1) to that of the "external world," with the particularity that he paired it with the idea of "immediate surroundings." His conception of what the environment is could be fleshed out as the "immediate effective environment"—that is, those factors present in the immediate surroundings of an organism that are deeply entangled with it (physiologically and structurally) so as to constitute a new unitary whole that is the bearer and enactor of "life."

In support of his outlook, Haldane advocated for an ontological understanding of mutual constitution (*ontological coconstitution*, in short), wherein the organism and its environment intertwine and give rise to a unified, interdependent entity that defies meaningful separation. The life of an organism, covering its interactions with both traditionally considered "internal components" and "external surroundings," endures as an active, cohesive, and inseparable whole. This persistent whole is not merely a rigidly demarcated physical structure or the sheer functioning of said structure; within it, there exists no distinct boundary that sharply separates a living organism from its environment.

Precisely because of this, Haldane identified a communal fault line that traverses different bio-theoretical positions, from mechanistic perspectives to organicist and vitalistic ones: organism and environment are always treated as distinct and separate entities. He proclaimed, "Vitalism in any form has the same fundamental defect as the mechanistic theory of life. It assumes that a living organism and its environment can be separated in observation and thought, when they cannot be separated" (Haldane 1931, 31). And in a different passage of his work, we read, "The mistake of the vitalists, and of the essentially similar 'organicist' or 'organismal' biologists, was, as I have repeatedly tried to point out in previous writings, to separate a living organism from its environment" (Haldane 1936, 45–46; see also Haldane 1917, 112). Although an early popularizer of the term "organicism," as we saw in chapter 2 (section 2.4.1), Haldane became more and more wary of the positions of his organicist colleagues as the years passed.

Haldane postulated that if the actions of an organism are influenced by its environment, and conversely, the actions of the environment are influenced by the organism, then the conventional notion of reciprocal causal action fails to capture their tight integration (Haldane 1929, 12). The interwoven nature of these factors resists a clear separation and a framing in terms of causal chains. Likewise, he rejected the possibility of reducing the organism-environment pairing as a "mere mechanical relation" (Haldane 1936, 35). Hence, Haldane reasoned, an alternative form of relationship needs to be proposed. In broad, unqualified terms, he labeled this relationship "reciprocity," although it can be best comprehended by what I have christened here "ontological coconstitution."

In early-twentieth-century biology, holistic thinkers, including Haldane and other figures such as Jakob von Uexküll, found inspiration for their conceptions of reciprocity in the writings of Immanuel Kant. For instance, Uexküll peered into the organism-environment relationship using a concept known as *Einpassung*. This term, fashioned in direct reference to Darwin's notion of adaptation (*Anpassung*), encapsulates the idea of "adaptive insertion." The prefix *ein-* accentuates the fusion between the animal and its surroundings, a reciprocal interdependence that brooks no gradations (Brentari 2015, 78). In opposition to the Darwinian notion of adaptation, Uexküll considered that organisms are not adapted to their surrounding conditions but are *adaptively inserted* in them: natural selection-based evolution could not occur given that organisms fit perfectly in their settings, and there is a subtending harmony in the connections running in both directions. *Einpassung* is thus another example of a view of (holistic) ontological coconstitution, even if an antievolutionary one. These interpretations usually started from, and stretched, certain insights furnished by Kant.

In his *Kritik der Urteilskraft*, Kant described reciprocity as the distinctive organizational pattern that governs organismal development. According to him, an organism emerges and thrives as a unified whole through the reciprocal interactions of its constituent parts. In the prevailing view, linear trajectories and connections impose themselves, capturing "series (of cause and effect)" that are "always descending" (5:372; Kant [1790] 2000, 244), but when we turn our attention to the inner components of an organism, we find that these intertwine. They are bound together "into a whole by being *reciprocally* the cause and effect of their form" (5:373; Kant [1790] 2000, 245; emphasis added). This reciprocal connection, distinct from the linear model of causes preceding effects, imbues the organism with the qualities of a "self-organizing being" (5:374), setting it apart from the realm of inanimate objects (Kant [1790] 2000, 245). As philosophers Laura Nuño de la Rosa and Arantza Etxeberria put it, "Unlike what happens with the parts of a watch, [for Kant] the parts of an organism are, reciprocally, cause and effect of its form. Organisms are *self*-producing totalities that do not require any external agent, since their parts produce themselves reciprocally" (Nuño de la Rosa and Etxeberria 2010, 191; emphasis in original; my translation). Within the web of causes that make up an organism's existence, each individual part exhibits profound interconnectedness: There seem to be no isolated parts, and these are rather perceived "as existing for the sake of the others and on account of the whole" (5:373–374; Kant [1790] 2000, 245; for careful exegeses of Kant's view of reciprocity, see Friebe 2008; Toepfer 2011c). The puzzle is that organisms seem to involve a sui generis form of *efficient* causation that bears no analogy to any other form of causality that we, as humans, have encountered or engendered (e.g., the causality involved in the assemblage of a machine). As we will see in chapter 6 (sections 6.2.4), for Kant it was precisely this internal causal structure that gave rise to the teleological judgment of organisms.

Influenced by Kant, many biologists appealed to reciprocity to refer to organismal development. For instance, the organicist embryologist Oscar Hertwig claimed, "The ovum is an organism which multiplies itself by division into many organisms similar to itself, and . . . it is through the reciprocal action [*Wechselwirkung*] of all these many elementary organisms at each stage of the development that the organism as a whole is gradually and progressively established" (Hertwig 1894, 85; my translation). And in another work, he also underscored the interrelationship between parts and their overall dependence on the organismal whole: "The parts of an organism develop in relation to each other or the development of a part depends on the development of the whole" (Hertwig 1892, 480; my translation). As another example of this, in his book *The Unity of the Organism, or the Organismal Conception of Life*, the American organicist William Emerson Ritter (1919) also argued extensively and vouched for the idea that the parts of the organism are integrated in their action and work together to produce unified results in the organismal whole. Equally, E. S. Russell held that in organismal development "the activities of the parts work together for the good of the whole" (Russell 1930, 168), and Lawrence Henderson asserted that the organism is "an autonomous unit in which every part is functionally related to every other and exists as the servant of the whole" (Henderson 1917a, 21). As we have seen, organicist authors were in agreement on this point.

Over the course of time, and especially in early-twentieth-century biology, the Kantian view of reciprocity between the organismal whole and its parts was expanded and applied to different relata, such as organisms and their environments.[3] Indeed, many organicists and holists saw a resemblance between organic part-whole reciprocity and organism-environment reciprocity. Needham, for instance, noted, "Parts are not entities having an existence in their own right, but only by virtue of their position and function in the organism of which they form parts. This relatedness or going-togetherness is the hallmark of an organism, and it may be noted that it can interlace with other organisms, *just as a living organism and its environment is inextricably intertwined*" (Needham 1928b, 34; emphasis added). Along those lines, Haldane stressed, "The maintained co-ordination is present, just as much in the relations between organism and environment as in the relations between the parts of an organism itself" (Haldane 1936, 45–46), and then again, "We can distinguish the same co-ordination between specific structure and activity in the relations between organism and environment as in the relations between the parts of an organism, so that organic wholeness is just as manifest in the relations between organism and environment" (Haldane 1927, 909). For many holistic thinkers, reciprocity ran through the dynamics of organismal parts and in the engagements that organisms have with their surroundings: "Every part of the organism must be conceived as actually or potentially acting on and being acted on by other parts of the environment, so as to form with them a self-conserving system" (Haldane 1884, 45). In a concise pronouncement,

Haldane declared, "An organism and its environment are one, just as the parts and activities of the organism are one" (Haldane 1917, 99).

Throughout the history of biology, this perspective of what reciprocity entails (i.e., a profound reciprocal coupling between the organism and its environment in the context of wholeness) emerged repeatedly in the wake of the Organicist Movement. However, as noted by Baedke (2019, 307), this shift in attributions of reciprocity also presented a challenge in determining the demarcation between the organism and its embedding environment. If holistic reciprocity is accepted, then the distinction between these entities becomes indescribable, leaving open a complex question of what precisely delineates their separation.

Similarly, we can see this poignant issue in a quote by Georges Canguilhem pondering on the link between the living and its milieu:

> *The relationship between the organism and the environment is the same as that between the parts and the whole of an organism.* The individuality of the living does not stop at its ectodermic borders any more than it begins at the cell. The biological relationship between the being and its milieu is a functional relationship, and thereby a mobile one; its terms successively exchange roles. The cell is a milieu for intracellular elements; it itself lives in an interior milieu, which is sometimes on the scale of the organ and sometimes of the organism; the organism itself lives in a milieu that, in a certain fashion, is to the organism what the organism is to its components. (Canguilhem [1952] 2008, 111)

We are now in a better position to understand why in Haldane's holistic viewpoint it is impossible "to distinguish separately the factors concerned," and life is regarded as an "integrated unity covering both organism and its environment" (Haldane 1935, 12). Ontological coconstitution is the outcome of embracing a Kantian understanding of reciprocity and extrapolating it to a new domain (i.e., the organism-environment relationship) under a Bernardian framework of constant regulatory changes aimed at the maintenance of the whole.

Nonetheless, as we will see, this way of approaching organism-environment reciprocity was not the only contender in early-twentieth-century biology. In the next subsection, I present a different understanding of reciprocity: mutual structural fitting.

5.2.3 Mutual Structural Fitting or the Fitness of the Environment

> The most recent biological science studies the living organism as a unit composed of the body and its particular environment: so that the vital process consists not only in an adaptation of the body to its environment, but also in the adaptation of the environment to its body.
> —Spanish philosopher and essayist José Ortega y Gasset (1918, 43; my translation).

Many authors of the Organicist Movement drew from physiological work to advance their claims, Haldane not being an exception on this front. At the same time, they complained about the prevalent lack of proper attention to the organism-environment bond in the discipline. For instance, theoretical biologist Joseph Henry Woodger protested that up until then

> [physiology] has been compelled to consider each organ separately, and has been so engrossed in its task that the organism as a whole has been lost sight of, and, in consequence, the very existence of a relation between the *organism as a whole* and the environment has easily been forgotten. To the physiologist the environment is nothing more than a place between which the organism exchanges of energy and material take place. (Woodger 1925, 675; emphasis in original)

In a retrospective assessment of early-twentieth-century physiology, Robert Perlman concurs with this limitation, which should also be understood as part of Bernard's justification for the autonomy of experimental physiology (Coleman 1985): "Concern with the internal workings of organisms has led physiologists to focus on the maintenance of homeostasis in the internal environment rather than on the interactions of organisms with their external environments" (Perlman 2000, 174). An exception to this state of affairs of environmental obliviousness was Lawrence Joseph Henderson, a leading American physiologist of the interwar generation.

Henderson, a distinguished figure at Harvard University, held the prestigious positions of Abbot and James Lawrence Professor of Chemistry, as well as chairman of the Society of Fellows. He obtained a bachelor's degree in biological chemistry at Harvard in 1898 and pursued further studies at Harvard Medical School, obtaining his MD in 1902. Seeking to expand his scientific horizons, Henderson conducted research in Franz Hofmeister's physicochemistry laboratory in Strassburg from 1902 to 1904. His return to Harvard in 1904 marked a turning point in his career as he assumed the role of lecturer in biological chemistry (for biographical sketches, see Ferry

1942; Parascandola 1971, 67–111; Fry 1996). It was during this time that he decided to investigate the phenomenon of acid-base equilibrium. Thus, the following decade he investigated the causal processes governing neutrality regulation within organisms through the theory of weak acid dissociation (Henderson 1913). One of his most celebrated contributions is the so-called Henderson-Hasselbalch equation for evaluating the buffering properties of weak acids and bases from their dissociation constant. Some of Henderson's endeavors led to the publication of the influential *Blood: A Study in General Physiology* (Henderson 1928). In 1920, he established the Laboratory of Physical Chemistry at the Medical School.

Like Haldane, Henderson drew inspiration from Bernard's exploration of the constancy of the internal environment to solidify his ideas about organismal regulation (Virtanen 1960, 84–85; Parascandola 1971, 90).[4] Of Bernard, he once stated, "Bichat and [von Baer] are the first of *the organicists*. Their successor is Claude Bernard" (Henderson 1917b, 74; emphasis in original). In Henderson's work, we find statements such as "The right working of physiological processes depends . . . upon accurate adjustment and preservation of physico-chemical conditions within the organism" and "Active protoplasm everywhere, as well as that which surrounds it—the environment and the *milieu interieur*—appear to be and to have been always of stable reaction" (Henderson 1913c, 389–391). In fact, Henderson, along with his colleague Walter B. Cannon at Harvard, who worked on stabilization factors of the internal environment and pioneered the concept of "homeostasis" (e.g., Cannon 1929), played a significant role in rekindling scholarly interest in Bernard's work during the interwar years (Cross and Albury 1987, 172).

This link to Bernard's thought is also important to showcase Henderson's organicist commitments (see Parascandola 1971). In the introduction to the English translation of Bernard's *An Introduction to The Study of Experimental Medicine*, he stressed,

> *The theory of organism is more than a philosophical generalization; it is a part of the working equipment of the physiologist,* fulfilling a purpose not unlike that of the second law of thermodynamics in the physical sciences. . . . The theory of the constancy of the internal environment, a related theory, we owe almost wholly to Claude Bernard himself. There is no better illustration of his penetrating intelligence. (Henderson 1927, viii; emphasis added).

Bernard had underlined the importance of the external environment for organismal life but retained the internal environment as the proper epistemic locus of physiology (see also Cannon 1929, 427), and this, as we will attest below, was a point of divergence for Henderson's organicist work:

> The conditions necessary to life are found neither in the organism nor in the outer environment, but in both at once. Indeed, if we suppress or disturb the organism, life ceases, even though the environment remains intact; if, on the other hand, we take away or vitiate the environment, life just as completely disappears, even though the organism has not been destroyed. . . . *Since the outer environment, on the other hand, infiltrates into the inner environment, knowing the latter teaches us the former's every influence.* Only by passing into the inner, can the influence of the outer environment reach us, whence it follows that knowing the outer environment cannot teach us the actions born in, and proper to, the inner environment. The general cosmic environment is common to living and to inorganic bodies; but the inner environment created by an organism is special to each living being. Now, *here is the true physiological environment; this it is which physiologists and physicians should study and know.* (Bernard 1927, 74–76; emphases added).

Of interest for the thoroughgoing discussion on distinct construals of organism-environment reciprocity is Henderson's 1913 book *The Fitness of the Environment*. While working on the physiological mechanisms involved in the maintenance of neutrality, Henderson was struck by the fact that these seemingly possessed "a remarkable and unsuspected degree of efficiency" (Henderson 1908, 448). This is what set him on the path to think about the "fitness" of environmental components. On the one hand, it became evident to him that the effectiveness of neutrality regulation relied, in part, on the unique properties exhibited by certain environmental compounds, such as the dissociation constants of carbonic acid and monosodium phosphate, as well as the gaseous nature of carbon dioxide, which facilitated its facile excretion. On the flip side, it was clear to Henderson that the delicate balance of acid-base equilibrium is knotted with various other physiological processes happening within an organism, all working together to uphold neutrality. These thoughts led him to concentrate on self-regulation and the underlying organization of biological systems (Parascandola 1971, 70–71; see also Needham 1928a, 85). For instance, in a paper about acid excretion, Henderson claimed that "phosphoric acid, one of the chief excretory products of the urine, possesses the very highest possible efficiency in the physiological process for regulating the ratio of acids to bases in the body fluids by means of renal activity. *Certainly there seems to be nothing in evolutionary theory to explain it*" (Henderson 1911, 417; emphasis added).

From these kinds of biochemical indications, Henderson made a generalizing inference about the properties of abiotic environmental components and what they afford for evolutionary processes: "*I saw that fitness is a reciprocal relation*, that adaptations in the Darwinian sense must be adaptations to something, and that complexity, stability, and intensity and diversity of metabolism in organisms could not have resulted through adaptation unless there were some sort of pattern in the properties of the environment that . . . is both intricate and highly singular" (Henderson quoted in Parascandola 1971, 73; emphasis added). For Henderson, there was something special about environments that made evolution possible in the first place, and for this reason he embarked on writing *The Fitness of the Environment*.

The opening statement lays down the main thesis of the book: "*Darwinian fitness is compounded of a mutual relationship between the organism and the environment*. Of this, fitness of environment is quite as essential a component as the fitness which arises in the process of organic evolution; and in fundamental characteristics the actual environment is the fittest possible abode of life" (Henderson 1913a, v; emphasis added; see also Henderson 1913b, 105). Along those lines, Henderson claimed that "the primary constituents of the environment . . . , the very substances which are placed upon a planet's surface by blind forces of cosmic evolution, serve with maximum efficiency to make stable, durable, and complex, both the living thing itself and the world around it" (Henderson 1913a, 5; for an analysis, see Sprenger 2019, chapter 3).

Historian and philosopher of science Iris Fry articulated the central question at the heart of Henderson's work: "How relevant are the properties of the material universe to biological evolution?" (Fry 1996, 160). In other words, how do the properties of the factors that make up environments contribute to the survival and evolution of organisms? If these properties are "fit" for life (i.e., strongly supporting its development and further evolution), can we consider this fitness to be mere coincidence given that deific intervention ought to be ruled out as a credible explanation? For Henderson, there was an overlooked *biological* significance of the properties of matter, as the subtitle of his book hints at, which should be given proper scientific and philosophical consideration (Henderson 1913a, 65). In a later work, Henderson said that the physicochemical conditions of environments are independent of the existence of biological systems, but "biology is nevertheless dependent upon them, for life can manifest itself only in active physico-chemical systems" (Henderson 1914, 525).

To address the complex issue of the environmental suitability for organismal life, he narrowed down his investigation to the role of a handful environmental components—namely, carbon dioxide, water, and their composing elements (i.e., carbon, oxygen, and hydrogen). He conducted comparisons between the properties of water and carbon dioxide and those of other substances, including ammonia and silicon compounds (Henderson 1913a, 65–67). For instance, he highlighted the significance of water's specific heat, which contributes to temperature stability in oceans, creating favorable settings for organisms. Additionally, he explored how organisms have adapted to leverage water's high heat of vaporization, enabling them to achieve high efficiency in manifold physiological processes (Henderson 1913a, 70–71, 86–88).

Similarly, carbon dioxide exhibits two significant properties that contribute to its suitability for sustaining living processes. First, its solubility in water, characterized by a high absorption coefficient and extensive mobility, allows for efficient interactions. Second, its unique acidic nature, attributed to its specific dissociation constant, enables its bicarbonates to help maintain neutrality within the organism. These properties, Henderson reasoned, enhance carbon dioxide's environmental fitness and its capacity to support life (Henderson 1913a, 133–152, 269).

Turning to a subset of biologically relevant chemical elements, Henderson remarked that hydrogen, carbon, and oxygen possess a multitude of properties that facilitate physicochemical exploration in evolutionary terms. These elements provide a broader spectrum of variables within organismal systems, making an allowance for increased complexity and stability.

Henderson accentuated the requirement of (a certain extent of) environmental stability for the survival of organisms. For instance, he stressed that the ocean, as an exemplary environment, offers optimal conditions conducive to life, where factors such as temperature, alkalinity, and ionic concentration are (most of the time) precisely regulated (Henderson 1913a, 183–190). This, in Henderson's bio-philosophical horizon, meant that environmental equilibrium was involved in satisfying the delicate circumstances that are necessary for organisms to thrive. In their evolution, marine warm-blooded animals, for instance, acquired their own distinct environments, *milieux intérieurs* like blood or lymph, which serve roles comparable to the stability provided by their external environment, the adjacent ocean that surrounds them (Henderson 1913a, 31–33, 186). Hence, specific properties of environmental components carry out major contributions in establishing and maintaining stable conditions that organisms encounter.

Henderson's concept of the environment encompasses the broadest range of universal physical parameters that are significant for organized systems. To recover a distinction that was advanced in chapter 3 (section 3.1), Henderson understood the environment as the "external world," an expansive entity independent of the presence of organisms, thus rejecting the relationality thesis (i.e., to every organism there is a corresponding environment). It is noteworthy that Henderson's understanding of the environment includes, with the same importance allotted to them, both its internal (Bernardian) and external aspects, with identical chemical constituents fulfilling similar functions in both domains, especially in the provision of matter and energy for the organism (Fry 1996, 165; see Henderson 1913a, 132).

Henderson saw *The Fitness of the Environment* as primarily an "*inquiry into the relation between the organism and the environment*" (Henderson 1913a, 524; emphasis added; see also Lillie 1913, 337). The relationship he thought he had uncovered was one of, as I label it here, "mutual structural fitting" that led to the emergence and subsequent adaptation of life. Mutual structural fitting between organism and environment can be understood as the thesis that claims that, besides considering organisms as structurally adapted to their surroundings as a general outcome of evolution by natural selection, the specific physicochemical properties of the (internal and external) environment should also be regarded as adapted, for they support the development of life: "*The fitness of the environment is one part of a reciprocal relationship of which the fitness of the organism is the other*" (Henderson 1913a, 271; emphasis added; see also Lillie 1913, 337). Stated differently, in this view the reciprocity between organism and environment is actually the reciprocity that results from the mutual structural adaptation (i.e., fit) of these relata. For Henderson, natural selection might have molded the characters and activities of organisms, but its prolonged action does not change the primary qualities of (abiotic) environmental components, which are important on an equal footing in the production of adaptive phenotypes—for instance, as targets of adaptations and facilitators of physiological processes (Henderson 1913a, 274–275). In this sense, organismal phenotypes inexorably have to work on, and evolve from, what the environment disposes of.

With his work, Henderson wanted to rehabilitate the neglected component in the reciprocal binomial (organism-environment) behind the adaptation of living forms. He mentioned that "although Darwin's fitness involves that which fits and that which is fitted, or more correctly *a reciprocal relationship*, it has been the habit of biologists since Darwin to consider only the adaptations of the living organism to the environment" (Henderson 1913a, 5; emphasis added). However, as Henderson dug deeper into the nature of environmental components, he also found himself compelled to give due consideration to the converse arrow of this reciprocal relationship: the organism that faces environing conditions and exhibits an intricate internal organization. Parascandola (1971, 80) hints that "Henderson's interest in the fitness of the environment forced him to consider the nature of the organism more carefully. In order to ask whether the environment is truly fit for life, he had to ask himself what the chief characteristics of the living organism were." That is why Henderson devoted so much space and effort to discussing biological organization in subsequent works—for instance, in his 1917 book *The Order of Nature* (Henderson 1917a; see also Henderson 1918, 572–576).

Henderson's *Fitness* was received with both praise and skepticism across diverse communities of biologists. On a positive note, plant physiologist Burton E. Livingston said, "Biologists have wondered at and tried to 'explain' the fitness of organisms to their surroundings, but the little volume before us appears to embody the first serious scientific attempt to study the adaptations of the surroundings to living things" (Livingston 1913, 315; see also Pearse 1922, 144). In contrast, organicist Ralph S. Lillie criticized the omission of nitrogen from the elements surveyed by Henderson and raised a common criticism that suggested that the facts Henderson presented could just as well be interpreted counterfactually, using J. D. Bernal's words, "as evidence that life had to make do with what it had, for if it had failed to do so it would not be there at all" (Lillie 1913, 339; see Bernal 1967, 169). Environmental components seem to be fit for life only in a post hoc reading of extant organism-environment interactions and physiological processes.

In any case, Henderson's musings proved to be influential, and he mobilized his biochemical and physiological views of the fitness of the environment to partake in certain evolutionary debates of the time—for example, concerning orthogenetic trends in evolution (Henderson 1922). Through Henderson, another view of organism-environment reciprocity had been put up in the market of ideas of early-twentieth-century biology.

Whereas the view of mutual structural fitting as a hallmark of reciprocity was exclusively defended by Henderson and some of his followers, in the next section I will go over another understanding of reciprocity that was shared by a larger group of early-twentieth-century biologists: reciprocal causation.

5.2.4 Reciprocal Causation between Organism and Environment[5]

At the beginning of this chapter (section 5.1), I mentioned that historian Georg Toepfer highlights a central distinction between interpretations of reciprocity that frame it in terms of the mutual dependence (*interdependence*) or the mutual influence (*interaction*) of the relata involved (Toepfer 2011b, 738). So far, I have discussed two understandings of organism-environment reciprocity that accord with the view of interdependence—namely, ontological coconstitution (ontic interdependence) and mutual structural fitting (structural interdependence). In contrast, in this section we will delve deeper into how the view of mutual (causal) interaction took root in early-twentieth-century biology.

As we have seen, Haldane and his holistic physiological standpoint epitomize the rejection of causal thinking to construe the organism-environment relationship. Summarizing Haldane's viewpoint, the Scottish organicist biologist and polymath D'Arcy W. Thompson (for historical context of his organicist thinking, see Esposito 2014) declared, "[Haldane contends that] we must sublate our view of the nature of life *from a mechanical or cause-and-effect category to a category of reciprocity*, and from that further to one in which the parts must be regarded as determined in relation to an idea of the whole; in other words that *organic processes cannot be reduced to series of causes and effects*" (D. W. Thompson 1884, 419; emphases added). Nevertheless, in contrast to Haldane's strictures, many early-twentieth-century biologists indeed believed that organism-environment reciprocity could effectively be spelled out in causal terms.[6]

The literature of these decades is rife with experiments, interpretations, and statements regarding how changes in environmental factors *causally affect* development, morphology, physiology, and evolution and how organisms do their proper share in these interactions. Recall all the ground that I covered in chapter 2 (sections 2.3 and 2.4.2). For instance, the British anatomist Gilbert C. Bourne asserted, "It has been proved by experiment that very small changes in the chemical and physical environment may and do produce specific form-changes in developing organisms, and in such experiments the consequence follows so regularly on the antecedent that *we can not doubt that we have true relations of cause and effect*" (Bourne 1910, 733; emphasis added; see also Child 1915, 41, 53–54).[7]

Adding an emphasis on the reciprocal bond between organism and environment, Burton Livingston, a plant physiologist mentioned earlier as one of the reviewers of Henderson's book, expanded on this issue in an article titled "Environments" published in *Science*. Livingston stated, "It is always true that many features of an organism are continuous with corresponding features of the environment; there is always interchange of material and energy between these two portions of a more complete system, and *organism influences environment while environment influences organism*" (Livingston 1934, 571; emphasis added).[8]

The idea of reciprocal interaction and the causal link between organism and environment was present as well in other disciplines and domains of the biological sciences during the interwar period, such as animal behavior studies and plant ecology (see, e.g., Salisbury 1930, 59; Munster Strom 1929, 106; for discussion of animal behavior studies, see De Bont 2010), and many authors derived important evolutionary repercussions out of it:

> *The organisms* produce vegetational change through their effects upon the environment, and upon each other through the environment; and by the production of new forms, varieties, species, through evolution. The *environment* brings about changes in the vegetation because it, too, is undergoing constant change, partly inherent in itself but in part caused by the action of the vegetation upon it. Through selection it is also of great importance in the process of evolution. (Cooper 1926, 398; emphases in original)

We could understand this general view of "reciprocal causation" as sequential feedback loops of causes and effects between two interacting, yet separate entities or processes. As Walsh explains, "According to the causal construal, organisms cause changes to the features of the environment external to them. . . . These modified physical conditions, in turn, redound upon the organisms" (Walsh 2015, 180; see also Casetta 2023, 76–77). Let me now explain further how this view of organism-environment reciprocity was mobilized in the theoretical views of two influential holistic and organicist biologists: Adolf Meyer-Abich and Conrad Hal Waddington.

In chapter 4 (section 4.4.1), I covered the philosophical and scientific background of the German theoretical biologist Adolf Meyer-Abich, including his reinterpretation of the "complementary principle" to renegotiate the traditional understanding of the segregation between internal and external factors in an organism. Here, it is important to emphasize how reciprocal causation featured in his theory of holobiosis.

As previously recounted, starting in the 1940s and continuing thereafter, Meyer-Abich put forth an evolutionary theory known as holobiosis (Baedke et al. 2020, 151). To grasp the significance of reciprocal causation within Meyer-Abich's theory of holobiosis, it is important to contextualize his influences and the reworking of ideas put forth by Jakob von Uexküll. As we saw in section 4.3.3, Uexküll championed the tight connection of organisms and their surrounding conditions. Uexküll formulated the concept of the *Umwelt* as a product of an organism's sensory exploration. According to him, each organism is complexly linked with its own surroundings, perceiving them subjectively and taking action accordingly. Therefore, the *Umwelt* comprises a "perception world" (*Merkwelt*), where sensory receptors capture stimuli and neural processing occurs, and an "effect world" (*Wirkwelt*), where the organism exerts causal influence through various traits and behaviors. The combined perceptions and actions of the organism establish a feedback loop, known as the "functional circle" (*Funktionskreis*), between the perception world and the effect world. For instance, when organisms venture into new environments, they strive to establish new functional circles by modifying their dietary habits. If these actions do not yield rewards (i.e., if a functional circle cannot be completed, such as when the new diet lacks nutritious items), the organism abandons this setting and reverts to its previous dietary patterns or explores alternative options. Throughout this process, the surrounding conditions prompt the organism to reconfigure neural networks and physiological receptor-reaction connections (Uexküll 1926, 126–129, 155–158).

For Uexküll, the organism passes through a so-called "environmental tunnel" (*Umwelttunnel*) throughout its ontogeny, in which the environment and the organism reciprocally act on one another: "By environmental tunnel I understand the life path of a subject, which is closed on all sides by objects, which serve the sense organs of the subject as feature carriers and its action organs as effect carriers" (Uexküll 1922, 142; my translation). According to the representation of the environmental tunnel, an organism can be depicted as a rolling cogwheel, and the environment as its underground. The wheels' joints represent the organism's receptive properties, and the pivots represent its action-executing features. Throughout the lifetime of an organism (i.e., the wheel rolls over the underground), it is affected by and affects the environment. Outgoing arrows in the wheel mark the beginning of activities of the organism in the environment; the ingoing arrows mark their ends and the perception of an environmental event (Uexküll 1922, 143). As we see, Uexküll's view of reciprocity conceptualizes the organism-environment relationship as highly individualized and mediated through experience: organisms, like isolated bubbles, live in their own *Umwelten*, their own worlds-as-perceived and worlds-as-acted-on. It is likely because of this, as we will discuss in the next section, that Uexküll's *Umweltlehre* was contested among biologists.

As an exception to this trend, Meyer-Abich's perspective integrated what he assumed to be formerly distinct "functional circles" through *reciprocal causation* between organisms. In his view, not only do organisms construct their own individual environments, as proposed by Uexküll, but they also collaboratively causally construct interconnected and shared environments for one another. Meyer-Abich argued that macroevolutionary change is driven by reciprocal processes between increasingly integrated organisms, which first constitute symbiotic partners, then holobionts, and finally systems of organs in a larger whole. For Meyer-Abich, holobionts exhibit the remarkable capacity to manifest novel traits and venture into uncharted environments of their own making. An exemplary illustration of this phenomenon can be found in lichen holobiosis. Over evolutionary time, as green algae and/or blue-green algae increasingly interact with fungi in protracted causation chains, a metabolic entity with distinct characteristics, such as novel metabolites and reproductive structures, arises. This new unit possesses an altered energy status, enabling access to previously unexplored ecological niches (see Meyer-Abich 1943, 1950). In essence, for

Meyer-Abich, during holobiosis, symbiotic organisms, through reciprocal causal interactions, engender new environments, which they both shape and encounter in unprecedented ways in the evolutionary process (Baedke et al. 2020, 2021).

Another evolutionary take on reciprocal causation, although a very different one, was developed by the British polymath Conrad Hal Waddington. Waddington received training in the disciplines of geology and paleontology during his time at Clifton College and later at Sidney Sussex College, University of Cambridge. Alongside these pursuits, he also did some philosophical studies, which provided balancing insights to his training as a natural scientist. In 1938, he went through graduate studies in paleontology, but his research interests soon shifted toward genetics and embryology (for biographical sketches, see Robertson 1977; Fabris 2019). Securing a research fellowship at the Strangeways Laboratory of the University of Cambridge, he focused on experimental embryology, particularly conducting tissue ablation and transplantation experiments. During his tenancy at Cambridge, he fostered fruitful collaborations with Joseph and Dorothy Needham as they investigated the biochemical underpinnings behind the so-called Spemann-Mangold organizer (see, e.g., Waddington et al. 1933; see also Peterson 2016, chapter 5). It was in response to the puzzlement generated by the large array of factors that induced the same axis duplication phenotype, mimicking the effects of grafting the vertebrate organizer, that Waddington and the Needhams spearheaded an initiative to assemble an interdisciplinary group of biologists, mathematicians, physicists, and philosophers (e.g., Joseph Henry Woodger, Dorothy M. Wrinch, J. D. Bernal, Karl Popper, Dorothy Crowfoot, and Lancelot Law Whyte). From 1932 to 1938, the members of the Theoretical Biology Club (TBC), as it later came to be known, convened regular meetings (Abir-Am 1987).[9] While their initial aim was to give answers to enigmatic embryological phenomena, these discussions later sought to establish a robust epistemology for an autonomous "theoretical biology" (Peterson 2016, chapters 5 and 6; see also Abir-Am 1991). The conversations held within the TBC revolved around the suite of possible concepts on which biological science at large could rest, such as the recognition of the organism as the primary ontological unit, along with the emphasis on the stratified organization of biological systems at various levels (Etxeberria and Umerez 2006; see also Nicholson and Gawne 2015,

360–367; Baedke and Fábregas-Tejeda 2023, 125)—two central organicist tenets, as we saw in chapter 2 (section 2.4.1).

After the TBC disbanded and the Organicist Movement lost its momentum (as we will see in section 5.3), Waddington moved to the University of Edinburgh. From 1947 until his passing in September 1975, Waddington held the position of Buchanan Professor of Animal Genetics there. It was during this period that he pursued some empirical and theoretical investigations aligned with the fundamental epistemological and ontological principles of organicism, albeit not always with resounding success (Peterson 2011). A notable example is the concept of "chreodes," referring to the open trajectories within the iconic epigenetic landscape (for an analysis, see Baedke 2013), which, as argued by Haraway (1976, 60–63), can be traced back to his earlier organicist framework.

Alfred North Whitehead's perspective on the organism-environment relationship held great sway over British organicists such as Waddington and Joseph Henry Woodger. These scholars found inspiration in the Whiteheadian "philosophy of organism," a process-oriented outlook that draws attention to the tight interconnectedness between an organism's developing parts and its surroundings. Whitehead (1929, 163) posited the existence of "two sides to the machinery" of evolution. On one side lies natural selection, where the externalist "givenness of the environment" reigns supreme. Yet Whitehead considered that there exists another side that had received less attention: "The other side of the evolutionary machinery, the neglected side, is expressed by the word creativeness. The organisms *can create their own environment*" (Whithead 1929, 163; emphasis added). Waddington, too, embraced this Whiteheadian notion of (what we could perhaps anachronistically call) "proto-niche construction" and developed it further in his own evolutionary theorizations.[10]

In Waddington's broad perspective, genetics and natural selection alone, either separately or in combination, fell short of providing a complete explanation for organismic evolution (Waddington 1957, 188–189). To address these purported explanatory gaps, he advocated for the integration of ontogenetic change and exploratory behaviors, especially what he termed the "exploitive system" (Waddington 1957, 104–108). This concept hinged on the seemingly simple, yet overlooked idea that

> animals . . . live in a highly heterogeneous "ambiance," from which they themselves select the particular habitat in which their life will be passed. Thus the animal by its behaviour contributes in a most important way to determining the nature and intensity of the selective pressures which will be exerted on it. Natural selection is very far from being as external a force as the conventional picture might lead one at first sight to believe. (Waddington 1959a, 1635–1636)

In other words, "before an organism's environment can exert natural selection on it, *the organism must select the environment to live in*" (Waddington 1961, 89; emphasis added; see also Waddington 1975, 273). Waddington recognized that this ability of organisms to "choose" their abodes wielded a profound influence as an ultimate cause in evolution. Indeed, Waddington delineated the exploitive system as one of the four subsystems, alongside the genetic system, the epigenetic system, and the natural selection system, all of which, for him, interacted with equal causal significance in the evolutionary process (for discussion, see Plotkin 1988, 4–10). For Waddington, the influence exerted by organisms on their chosen environments leads to feedback loops with selection pressures as generations pass.

For instance, he subverted the neo-Darwinian textbook exemplar of adaptation and directional selection in action of the peppered moth (*Biston betularia*) in the industrially darkened woodlands of Birmingham. In Waddington's hands, the industrial melanism case defied the belief that natural selection alone governs the fate of heritable phenotypic variants based on their contributions to fitness differences. Instead, he emphasized the nontrivial, evolutionary-consequential role of organisms in selecting their own environment: "The effective environment in which they are subjected to natural selection is, in fact, the darkened bark which they themselves choose; it is not something completely external, but is a combination of the outside world and the moth's own behavior" (Waddington 1961, 90).

Against this backdrop, Waddington urged biologists to abandon two commonly held views on the organism-environment relationship. First, he challenged biologists to reevaluate their conviction in thinking of natural selection as a purely external force in the face of recognizing that an animal's behavior actively shapes the nature and intensity of the selective pressures it experiences (Waddington 1959a, 1636). Second, rather than relying on a unidirectional causation framework, he called for a recognition of causal reciprocity and the interdependence of different subsystems. In doing so, he made emphatic the need to move beyond a simplistic view of evolution and embrace the complex dynamics and feedback loops that exist within and across biological systems. Waddington (1959b) asserted that *"we have to think in terms of circular and not merely unidirectional causal sequences"* (400; emphasis added). As we will see below, it was these two views, namely, constructivism instead of externalism—with the emphasis on organisms experiencing their environments—and reciprocal causation instead of unidirectional causality, that came under attack in evolutionary biology and, among other factors, led to the temporary demise of organism-environment reciprocity.

5.3 The Rise and Fall of Organism-Environment Reciprocity in Postwar Biology

Understanding the factors underlying the waning theorizations about organism-environment reciprocity in twentieth-century biology not only illuminates its resurgent relevance in present-day debates but also reiterates its significance as a multifaceted subject that faces several challenges for scientists and philosophers.[11] This downward trajectory was entwined with the broader fate endured by organicist and holistic perspectives in biological science, as they gradually faded into the background after World War II ended.

In recent times, historians of biology have investigated the factors contributing to the decline of organicist and holistic positions from discussions, theories, and biological practices during the latter half of the twentieth century. In the British context, Donna Haraway proposed that the organicist perspective within the TBC lost its prominence due to a lack of proper institutional and disciplinary support necessary to sustain it:

> Needham tried to construct an institute . . . but was unable to obtain needed financing. Beginning in 1934 he corresponded with Dr. Tisdale of the Rockefeller Foundation, which was then interested in fostering study on the borderlines of traditional disciplines. . . . Needham submitted . . . a plan for an Institute for Physico-chemical Morphology. . . . By 1938 the idea was dead. . . . The reasons are controversial and complex, but the success of Needham's institute certainly would have altered the course of biological investigation in England after the 1930s. Instead, factors combined to break up the collaboration of members of the paradigm community, and World War II finally sealed the issue. (Haraway 1976, 134; see also Peterson 2016, chapter 7)

In their initial plan for an organicist research institute, Needham proposed a lineup of key figures to lead various sections: Conrad Hal Waddington was designated to head experimental embryology, Joseph Henry Woodger was chosen for theoretical embryology, and the Russian geneticist Theodosius Dobzhansky was initially suggested as the head of genetic embryology. As the institute never came into fruition, Dobzhansky eventually joined the ranks of Thomas Hunt Morgan's Californian fly group and became instrumental in the Modern Synthesis and evolutionary genetics thereafter (Peterson 2016, 118).

As Pnina Abir-Am (1982, 341) asserts, the decisions made in funding policies played a crucial role in shaping the trajectory of emerging biological disciplines, if not dictating their directions entirely, in the second half of the twentieth century. While the Rockefeller Foundation withheld its support for the organicist pursuits of Needham and his colleagues, they redirected their focus toward what would later be known as "molecular biology." The postwar era saw funding policies in the life sciences favoring reductionist but tractable and profitable fields like molecular biology over holistic (and hard to operationalize) research championed by, among others, organicists in diverse corners of the world. This shift toward the "molecular view of life," as it was famously called by historian Lily Kay (1993), overshadowed the organicist perspective. As Brooks (2019, 2) remarks, "It was, as the story goes, the politics of research funding that seemed to doom organicism: With the molecular revolution just around the corner, it seemed simply the wrong place and time for the movement to take root." Efforts to uphold the organicist perspective over the molecular one were occasionally discussed in prominent journals like *Science*, but they did not yield the desired outcome of keeping organicism afloat (see, e.g., Schaffner 1967).

Like their British counterparts, German holists also had plans to establish a center dedicated to holistic and organism-centered research. In 1942, Adolf Meyer-Abich and particle physicist Pascual Jordan founded the journal *Physis: Beiträge zur naturwissenschaftlichen Synthese*, announcing their intent to create a research institute exploring organicist and holistic themes across multiple scientific disciplines (Beyler 1996, 268–269). However, these aspirations never materialized, and the journal *Physis* failed to gain traction (Dahn 2019). The project to institutionalize German holism, complete with research institutes, journals, and specialized communities, also failed to flourish (Beyler 1996, 267–270; see also Amidon 2008, 366). The discredit of German holistic biology after World War II further hindered its progress. Due to its association with Nazi ideology, biologists from other regions regarded these theories as anathema, deliberately avoiding references to German-speaking holistic and organicist authors (Wise 1994, 244; for discussion, see Harrington 1996; Rieppel 2016; for discussion on weak continuities after 1945, see Reiß 2022, 390–392).

Furthermore, during this period, many prominent figures in the international Organicist Movement passed away. The deaths of Haldane in 1936, Wheeler in 1937, Henderson in 1942, Schaxel in 1943, and Ritter, Dürken, and Uexküll in 1944 marked the loss of important members. What is more, during the Nazi upsurge in Austria, organicist scientists based at the Vivarium were expelled from their workplace and barred from entering, and some, such as its director Hans Przibram, were transported to, and later executed in, concentration camps (Taschwer 2014, 2020). Other organicists shifted their research interests to new areas. For instance, Woodger turned to logic, Needham became a sinologist, and Bertalanffy focused on "systems," rather than organisms and pushed for a "General Systems Theory" (Nicholson and Gawne 2015, 372; Peterson 2016, chapter 9). Some organicists engaged in politics or turned to popular science writing, such as Schaxel and Bernal. For many German-speaking advocates of holistic ideas, the end of World War II signified a definitive caesura with the past. In the 1940s and 1950s, the number of monographs and papers discussing organisms (and their relations with their environments) as central biological units notably declined (Baedke 2019, 297).

Furthering this decline, the disbandment of the TBC was accompanied by a chorus of critical voices on organicism, including that of British immunologist Peter Medawar. Despite his early tutelage under the guidance of Woodger and Waddington, Medawar turned into a zealous detractor of organicism. He vehemently denounced it as an archaic, speculative, and ineffectual approach to biology. Medawar's ascent to influential positions provided him with a platform to publicly decry the organicist framework (Peterson 2016, 156–158). For example, in his review of *Problems of Life* by Ludwig von Bertalanffy, he shoveled a coarse attack to organicism and the strawmen it allegedly built:

> The theories and attitudes of mind which the "organismic conception" supersedes are so conspicuously foolish that one wonders who can have had the temerity to propound them: and so we wait in ghoulish pleasure, page after

> page, for quotations from or references to the authors whose fatuous views they represent. But we wait in vain: expectation gives way to doubt and then at last to the full exactions of hope confounded: there are no such views and no such people, and I submit that it was not necessary to invent them. The "analytical-and-summative mechanist" is merely a sort of lay devil invented to allow nature-philosophers the pleasures of indulging in the rites of exorcism. (Medawar 1954, 105–106)

Simultaneously, the leading evolutionary theorists Ernst Mayr and Theodosius Dobzhansky played a role in discrediting the Organicist Movement. They branded it as "(neo-)Lamarckism," and Mayr even went so far as to suggest that Waddington was a "Lysenkist" (see Robison 2018; see also Alexandrov and Aronova 2004, 46). Such stigmatizing labels, though misaligned and certainly undeserving, contributed to the dismissal of organicist scholarship in the latter half of the twentieth century (Peterson 2016, chapter 11).

Within this postwar juncture, the Organicist Movement receded from the international stage. Concurrently, the Modern Synthesis, a movement seeking theoretical unification and consilience in evolutionary research, and especially discipline building and programmatic reforms in the social-professional terrain, gained ascendancy (Smocovitis 1992, 1994). The emergence of molecular biology and subsequent trends in evolutionary biology, which became more and more gene-focused as decades passed, further shaped the scientific landscape (Suárez-Díaz 2017; see also Hein 1969, 238–253; Cortés-García and Etxeberria Agiriano 2023).[12]

Fox Keller (1990) has contended that the institutionalization of molecular biology brought several conceptual shifts to biological science writ large, one of them being the "*relocation of the essence (or basis) of life*" (391; emphasis in original). "The locus of vital activity was now to be sought," Fox Keller argued, "not in the physicochemical *interactions and structures of the organism-niche complex*, of the organism itself . . . but rather, in the physicochemical structure of one particular component of the cell: namely, in the genetic material, or, more exactly, in the gene" (Fox Keller 1990, 391; emphasis added). These transformative currents affected the explanatory standards within evolutionary research, often downplaying, neglecting, or abstracting away the central ontological and epistemic functions ascribed to organisms by the Organicist Movement, including the importance of organism-environment reciprocity.[13]

The extrication of genes from their organismal context, even if sometimes epistemically justified, constituted a significant explanatory shift. Gawne et al. (2018) have argued that the dominant perspective among evolutionary biologists in the latter half of the twentieth century onward gravitated toward a much-simplified view of the genotype-phenotype map, wherein genes assume the unquestionable role of primary determinants of phenotypic traits. Regrettably, this narrowed focus obscured the organicist insight that a comprehensive explanation of phenotypic traits requires the integration of findings across all levels of organismal organization.

While molecular approaches offered a powerful research program that, to a great extent, achieved heuristic and explanatory success precisely by abstracting from the context of the entire organism, population geneticists concentrated their efforts on theorizing and measuring allele transmission and population dynamics. This is captured in Dobzhansky's famous characterization of proper evolutionary science: "Since evolution is a change in the genetic composition of populations, the mechanisms of evolution constitute problems of population genetics" (Dobzhansky 1937, 11). As a result, the developing organism, once considered a locus of concrete, dynamic material interactions with its environment, was relegated, at best, to the status of a secondary performer in the playground of evolution (for analysis, see Cortés-García and Etxeberria Agiriano 2023), or at worst, it was treated as an epiphenomenon or a mere vessel for the replication of genetic programs (see Walsh 2019). Although there were valid scientific justifications for adopting this approach, and certain proponents of the Modern Synthesis retained an organismic-ontogenetic perspective to a certain degree, a consequence was the marginalization of the organism as a central ontological and theoretical unit within biology.[14]

In contrast, populations were enshrined as the proper units of evolutionary change and analysis. As philosopher Thomas A. Goudge noted, "Biologists have recently found it useful to give a prominent place in their theories to the concept of a plant or animal population. Genetics, ecology, paleontology, and especially the theory of evolution have been able to make notable advances as a result of employing the concept in their interpretations and inferences" (Goudge 1955, 272). By the 1960s, the field of population genetics had amassed a large corpus of mathematical models, and it had garnered a wealth of empirical data through laboratory and field experiments exploring variations in fitness components and quantitative traits. For instance, studies examining visible and chromosomal polymorphisms in natural populations, as well as investigations into human blood groups and select biochemical polymorphisms (e.g., the sickle cell hemoglobin variant maintained by heterozygote advantage), provided valuable insights for the discipline during this period (for an overview, see Lewontin 1974; see also Allison 1964).

However, besides the copious epistemic import that genetic analyses and the concept of (genetic) population brought to evolutionary biology, authors such as Mayr also propelled the narrative that the swing away from developing organisms was inevitable to leave behind confounded views of evolution—what in his opinion were a broad family of essentialist and wrong-headed *typological views*. Mayr claimed to trace the historical roots of "typological thinking," the supposed benchmark of organicism and other unlawful approaches such as neo-Lamarckism and orthogenesis, back to Plato, while simultaneously taking Charles Darwin to be the eminent champion of "population thinking" (Mayr 1959a, 2–4). To Mayr's dismay, historians had left a gaping void by consistently overlooking this decisive dimension of the Darwinian revolution (Mayr 1959b, 215–216). By the early 1960s, Mayr relished portraying the transition from typological to population thinking as "perhaps the greatest revolution that has taken place in biology" (Mayr 1963, 5). However, in Mayr's historiography, which has been extensively challenged by recent scholarly work (e.g., Amundson 2005; Winsor 2006; Witteveen 2015, 2016), Darwin's eradication of typological thinking had not been an unqualified success. According to Mayr's account, the remnants of this mode of thought persisted, although tenuously, until the advent of the Modern Synthesis, in which he himself played a leading role. This synthesis, for Mayr, expedited the expulsion of typological (organismal) thinking from biology by integrating genetics and natural history insights through a common Darwinian framework (see Mayr 1972, 1982, 1993a). When facing pushback from his contemporaries, Mayr alleged that this self-aware gross distortion of the history of evolutionary thinking in general was "a temporary oversimplification," as he once stated (Mayr quoted in Witteveen 2015, 20); however, he never backed down, and he strongly believed that the potential risk of oversimplification was compensated for by the rhetorical advantage of having a clear-cut distinction that could effectively address numerous debates in both historical and modern biology while stressing the importance of Darwinian thinking.

Alas, this is the unquestionable story, stemming from a biased historiography, that many biologists still faithfully repeat to this day.

In contemplating the ramifications of partially forsaking the organism and fully embracing populations and genes, one must inquire about what happened to the notion of reciprocity. What consequences did this abandonment engender, and how did it shape the scientific understanding of reciprocity within the biological sciences?

5.3.1 The Decline of Organism-Environment Reciprocity and Impoverished Views of the Organismal Environment

The epistemic emphasis put on organism-environment reciprocity significantly diminished with the institutionalization of the Modern Synthesis and subsequent developments in evolutionary biology. Throughout the latter half of the twentieth century, perspectives on organism-environment reciprocity were progressively marginalized (Baedke et al. 2021, 6–8; Baedke and Fábregas-Tejeda 2023).

In 1953, for instance, paleontologist George Gaylord Simpson stressed, "Organism and environment obviously interact and obviously are closely fitted, that is, adapted to each other. Yet . . . it is improbable (to say the least) that the effects of the interaction can become heritable directly and in the same form" (Simpson 1953, 110). He reasoned that given that organism-environment interactions during ontogenetic time are not genetically inherited, they can be safely left out of (population and gene-based) evolutionary biology. Likewise, the examination of phenotypic plasticity underwent a notable recontextualization within a populational framework: organism-environment interactions became encompassed by the framework of "norms of reaction," which viewed plasticity as an adaptive phenomenon occurring at the population level with a clear (selectable) genetic basis (see, e.g., Dobzhansky 1955; Dobzhansky and Spassky 1963; for an analysis of this changeover, see Sarkar 1999).

Along those lines, Dobzhansky advanced influential views on how to take in the relationship between evolution and environment and how allelic differences were the effective mediators of adaptation:

> The relationships between environment and evolution are subtle enough to make them often misunderstood. *Evolution is controlled by the environment on two levels, and yet the environment cannot be said to impose changes on living species. The first*

> *level is that of mutation. The second is that of natural selection.* The most accurate, although metaphorical, way of describing the dependence between evolution and environment is to say that *environment provides the challenges to which* [populations of] *organisms may or may not respond by adaptive modifications.* Mutations are, in the last analysis, physicochemical alterations in the genes or chromosomes; hence they cannot be wholly independent of the environment. And yet mutations are described as random and undirected changes. . . . To suppose otherwise is to believe in magic. (Dobzhansky 1950b, 165; emphases added)

Dobzhansky viewed the environment as having a limited but significant evolutionary role, primarily by bestowing challenges for adaptation. For him, the response of populations to these challenges was solely determined by genotypic differences, which could fully account for whether organisms adapted to the environmental conditions they encountered.[15] This view was shared by many other influential evolutionary biologists, such as the American botanist and Modern Synthesis architect G. Ledyard Stebbins.[16]

As an additional outcome of this perspective, while the environment was considered an external causal factor that generates selection pressures on populations, it was also perceived by scientists as a "source of error that reduces precision in genetic studies" (Falconer 1960, 140). For this reason, biologists reasoned that the environment had to be controlled for and reduced to the greatest extent possible in experimental settings and explanatory frameworks. In response to this limiting attitude, Waddington raised objections to the status quo view of the environment present in his scientific milieu:

> Darwin's theory was interpreted to mean that all living things . . . had been brought into being by the collocation of two entirely independent factors, on the one hand the occurrence of mutations whose nature was totally unconnected with any ambient circumstances, and on the other a sieving process in which the environment merely selected from among organisms which were offered to it ready-made as units of being, not in any way of potentiality. *Any further influence which the environment might have was degraded to the status of mere "noise" in the system of genetic determination.* (Waddington 1957, 188–189; emphasis added)

As an example of the noise-view of the environment, the British entomologist Vincent Brian Wigglesworth recounted in 1961 how skeptical many geneticists were of environmentally triggered phenotypic changes, as these went against their theoretical expectations: "I have found that some geneticists are inclined to be apologetic about these phenomena, and to regard them as examples of unfortunate defects in the delicate genetic machinery: as R. A. Fisher once said to me, it is not surprising that such elaborate machinery should sometimes go wrong" (Wigglesworth 1961, 107). It was not uncommon for evolutionary biologists to think that plastic phenotypes were defects of naturally selected genetic programs. Along those lines, for example, the British evolutionary ecologist Anthony David Bradshaw pointed out the derisive attitude of plant evolutionary biologists in the mid-1960s towards phenotypic plasticity: "Much of the recent work on plant adaptation has eschewed very carefully any consideration of plasticity. *Any modifications induced by the environment during the course of an experiment are usually considered only an embarrassment*" (Bradshaw 1965, 148; emphasis added).

Waddington's critique was indeed not off target. As another example of an impoverished view of the action of the environment in evolutionary processes, Mayr claimed that "the *true role of the environment in evolution* could not be understood until the nature of small mutations and of selection was fully comprehended" (Mayr 1970, 2; emphasis added). Indeed, it was Mayr's and several evolutionists' conviction that recent work in evolutionary biology had "restored" the environment "to the place of one of the most important evolutionary factors but in a drastically different role than it held in the various 'Lamarckian' theories. The new role of the environment is to serve as principal agent of natural selection" (Mayr 1963, 7; see also Mayr 1963, 310). For them, this was not a depleted construal of the environment but, rather, an advantageous clarification of its proper role in evolutionary processes.

This stance, coupled with Mayr's highly influential distinction between *proximate* and *ultimate causes* in biology, steered research away from the exploration of organism-environment reciprocity in development (Mayr 1961; see also Mayr 1982, 67–68). Under this dichotomy, the organism is regarded primarily as a developmental unit. Proximate causes manifest through the decoding of a genetic program. Conversely, the environment only assumes evolutionary significance as the reservoir and source of selective pressures, serving as ultimate causes that shape the composition and frequency of the genetic programs harbored by individuals in organismal populations. As briefly mentioned in chapter 1 (section 1.1), aligned with these perspectives, mainstream evolutionary biology increasingly adopted an asymmetric and unidirectional view of the organism-environment relationship—which reduced it to be synonymous with evolutionary "adaptation"—as exemplified by Williams and others (e.g., Williams 1992, 484; for a recent defense, see Fromhage and Houston 2022).

Authors like George Gaylord Simpson tried to argue that adaptation ultimately is the outcome of the organism-environment dyad, so both terms should be discussed in evolutionary research:

> Pittendrigh goes on to point out that adaptation is not reciprocal between organism and environment; the organism is adapted to the environment and not vice versa. So far, and beyond possible quibbles as to the wording, I agree, but not so entirely when he goes on to say that "The word 'adaptation' should be restricted to discussion and description of the organism," and especially not when he equates organization with adaptation. No more satisfactory is the frequent opposite attitude . . . that because what is adapted *to* is the environment (broadly speaking) adaptation is basically controlled by or uniquely relative to the environment. (Simpson 1958, 520; emphasis in original)

Indeed, Simpson continuously stressed the importance of the organism-environment relationship, even devoting a full chapter (chapter 6, "Organism and Environment") to the subject in his book *Tempo and Mode in Evolution*, but always through the lens of adaptation. "How necessary it is to consider evolution," he asserted, "in terms of interaction in organism and environment. In this system *adaptation is the crucial element*" (Simpson 1944, 180; emphasis added).

In general, evolutionary biologists came to a consensus. They recognized the influential (selective) causal arrow originating from the environment and impinging on organisms; however, they considered that organisms do not reciprocate in a manner that carries evolutionary significance within these processes. Thus, organismal action does not deserve to be included explicitly in theoretical constructs, models, and explanations with an evolutionary bent. In particular, reciprocity was found elsewhere by population biologists, as will we see in the next subsection.

5.3.2 Reciprocity in Other Relata: Populations, Genes, and Environments

The question of reciprocity qua biological relation and its broader implications continued to captivate the attention of ecologists and evolutionary biologists in the second half of the twentieth century. Rather than receding from the scientific agenda in general, it underwent a shift in focus, with numerous studies exploring the reciprocity among various relata beyond the traditional domain of organism-environment reciprocal couplings. In particular, *gene-environment reciprocity* was increasingly studied in population and quantitative genetics and closely allied areas (see, e.g., Haldane 1946; Lerner 1950; Falconer 1952; for discussion, see Loison 2024). Novel mathematical tools were developed and mustered to study these reciprocities; for instance, the method of "path analysis" pioneered by Sewall Wright was used to trace reciprocal interactions between variables with or without time lags in responses (Wright 1960). Out of these and other studies grew the idea of "genotype-environment interaction," commonly denoted as G×E, which models situations in which distinct genotypic groups exhibit differential responses when exposed to, mutatis mutandis, identical environmental conditions (for historical background, see Tabery 2008). These phenotypic variations are frequently illustrated with reaction norm graphs, which visually represent the range of responses exhibited by different genotypes across various environmental contexts.

Some influential authors, such as Dobzhansky, believed that studying dynamic gene-environment interactions would capture the reciprocal essence of life and its evolution: "The 'stability' of the genes is peculiarly dynamic—they change to produce their own copies. Self-reproduction is the fundamental quality of life that distinguishes it from inanimate nature. This explains the apparently paradoxical nature of life: *life changes the environment and is changed by the environment*, and yet it preserves an inner continuity which is, in fact, its basic property" (Dobzhansky 1950b, 163; emphasis added).

In the second half of the twentieth century, many models of reciprocal relations between genes and populations, as well as genes and environments, addressed population regulation by genetic feedbacks, positive and negative frequency-dependent selection, and distinct kinds of eco-evolutionary dynamics (e.g., Fisher 1930; Lewis 1954; J. B. S. Haldane 1956; Mather and Jones 1958; Jones and Mather 1958; Pimentel 1961, 1968; Wright 1966; Charlesworth 1971; Kirkpatrick 1982; J. N. Thompson 1998 and references therein; for a recent overview, see Svensson 2018). Their explanatory and heuristic importance notwithstanding, the vast majority of these models did not encompass organism-environment reciprocity but focused on other relata (e.g., gene-population, gene-environment). As population ecologist Juha Tuomi diagnosed, "When the organism-environment relation is omitted, this does not hurt the logic of population genetic models which analyse the consequences of selection pressures on gene frequencies" (Tuomi 1981,

27; for discussion on this issue within quantitative genetics, see Fábregas-Tejeda and Baedke 2023a). In these developments, we see again another instance of how the organism lost its previous explanatory function as a causal agent that both constructs its environment and it is affected by it in evolutionary dynamics.

The study of reciprocity of different (nonorganismal) relata was not exclusive to population and quantitative genetics. Let us now briefly look at the case of *population-environment reciprocity* in population ecology in some detail before moving to the contemporary biological landscape in section 5.4. This episode showcased an interesting debate regarding the proper relata of reciprocal interactions—namely, organism-environment or population-environment—and between the notions of organismal environment versus population environment.

5.3.3 Population and Environment as Relata

Ecological studies of population growth started to increase their pace in the 1920s. For instance, the American biologist Raymond Pearl conducted experiments using cultures of *Drosophila* populations in controlled environments of known size. Through his observations, Pearl drew some conclusions regarding the behavior of populations under these conditions. His findings indicated that populations often exhibit a tendency to approach a saturation point specific to their respective environments (Pearl 1925, 25–44). Likewise, researchers such as the Canadian entomologist William Robin Thompson, the American physical biologist Alfred Lotka, and the Italian mathematician Vito Volterra made advancements by quantitatively investigating the factors that contribute to population stabilization through mathematical calculations of theoretical rates of reproduction (W. R. Thompson 1928; Lotka 1925; Volterra 1926). As historian Sharon Kingsland (1986) has shown, this mathematization trend in population ecology consolidated in the 1930s and 1940s (see, e.g., W. R. Thompson 1939).

In this scientific landscape, population and environment began to be treated as relata in interaction. The German-born Israeli entomologist Friedrich Bodenheimer, for instance, argued that environmental factors, and not internal factors of organismal competition and struggle for life, caused fluctuations in animal populations (Bodenheimer 1928). Bodenheimer was trying to conceptually articulate a relationship between environment and population, without considering the mediation of individual organism-organismal environment connections. Another example of this conceptual articulation can be found in the work of Russian population ecologist Georgii Frantsevich Gause. Some of his experiments consisted in taking a few select populations of protozoans, confining them within a controlled and uncomplicated test-tube environment, and closely observing the outcomes of their interactions. The questions at hand were whether these populations could coexist harmoniously or face extinction. Additionally, Gause (1937) sought to determine whether population oscillations, as anticipated by the existing mathematical models of population ecology, could be generated within this laboratory setup. Gause encountered difficulties replicating the oscillations predicted by the Lokta-Volterra predator-prey equations. Instead, what occurred was the predator species eradicating all the prey and subsequently succumbing to its own demise. However, Gause (1937) managed to achieve a semblance of oscillatory patterns in the populations by skillfully manipulating the environmental conditions.

In this kind of investigation, the notion of "regulation" became a theoretical cornerstone. Hagen (2021) has provided extensive evidence to document how dominant strands of ecology and evolutionary biology relied on and reimagined the concept of "homeostasis" and "regulation," two ideas that had been very important for organicist and holistic thinkers such as Henderson (see section 5.2.4). These permeated various aspects of their frameworks and theories. The notion of "population regulation" is just one example of this conceptual ingress from early twentieth-century physiology (see also Wilbert 1970).

Throughout the 1930s and 1950s, the research surrounding population regulation witnessed a gradual division (for discussion, see Waddington 1965; Collins 1986). "Density-independent" and "density-dependent" forms of regulation were pitted against each other. The Australian entomologist Alexander John Nicholson, in particular, was steadfast in maintaining that an environmental stressor could engender alterations in population density, subsequently influencing the degree of competition within the population. As the strength of competition fluctuated in response to these changes, the population would successively strive to reach a state of density equilibrium with the environmental stressor. Population biologists thought this revealed complex interactions established between ecological factors, competition dynamics, and the drive of populations to attain a balance within their environments. In other words, compensatory reactions counteract to some degree the adverse effects of the stresses to which the populations are subjected, and thus, population and environment are entangled in a reciprocal causal loop (Nicholson 1933).[17] This reciprocity, for Nicholson, was in fact natural selection acting on populations (Nicholson 1960; see also Gause 1934, chapters 2 and 3).

The density-independent faction was personified by William Thompson and his followers. They argued that populations were under the sway of extrinsic factors, such as climatic conditions, which operated independently of population density (W. R. Thompson 1956). At its core, this debate was entwined with the very definition of a *population*. Proponents who ascribed significance to density-dependent factors viewed the population as an entity with its own distinct identity. Therefore, the concept of self-regulation implied a high degree of cohesiveness. In contrast, advocates of density-independent factors viewed populations as aggregative sums of individual organisms whose existence was the focal point.

In this sense, not all population ecologists agreed with the central displacement of the organism as an environmentally embedded unit and evolutionary meaningful relatum. For instance, Thompson, a contributor to population ecology himself, launched a pointed critique against Nicholson, accusing him of misconstruing the fundamental tenets of evolutionary theory. At the crux of Thompson's disagreement was Nicholson's contention that the balance and the regulatory mechanisms within a population suggested the actuality of a cohesive natural unit (i.e., the population as a singular counter relatum to an environment). Thompson, however, dismissed this notion as bordering on the supernatural, asserting that populations lacked independent existence and, consequently, could not possess self-regulatory properties. For him, only individual organisms existed de facto, and it was through understanding the physiology and behavior of these individuals that one could elucidate downstream population control. Thompson took issue with the bio-mathematicians' presumption that populations behaved as a unified entity governed by mathematical laws (Kingsland 1986, 250). He sought to debunk the notion of "population" as an autonomous, self-individuated entity, redirecting the focus toward individual organisms as the primary agents of population regulation (Kingsland 1986, 251).

A continuing controversy involving the uncertain ontological status of biological populations kept resurfacing for the next decades (see, e.g., Jonckers 1973 and references therein). In any case, it is undeniable that mid-twentieth-century biologists conceptualized population and environment as relata that could be modeled and scientifically appealed to, and whose interactions could be measured and described. This was the case in a diverse range of biological disciplines, from systematics to population ecology and quantitative and population genetics (see, e.g., Simpson 1961, 154; Davidson and Dunn 1967). Views of population-environment reciprocity became more and more prominent in the 1950s and 1960s (see, e.g., Wilbert 1961; Bodenheimer and Schiffer 1952; Nicholson 1954a, 1954b). An example of this comes from the work of Stebbins:

> The technical difficulties involved in providing a complete chain of information from molecular structure through population dynamics to *the identification of specific population-environment interactions* are admittedly great. Although many years may elapse before this complete chain of knowledge has been synthesized, nevertheless the way toward its accomplishment has been shown by the exploratory research that has already been accomplished. (Stebbins 1972, 458)

The epistemic move to consider population as relatum was only possible because a parallel concept sedimented in the minds of biologists: that of *population environment*, something distinct from an organismal environment. The idea of "population environment" is now common conceptual currency across ecology and evolutionary biology. Along those lines, Brandon (1992) has argued that the only concept of environment that fits the explanatory needs of the theory of evolution by natural selection, one of the conceptual keystones of population and quantitative genetics, is that of a (population) "selective environment" (i.e., factors external to a population that affect its members' relative reproductive success). He contends that although no two organisms will ever interact with the exact same set of environmental factors, they need to be seen as members of a homogeneous reference class, namely, a "selective neighborhood." Explaining or predicting changes of frequency of one genotype over another due to the action of natural selection requires that the two are related to a homogeneous selective environment that encompasses the whole population and that one is better suited to it than the other (Brandon 1992, 2012). This had already been anticipated by many early-twentieth-century commentators, such as the organicist E. S. Russell (1909b, 77).

As Antonovics, Ellstrand, and Brandon stated in another article regarding the population environment, "For the theory of natural selection to have explanatory power with regard to how adaptations originate, the concept of environment is as important as that of fitness" (Antonovics et al. 1988, 280; see also Abrams 2009; for a critique, see Walsh 2022).[18] The idea of the population (selective) environment was also important in debates in the 1950s about how to empirically assess and distinguish natural selection outcomes from other (purportedly) population-based evolutionary processes such as drift (see, e.g., Millstein 2008).

Similar to the relationality desideratum explored in chapter 3 (section 3.1.1), philosophers of biology usually contend that the problem of defining a "population environment" requires that we start with a situated population in the first place (but see Glymour 2011). They reason that every biological population is embedded in an environment and that environments are only predicated of particular populations with which they hold a set of relations. In particular, Millstein (2014) has argued that the problem of situating populations can be solved if we construe populations, in a causal interactionist sense, as *individuals*, and pay attention to their boundaries. I won't pronounce myself on the issue of whether the isomorphism between population (qua individual) and (individual) organism is warranted or not, but it is interesting to note that both views lead to carving out different environments from relations with the same physical medium, and organisms and populations are said to be crucial for determining the nature of their effective or selective environments, respectively (see also Brandon 1992).

In a similar vein to the controversy surrounding the ontic status of population, the characterization of the "population environment" was a subject of disagreement among many biologists during the mid-twentieth century. Milne and many of his colleagues reasoned that the environment that affects individual organisms held greater significance than its posited populational counterpart: "In population dynamics, *the effective environment must be defined with respect to the individual because the population is itself an environmental factor having effect on the individual according to density*. This presents no difficulty since *what happens to a population is simply the sum of what happens to each of its individuals*" (Milne 1962, 2; emphases added; see also Milne 1957).

In contrast, the concept of the population environment served as a significant convergence point for population ecologists and population geneticists. For instance, at the 1957 Symposium on Quantitative Biology at Cold Spring Harbor, entitled "Population Studies: Animal Ecology and Demography," organized by L. C. Birch, Dobzhansky, Frank Lorimer, and Bruce Wallace, many scholars convened to discuss the importance of what could be said to be "population biology" writ large (*Population Studies* 1957). Participants included, among many others out of a list of 150 biologists, population geneticist Luigi L. Cavalli-Sforza, geneticist Natalia Dobzhansky née Sivertzev, geneticist Leslie C. Dunn, ecologist G. Evelyn Hutchinson, population geneticist and theoretical ecologist Richard Levins, evolutionary biologist Richard Lewontin, population ecologist A. Milne, population ecologist Alexander J. Nicholson, entomologist and ecologist David Pimentel, evolutionary biologist Leigh Van Valen, and developmental geneticist C. H. Waddington, many authors we have already mentioned in this chapter. One of the major debates revolved around determining the most appropriate understanding of the biological environment. As Solomon recounts in his review of the symposium, "There [was] some discussion of the concept of 'environment.' H. G. Andrewartha re-affirms his preference for thinking exclusively of the individual's environment. Th. Dobzhansky, A. J. Nicholson and Hutchinson oppose any such restriction and uphold the common and convenient use of the term with reference either to an individual or to a population" (Solomon 1959, 325).

Dobzhansky exemplifies a figure that supported the idea of both organismal environments and population environments within biological research. For instance, in 1957 he engaged in a debate with philosopher Thomas Morrill in the pages of *The American Naturalist* on how definitions of "environment" matter for conceptualizing or misconstruing adaptive evolution. Morrill argued that in evolutionary biology there was a semantic ambiguity regarding the environment concept that had consequences for how adaptive evolutionary change should be understood:

> In the article "What is an adaptive trait?" . . . , Professor Dobzhansky wrote: "The basic postulate of the modern biological theory of evolution is that adaptation to environment is the guiding force of evolutionary change." It seems to the writer that before we can discuss adaptive traits we must decide what is being adapted to. The term "environment" . . . is a big word, subject to such enormously different interpretations as to be useless without definition. (Morrill 1957, 269)

In response to Morrill, Dobzhansky, hastily covering many topics, replied,

> Mr. Morrill asks for a definition of the "environment" to which organisms become adapted in the process of evolution. As with some other fundamental concepts of biology (life, gene, individual, species), an attempt to produce a formal definition would land us in a morass where no simple idea can be conveyed without endless quibbling and hair-splitting. Fortunately, the situation is not as bad as some strict logicians imagine. . . . In the broadest sense, the environment is the whole universe. Such a broad concept would, of course, be worthless in biology; all organisms inhabit the same universe, but they often stand to it in different relations, even if they are close neighbors in the spatio-temporal continuum. Thus, *Drosophila pseudoobscura* and *Drosophila persinilis* fly together in the Transition Zone of the Sierra Nevada of California, but the former species is relatively more active in the evening and the latter in the morning. Their environments are, accordingly, not wholly identical. . . . Interactions *between organisms and environments are often reciprocal*; a culture medium inhabited by a clone of bacteria or by a strain of Drosophila is evidently not like this medium was before the inhabitants were introduced. *The only sense in which environment and natural selection may be "ending at the organism's skin" is that the so-called internal environment and genotypic environment are really different phenomena from the "external" environment*, and their names are metaphors which become misleading if taken too literally. Otherwise, the environment may be what the organism chooses it to be, within limits. (Dobzhansky 1957b, 269–270; emphases added)

In a close reading of the above passage, we can see that, even though Dobzhansky explicitly mentioned reciprocal interactions between organisms and environments, he was actually thinking about *populations* and not *token* organisms (e.g., the mention of different species of *Drosophila* in the Sierra Nevada of California). In fact, he ambiguously transitioned to the population level in the span of the same sentence: "Interactions between *organisms* and environments are often reciprocal; a culture medium inhabited by *a clone of bacteria or by a strain of Drosophila* is evidently not like this medium was before *the inhabitants* were introduced" (Dobzhansky 1957b, 269–270; emphases added). For Dobzhansky, even though organismal environments and population environments are valid biological construals and go together, the latter supply the evolutionary consequences of note (see, e.g., Dobzhansky and Wallace 1953).

Dobzhanksy came to accept the reciprocity of organisms and environments and populations and environments as he was influenced by the work of W. C. Allee and the ecologists of the Chicago School. They had stressed certain commonalities of organisms and populations qua relata (see Allee et al. 1949), and Dobzhansky approvingly cited their characterization in the revised third edition of *Genetics and the Origins of Species*, published in 1951:

> Allee et al. argue that a population is not a group concept but a spatiotemporal entity which possesses the following five organismic attributes. (1) A definite structure and composition. (2) The population is ontogenetic. It exhibits (as does the organism) growth, differentiation and division of labor, maintenance, senescence, and death. (3) The population has a heredity. (4) The population is integrated by both genetic and ecologic factors that operate as interdependent mechanisms. (5) *Like the organism, the population is a unit that meets the impact of the environment. This is a reciprocal phenomenon, since the population is altered as a consequence of this impact, and, in time, it alters its effective environment.* (Dobzhansky 1951, 15; emphasis added)

Dobzhansky's embrace of both concepts of environment was not universally shared among biologists, and the clash between supporters of organismal environments and population environments lasted for some years. In 1965, Maelzer commented on the persistent opposition between conceptions of organismal environments versus population environments in ecology:

> Attitudes to ecological experimentation and ecological theories are influenced by the acceptance of particular concepts of environment and of its influence on an organism. Ecologists today accept two concepts of environment, namely, the environment of an individual (e.g. Andrewartha & Birch, 1954 and others) and the environment of a population (e.g. Solomon, 1949, 1959). Proponents of each concept have failed to sway the opinion of those who adhere to the other view. (Maelzer 1965b, 395)

Interestingly, in the camp that was skeptical of population environments, some scientists argued that this notion was ultimately an abstraction, a useful shorthand that nevertheless carried the risk of pernicious reification for biological research (see also section 3.3.1). For instance, Andrewartha and Birch, in their famous work *The Distribution and Abundance of Animals*, claimed,

> The difficulty of thinking of the "environment of a population" is that it leaves out half the picture. *The phrase itself represents a false abstraction.* It is an example of what Whitehead . . . called "the fallacy of misplaced concreteness." It tends to confuse rather than to clarify. For this reason we speak of the "environment" of the individual, regarding the population as part of the environment rather than as itself having an environment. (Andrewartha and Birch 1954, 13; emphasis added; see also Mason and Langenheim 1957, 331)

Dobzhansky took the challenge of replying to Andrewartha and Birch and defended the legitimacy of speaking of population environments:

> Among the most important components of the environment which an individual has to face are other members of the Mendelian population of which this individual is part. But it can be argued that the following proposition is equally true, namely that *a Mendelian population is an organic system which encounters and reacts to physical and biotic factors in its environment.* Andrewartha and Birch have rightly pointed out that such a *dualistic concept of the organism-environment relation* incurs methodological difficulties. (Dobzhansky 1957a, 285; emphasis added)

The "dualistic concept of the organism-environment relation" that Dobzhansky is referring to in that passage is the theoretical coexistence of the dyads token organism-environment and population-environment. The interconnectivity within a Mendelian population (on this notion, see Dobzhansky 1950c; for discussion, see also Loison 2024), Dobzhansky argued, reveals that each individual—as a bearer of a unique genotype—constitutes an integral part of the environment for others. For him, this notion found compelling support in investigations conducted by Richard Lewontin, himself, Howard Levene, and Olga Pavlovsky. In a coauthored study by Levene, Pavlovsky, and Dobzhansky, the adaptive values of karyotypes in experimental *Drosophila pseudoobscura* populations were revealed to be influenced by factors beyond the physical environment and nutritional conditions. In fact, the presence of other karyotypes within the same population had an effect in shaping their adaptive characteristics (Levene et al. 1954). For Dobzhansky, this discovery emphasized the necessity of considering the genetic composition of a population when assessing its adaptive potential. In addition, Lewontin compared twenty-two strains of *Drosophila melanogaster* with respect to their larval viabilities and came to the conclusion that the viability of a genotype is related to the presence and interaction of other coexisting genotypes. The outcome of a specific genotypic combination cannot be reliably predicted based uniquely on the viability of the individual genotypes when weighed in isolation. Instead, the collective dynamics and back-and-forth between genotypes within a population exert an influence on their overall viability (Lewontin 1955). These studies made Dobzhansky think about the implications of how the presence of diverse genotypes can influence adaptive dynamics within a population (Dobzhansky 1957, 386–392).

As we have seen, in the second half of the twentieth century, genetic and population research was rife with (epistemically productive) abstractions and not a few unfortunate reifications. The shift away from organism-environment reciprocity was also driven by the desire of scientists to establish clear demarcations between individual organisms and their environments as a methodological necessity for fruitful genetic research, something that becomes quite complicated if organism-environment interactions are taken seriously throughout ontogeny (see section 4.5.1). As population geneticist John Burdon Sanderson Haldane, the son of the holist John Scott, succinctly expressed, a clear demarcation between organism and environment serves as a "practically and theoretically valuable abstraction" in genetic investigations (Haldane 1936, 349).[19]

Moreover, in the work of population geneticists and ecologists, environments were held constant in experimental interventions or considered homogenous in field settings across the individual organisms making up a population for the sake of epistemic simplicity and tractability. Waddington and his colleagues tried to point out this limitation of population models:

> In the development of the mathematical theory of evolution . . . , the environment is normally treated as uniform and the behaviour of various mutants in respect to it is discussed simply in terms of a coefficient of selection by which each mutant is characterised. In practice it is, of course, clear that the majority of species live under conditions in which the environment is not uniform but presents many minor variations. The variant types present in the population will react with these variations of the environment in much more complicated ways than are contemplated by the simple mathematical theory. (Waddington et al. 1954, 89)

Lewontin was aware of these limitations and invited his population-oriented colleagues to overcome them:

> Population genetics and population ecology are both at a critical juncture in their development as scientific disciplines. As with all other sciences, they have passed through a stage of over-simplification necessary for the sorting out of the various elementary processes influencing the observed phenomena with which they deal. . . . The Lotka-Volterra model of population increase . . . and Fisher's classical Fundamental Theorem of Natural Selection (1930) are attempts to predict the future status of a population either in terms of size or in genetic constitution, from a few basic parameters. The success of these theories and the applicability, to natural populations, of experiments constructed with these theories in mind are dependent upon two assumptions. First it must be assumed that the forces

> operating those determining birth and death rate and the direction of natural selection, are constant throughout the life of the population. Second, it must be supposed that rare or unique events intrinsic to the population play no decisive role in its future history. *The assumption of constant extrinsic forces is really an assumption of environmental constancy.* The force which natural selection exerts upon a gene is a function of the total environment, biotic as well as physical, in which the population reproduces. The same is obviously true of the population rate of increase. *Clearly the assumption of a constant environment cannot stand in nature, so that these deterministic theories must ultimately be modified or discarded.* (Lewontin 1957, 395; emphases added)

The second assumption Lewontin mentioned had to do with how organisms within a population interacted with one another and affected their surroundings with important evolutionary consequences. In other words, it involved the neglect of organism-environment reciprocity and the causal arrow that stemmed from organismal activities. As we will see in the next section, it took a while before the homogeneity of the (population) environment began to be questioned again (but see Levins 1968) and the importance of organism-environment interactions was accentuated with renewed strength in the agenda of ecology and evolutionary biology.

5.3.4 Theoretical Challenges of Organism-Environment Reciprocity

Against this background, we can start to realize that, from a theoretical standpoint, the transition from organism-environment reciprocity to alternative frameworks such as population-environment reciprocity is not entirely surprising, given the conceptual hurdles that the former encounters and the epistemic payoffs the latter has shown to successfully convey in ecological and evolutionary research (see above).[20] In particular, traditional approaches to organism-environment reciprocity (such as those surveyed in sections 5.2.3–5.2.5) encountered two significant challenges, which were already recognized by some scientists at the time and contributed to their decline in the latter half of the twentieth century, in addition to the funding-related issues and the agenda shifts in the life sciences described in sections 5.3 and 5.3.1. These challenges can be summarized as follows:

1. As I recounted in chapter 4 (section 4.2), some organicist and holistic authors struggled to establish meaningful boundaries between organisms and their environments. As we saw, this is an epistemic precondition for genetical and population research. These difficulties often led to

perspectives of reciprocity where organism and environment were intertwined, but how this could be probed through empirical study was not entirely clear. During the early twentieth century, many authors embraced holistic perspectives that viewed life as an interconnected unity, blurring the ability to discern independent or partially independent factors (sections 4.3.1 and 5.2.3 on ontological coconstitution). However, the standpoint that regarded the mutual (reciprocal) dependencies between organisms and their environment as forming a single inseparable system faced critique. As we saw in chapter 4 (section 4.3.2), Needham, for instance, challenged Haldane's stance and emphasized the methodological impossibility of unifying the organism and its surroundings. Needham argued that biologists needed individuation criteria to distinguish organismal systems from relevant environmental features, as the complete merging of boundaries posed significant challenges (e.g., how to determine where immediate or far-off surroundings end). In the pages of *Science*, the ecologist Amos H. Hawley also criticized the holistic approach to the organism-environment relationship, as this "when held too rigidly, leads to a disappearance of variables; every property tends to be seen as an aspect of another property. On this score, too, expediency must be consulted" (Hawley 1973, 1196). The blurring of boundaries and of clear cause-and-effect explananda and explanantia variables was a common issue in strong conceptions of organism-environment reciprocity. As a result, many biologists considered this idea intractable or cumbersome to investigate empirically.

2. Another significant problem lay in the main conceptions of the organismal environment proposed by reciprocity theories, which was deemed inadequate by many biologists. In particular, Uexküll's characterization of an organism's *Umwelt* (see section 5.2.5) was considered too narrow in scope. It was argued that the *experienced environment*, as perceived by an organism through its sensory receptors and specialized organs, did not encompass the entirety of its relevant surroundings (see, e.g., Goldstein [1934] 1995, 364; Hartmann 1950; for discussion, see Köchy 2019). Authors claimed that there are physical, external environmental factors that, although not directly experienced, can have causal consequences on organisms and shape their ecological and evolutionary trajectories. Moreover, the notion of "experience" itself became a subject of intense debate, with numerous researchers adamantly opposing any

investigation into aspects resembling subjectivity (see, e.g., Bierens de Haan 1947; Tinbergen 1963).

Theories of organism-environment reciprocity fell short of providing a clear understanding of how experienced and physical environments were interconnected and how they could be integrated within a unified conceptual framework. The limitations of the existing frameworks prompted biologists to seek alternative approaches that could account for the interplay between organisms and their many-sided environments in more scientifically tractable ways (e.g., a shift toward studying population-environment reciprocity or gene-environment reciprocity).

Related to item 2, an overt "externalist logic" in evolutionary explanations reigned supreme from the 1960s onward, and the main concepts of environment employed in the field attest to that shift (see Antonovics et al. 1988, 279–281; Godfrey-Smith 1996a, chapter 2; Walsh 2015). The notion of "individual (organismal) environments," such as that proposed by many organicists, was superseded by populational vantage points (see above), in spite of certain prominent authors such as Dobzhansky vouching for both construals.[21] As we saw in section 5.3.1, this externalist logic impoverished the diverse roles previously granted to the environment in development and evolution. Evolutionists commonly overestimated the causal influence of the selective environment on organisms and increasingly downgraded the influence of the nonselective environment.

The challenges concerning the delineation of boundaries between organisms and their environments, as well as the integration of experiential and physical environmental perspectives, are once again surfacing in contemporary discussions surrounding organism-environment reciprocity within fields such as niche construction theory. Addressing and contextualizing these long-standing historical issues was important to start making progress in our comprehension of the dynamic interactions between organisms and their environments as they are discussed today.

5.4 Organism-Environment Reciprocity in Contemporary Debates

The notion of organism-environment reciprocity has resurfaced and gained considerable traction within contemporary biological parlance, particularly in discussions surrounding the purported reciprocal nature of various developmental and evolutionary processes (see, e.g., Laland et al. 2011, 2013; Schwab et al. 2019; Uller and Laland 2019).[22] In particular, new positions vouching for both reciprocal causation and ontological coconstitution (as we will see below) are making a comeback on the horizon of biological and philosophical research.[23] This resurgence can be attributed to a growing inclination to reorient evolutionary reasoning around the organism itself, as evidenced by the works of many scholars (see, e.g., Nicholson 2014; Sultan 2015; Walsh 2015; Moczek 2015). These researchers have directed attention toward the role of organisms' phenotypic plasticity and their niche construction capacities, both of which, operating as feedback loops, actively modify their environments, thus affecting, for instance, the selective pressures that act on them.

Some biologists contend that the study of complex reciprocal interactions between developing organisms and their environments brings about important reconfigurations and expanded explanatory power to evolutionary biology (Laland et al. 2015; for discussion, see Baedke et al. 2020). At the same time, a philosophical debate surrounding the (alleged) novelty, limits, and scope of the theoretical tenet of "reciprocal causation" has emerged, primarily evaluating the concept's epistemological, explanatory, and heuristic roles in biological practice (e.g., Dickins and Barton 2013; Svensson 2017; Fábregas-Tejeda and Vergara-Silva 2018b; Buskell 2019; Hazelwood 2023). From the scientific side, niche construction theory (NCT) has stood at the forefront of this debate, but allied fields and research areas such as ecological evolutionary developmental biology (eco-evo-devo), the study of phenotypic plasticity, and developmental systems theory (DST) have also contributed (see Odling-Smee et al. 2003; Uller and Helanterä 2019; Baedke and Gilbert 2021).

In what follows, I will direct my attention to NCT, as it serves as a platform for uncovering the main concerns surrounding organism-environment reciprocity. Markedly, within the latest expansions of this conceptual framework, we observe the resurgence of recurring challenges that had previously impacted organicist and holistic research during the first half of the twentieth century (see section 5.3.4). By exploring these issues within the context of NCT, we can gain valuable insights and engage in an informed discussion that will help us to unravel the nature of the organism-environment relationship in biology.

5.4.1 Niche Construction and Reciprocal Causation

NCT challenges the notion of developing organisms as mere vessels of genetic information, instead emphasizing their active role in modifying their surrounding conditions through metabolic processes, activities, and choices. These modifications can have significant ecological and evolutionary consequences (see an overview in Odling-Smee et al. 2013). In this framework, organisms are recognized as exerting causal influence over their own evolution, as niche construction leads to nonrandom modifications and stabilization of environmental states, thereby imposing systematic biases on selective pressures affecting themselves and other species (Laland et al. 2016). Proponents of NCT see niche construction as a developmental process, and "the niche-construction perspective in evolutionary biology is all about exploring the evolutionary ramifications of coupling this particular developmental process with natural selection" (Laland and O'Brien 2011, 193).

NCT has tried to address a common shortcoming in traditional (populational) evolutionary models: "Except in coevolutionary models, it is usually assumed that there are no reciprocal effects of the evolving organism on the environment, although we know that it is common for organisms to do so. If organisms affect their environment, they will affect the conditions for natural selection for their offspring if not themselves, and this will affect the subsequent direction and rate of evolution" (Endler 1989, 330). Niche construction is said to have a codirective effect on adaptive evolution by imposing consistent statistical biases on selection pressures (Laland et al. 2017).

Organisms routinely engage in the modification of diverse physical and chemical conditions that constitute environmental parameters (e.g., by fabricating artifacts or altering nutrient cycling). By expanding these examples, Clark and colleagues have amassed compelling evidence showcasing the impact of niche construction on the variability and intensity of natural selection. Their models distinguish between environmental sources of selection that are influenced by the organisms' constructive activities and those that are not (Clark et al. 2020). Furthermore, niche construction has been shown to influence the rates of evolution and produce temporal lags in responses to selection, as well as generate cyclical dynamics (Laland et al. 2016).

In the context of NCT, organisms inherit two legacies from their ancestors: a set of alleles and a modified environment that harbors associated selective pressures shaped by previous instances of niche construction; the latter is referred to as "ecological inheritance." Proponents of NCT have argued that ecological inheritance differs from genetic inheritance in several key ways (Odling-Smee 1994, 2009; Odling-Smee and Laland 2011): (1) it is transmitted through environmental modification, not reproduction; (2) it relies on sustained niche construction, not replication; (3) continuous transmission from multiple organisms to others, both within and across generations, is possible; and (4) it is not dependent on genetic relatedness and thus can be vertical, horizontal, or oblique. Moreover, ecological inheritance influences parent-offspring resemblance by reconstituting developmental niches, which is crucial for trait recurrence across generations in many multicellular organisms (see Badyaev and Uller 2009).

Beyond development and inheritance, niche construction yields ecological consequences that extend across multiple domains, ranging from the distribution and abundance of organisms to the dynamics of trophic relationships and the modulation of matter-energy transfer. The implications of sustained rounds of niche construction extend to macroecology, where its effects become apparent through the establishment of "engineering webs" within communities and ecosystems (Laland et al. 1999; see also Laland et al. 2017). Furthermore, Fábregas-Tejeda and Ramsey (2024) have recently argued that niche-constructing activities can also influence drift probabilities by altering population size and individual variance in possible reproductive outcomes, known as "driftability."

In recent years, partially due to the impetus conferred by NCT-inspired research, there has been a growing recognition among evolutionary biologists of the centrality of reciprocal causation as both a theoretical principle and a modeling framework.[24] Advocates of this perspective argue that it can, for instance, account for the presence of stable, biased selection pressures acting on organisms. However, the concept of reciprocal causation remains somewhat ambiguous and calls for further philosophical clarification. This exploration is indispensable, I think, to fully grasp the implications and potential of organism-environment reciprocity in evolutionary biology. By addressing its ambiguities and clarifying its epistemological aspects, we can enhance our understanding of this concept and its role in advancing biological theorization.

In order to explain the feedback between constructing organisms and environments, as well as the developmental effects on evolutionary trajectories (and vice versa), some evolutionary biologists argue that the traditional dichotomy Mayr advanced between ultimate and proximate causes (see section 5.3.1) should be replaced by a framework of "reciprocal causation" (for a counterpoint, see Hazelwood 2023). In previous publications, Kevin Laland (now Kevin Lala) referred to the codetermination and mutual adjustment of some properties of organisms and their environments as "cyclical causation" (Laland 2004). According to this view, developmental processes coconstruct with natural selection the complementarity between the organism and its environment and affect the direction and rates of evolution. In other words, developing organisms are not only products but also causes of evolution and active starting points of evolutionary trajectories. Therefore, the proximate causes of developmental processes should not be strictly isolated from ultimate causes of evolutionary dynamics. Instead, organismal activities feed back to affect the environmental interfaces in which phenotypic traits are expressed, and thus the pace and/or direction of adaptive evolution can be modified. As a result, investigations into ontogenetic causal processes, ranging from differential gene expression and tissue differentiation to the constructive actions of organisms within their environments, offer valuable explanatory insights into the processes by which organismal evolution happens (Laland et al. 2015).

In tracing the origins of the notion of causal reciprocity between organisms and their environment, many authors refer back to Levins and Lewontin's *The Dialectical Biologist* (1985) as a key source. Others highlight Waddington's perspective on the "exploitive system" to underscore his "proto-niche construction" ideas (see section 5.2.5). However, as I have shown in this chapter, what these authors may not be aware of is that Levins and Lewontin are late successors of the broader Organicist Movement that embraced organism-environment reciprocity as a foundational notion for biological reasoning. Importantly, as we have seen (section 2.4), this organicist and holistic movement predates, and therefore did not react to or build on, the Modern Synthesis in evolutionary biology.

Odling-Smee and colleagues have recognized Waddington as someone who reconceptualized the relationship between organisms and their environment in directions that resonate with their constructivist framework. In their historiography, they position Waddington as a "precursor" to their own theoretical perspective, as he was pondering about "the many ways in which organisms modify their own selective environments throughout their lives, by choosing and changing their own environmental niches" (Odling-Smee et al. 2003, 29). Waddington's influence is further evident in the taxonomic classification of the two distinct forms of niche construction that Odling-Smee and colleagues originally identified: "*Perturbation* occurs if organisms actively change one or more factors in their environments at specified locations and times by physically changing them. . . . *Relocation* refers to cases in which organisms actively move in space . . . [exposing] themselves to alternative habitats, at different times, and thus to different environmental factors" (Odling-Smee et al. 2003, 44–45; emphases added).

The concept of "relocation" within NCT thus represents a reformulation of the Waddingtonian exploitive system. In fact, Odling-Smee discussed Waddington's ideas in his earlier studies on animal learning and evolution in the late 1970s and early 1980s, as well as in the seminal text where he introduced the term "niche construction:"

> Waddington realized that organisms don't just sit on the receiving end of naturally selecting inputs from their environments. On the contrary, they select . . . their own habitats, select their own mates, select and consume resources, generate detritus, and even construct parts of their own environments. . . . In addition, many organisms select, protect, and provision various "nursery" environments for their offspring. Therefore, Waddington argued, *evolution cannot just depend on naturally selecting inputs from environments to organisms; it must also depend on the outputs emitted by active organisms back to their environments.* (Odling-Smee 1988, 76; emphasis added)

The conceptual framework of NCT appears to align seamlessly with the notion of organism-environment reciprocity, particularly in its framing of reciprocal causation between two separate yet interacting relata. This alignment is evident not only in the aforementioned assertions on the need to overcome Mayr's entrenched dichotomy but also in certain historical antecedents of this conceptual framework. Nevertheless, recent expansions of NCT have broadened its scope to encompass a wider range of developmental processes that surpasses the original taxonomy of "perturbation" and "relocation." Within these new developments, there is a simultaneous endorsement of ontological coconstitution, instead of reciprocal causation, as a valid construct for understanding organism-environment reciprocity. However, in these novel directions, we encounter familiar challenges that have resurfaced, echoing past concerns that we uncovered in the historical arc of this chapter.

5.4.2 Returning Challenges of Organism-Environment Reciprocity Within Niche Construction Theory

Increasing philosophical interest in NCT has led to several different taxonomies distinguishing diverse kinds of reciprocal processes of niche construction that occur during organismal development. Aaby and Ramsey (2022) propose that within the framework of niche construction, organisms engage in three distinct ways of altering the relationship between themselves and their environments. These modifications occur through the (1) alteration of environmental factors, (2) changes in organismic features, or (3) modifications in the relations between organismic features and environmental factors. They call these three types *external*, *constitutive*, and *relational niche construction*, respectively. In contrast, with a different nomenclature, Chiu (2019) distinguishes between *physical niche construction* and *experiential niche construction*.[25] Physical niche construction, which could be taken to be synonymous with perturbational or external niche construction, involves organisms altering the physical properties of their environment, while experiential niche construction focuses on the way organisms perceive their surroundings, without necessarily causing physical changes to it. The concept of experiential (mediational) niche construction is usually traced back to the work of Richard Lewontin, who expressed that an organism's experience of its environment had bountiful effects for evolutionary dynamics:

> *If it is true that the life activities of an organism enter into the specification of its environment, then different phenotypes will make different environments.* Because evolution occurs by a change in the frequencies of heritable types in a population, it follows that during the evolution of a change in the frequencies of heritable types in a population the environments experienced by the population are changing. But this, in turn, means that selective forces, which are a consequence of relations between organisms and their environments, are also changing in a way that is sensitive to the change in population composition. (Lewontin 2000, 56–57; emphasis added)

Within Chiu's (2019) classification, mediational niche construction and relocational niche construction are regarded as subcategories of experiential niche construction. In yet another taxonomical undertaking, Trappes et al. (2022) have presented a conceptual framework that carves out three (purported) distinct mechanisms underlying organism-environment interactions: niche construction, niche choice, and niche conformance. Each of these mechanisms operates at the organismal level, influencing the alignment between an individual's phenotype and its environment, its fitness, and the specification of its *individualized niche*—defined as the environmental conditions required for the individual's survival and reproductive success.[26] In their terminology, niche construction sensu stricto involves the active modification of the (physical) environment by individual organisms, instigating changes that can influence their own phenotype-environment match and overall fitness. This is equivalent to "perturbation" in the original taxonomy of Odling-Smee and colleagues. In contrast, niche choice pertains to the active selection of specific environmental conditions by individuals—a rephrasing of relocation, habitat choice, or the Waddingtonian exploitive system. Through this mechanism, individuals exhibit a preference for particular environmental settings that align with their adaptive needs and characteristics. Finally, niche conformance refers to the capacity of individuals to adjust their phenotypes in response to fluctuating environmental conditions (adaptive phenotypic plasticity), a notion similar to constitutive niche construction sensu Aaby and Ramsey (2022).

These taxonomies of niche construction have broadened our understanding of the varied roster of reciprocal connections that exist between organisms and their environments in the context of development, ecology, and evolution. However, the current debate surrounding different types of niche construction continues to grapple with persistent challenges that theories of reciprocity have long faced, particularly in the early twentieth century (section 5.3.4).

One issue within this debate is the lack of explicit delineation, even if only in an epistemically expedient way, of meaningful boundaries between organisms and environments, which are crucial for research purposes. Scholars often fail to articulate the grounds on which these boundaries can be established and effectively exploited. Another issue is a dearth of theoretical guidance on how to integrate the experiential and physical dimensions of organism-environment reciprocity.

Regarding the first issue, initially, early NCT adopted an externalist perspective, treating organisms and environments as clearly distinct entities (Odling-Smee et al. 1996, 2003; see also Trappes 2021). NCT was built on the "factor-feature" conception of a niche originally proposed by Bock (1980). According to this perspective, a niche comprises the assemblage of selection pressures that act on organisms, with these selection pressures being environmental variables, referred to as "factors," that shape organismal traits, known as "features." With the exception of relocation, every instance of niche construction was primarily viewed as an alteration of the physical properties of the environment. The active role of organismal experience as a causal factor in reciprocal interactions did not hold a central position in early NCT. Subsequently, influenced by DST and its proposition that there is no distinction to be made between organism and environment (see section 4.5.1), proponents of NCT embraced more constructivist perspectives of the environment. Resembling earlier ideas of authors such as J. S. Haldane and Uexküll, contemporary strands of NCT emphasize the organism's central role in evolution while merging the boundaries between organisms and environments. They argue, more broadly, that organisms and environments are intricately engaged in reciprocal couplings and are intertwined in "cycles of contingency," as phrased by DST (see Laland and O'Brien 2011, 193). According to Sultan (2015), it is difficult to establish a meaningful boundary between organisms and their environment, as individual phenotypes inevitably influence both the external environment and the organism's experience of that environment (44–45). By acknowledging the reciprocal connections between an organism's experience and the external environment, Sultan suggests the inseparability and ontological coconstitution of organisms and their surroundings within the context of NCT.

In addition to the challenge of organism-environment inseparability, another obstacle arises in contemporary discussions of NCT: the lack of integration between approaches focusing on physical and experiential forms of niche construction. Although Lewontin reintroduced the concept of experiential niche construction to evolutionary biology, it has not gained substantial traction to date (for an analysis, see Chiu 2019). On this topic, Lewontin had originally stated,

> Organisms . . . transduce one physical signal into quite a different one, and it is the result of the transduction that is perceived by the organism's functions as an environmental variable. . . . *The biology . . . of an organism . . . determines its effective environment,* by establishing the way in which external physical signals become incorporated into its reactions. *The common external phenomena of the physical and biotic world pass through a transforming filter created by the peculiar biology of each species, and it is the output of this transformation that reaches the organism and is relevant to it.* (Lewontin 2000, 63–64; emphases added; see also Levins and Lewontin 1985, 97–106)

Critics of Lewontin's view of experienced environments have argued, for instance, that the notion of experiential or mediational niche construction simply represents a phenotypic response to selective environmental pressures. They caution about conflating changes in environmental experiences with changes in the environment itself. For instance, Brandon (1992) has stated that an organism's viewpoint and experiences are irrelevant when dealing with selective environments (68). In a more nuanced critique of Lewontin's vantage point, Peter Godfrey-Smith has argued that we should distinguish niche construction from organismal accommodation—namely, cases of an organism (physically) constructing an environment and cases where an organism developmentally accommodates to an environment by altering its features (Godfrey-Smith 1996b, 460). Such conflation, critics hold, may lead to the defense of antirealist and irresolvable holistic positions, where organisms subjectively construct their environments without any anchorage to the world.

While recent years have witnessed attempts to bolster the significance of experiential mediation of the environment in organismal evolution (see, e.g., Sultan 2015; Walsh 2015, 2022; Chiu and Gilbert 2020), some of these approaches still adhere to the Uexküllian perspective, wherein the entire environment of an organism is interpreted solely as the experienced environment. For those seeking to avoid this stance, compelling arguments must be presented to mark off the evolutionarily relevant physical environment of organisms from their experienced environment. One needs to answer why and how certain downstream physical effects on the surrounding conditions can be discerned and excluded from the processes of experiential niche construction, and how closely experience influences evolutionary processes, such as population dynamics, in a distinct manner from the physical effects on selection pressures (i.e., perturbational or external niche construction). For example, it remains largely unclear how shifts in an organism's experiences can alter the selection pressures acting on itself and conspecifics or establish individualized niches (but see Müller et al. 2020). I believe that this lack of integration is partly due to the fact that proponents of experiential niche construction often embrace a perspective of organism-environment inseparability and organism-environment reciprocity as ontological coconstitution, which can present challenges when it comes to methodological implementation. These challenges, coupled with longstanding objections against intractable holism, contribute to the ongoing tension between approaches addressing physical/external and experiential niche construction, as well as the broader application of experiential niche construction within evolutionary theory.

Against this background, we could ask, What is the best construal to understand organism-environment reciprocity? Is it reciprocal causation or ontological coconstitution? Are there ontological and epistemic or methodological reasons to prefer one over the other? In what follows (section 5.5.1), I will argue that reciprocal causation is the most suitable understanding of organism-environment reciprocity, given what we can say about the metaphysical underpinnings of organismal environments (chapter 3) and the problem of organism-environment separation (chapter 4). Moreover, I will argue that also in the epistemological domain there are several advantages for favoring reciprocal causation over ontological coconstitution (section 5.5.2) as a construal of organism-environment reciprocity.

5.5 Adjudicating between Reciprocal Causation and Ontological Coconstitution

5.5.1 Reciprocal Causation vis-à-vis Ontological Coconstitution: Metaphysical Remarks

In the beginning of this chapter, reference was made to Toepfer's (2011) characterization of reciprocity, where he suggests that this notion, when broadly interpreted, pertains to the interconnectedness of different components within a system. This interconnectedness can manifest as mutual dependence or as mutual influence. If we follow Toepfer's characterization, ontological coconstitution aligns with a conception of reciprocity that underscores the close-fitting interdependence of the relata involved. In contrast, reciprocal causation asserts the bidirectional interactions between organisms and their environments. Now it is time to go deeper into the philosophical analysis of these two positions.

We have already seen that reciprocal causation is typically portrayed as the back-and-forth of feedback loops between two distinct entities: organisms and their environments. Organisms affect the environments surrounding them and, conversely, environmental factors exert changes on organisms, prompting them to respond one more time by modifying specific components of their environment. These alterations, in turn, reverberate back to impact the organisms, thereby initiating a spiraling pattern of reciprocal causation (for an analysis, see Walsh 2015, 180).

Ballard (1955) has argued that we need to distinguish between *linear causal relations*, and *cyclical, reciprocal causal relations*. Systems with cyclical relations are characterized by the presence of events that continuously engage in reciprocal causality. In a relatively isolated system with events A and B, if A causes an effect in B, A in turn, undergoes a change as a consequence. This type of relationship is referred to as *reciprocal causation* simpliciter by Ballard (1955). However, he argues, when the relation between A and B is *continuously reciprocal*, a sort of unique and transformative dynamic ensues. The change in B reverberates back to alter A, which, in its modified state, reacts differently to the changed B, and the cycle continues. Thus, a coupling emerges, comprised of codependent and evolving events or entities—a cyclical, reciprocal system. Ballard further suggested that organism-environment interactions correspond to this scheme of a continuously reciprocal dynamism: "The developing mutual adaptation of organism and environment is an obvious illustration of such a system. . . . As an animal lives in its environment, it changes that environment and becomes changed by this activity, with the result that animal and environment approach a state of mutual adaptation" (Ballard 1955, 208, 210). In this view, reciprocal causation is a diachronic relationship that extends over time (figure 5.1, panel I).

We should distinguish this particular understanding of reciprocal causation from what in recent literature has been categorized as "causal nonseparability." Walsh (2022) gives a nice overview of this notion: two causes, labeled *x* and *y*, are considered nonseparable when the impact of *x* on *y* at a given time *t* is influenced by the effect of *y* on *x* at the same time *t*.

I. Reciprocal causation

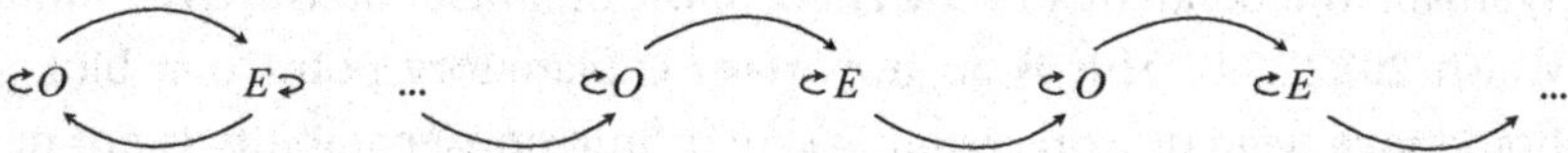

II. Ontological co-constitution

$$Œ \quad \cdots \longrightarrow Œ_{n-1} \longrightarrow Œ_{n} \longrightarrow Œ_{n+1} \longrightarrow \cdots$$

Figure 5.1
Two competing views of organism-environment reciprocity: reciprocal causation (I) and ontological coconstitution (II). See text for explanations. Source: Illustration by Guido I. Prieto.

While synchronic reciprocal causes can be identified separately, their causal dynamics cannot be fully comprehended in isolation. To entirely explain the effect of *x* on *y*, one must also consider the influence of *y* on *x*, thereby necessitating an understanding of the synchronic reciprocal relationship between the two and the back-and-forth of their effects (Walsh 2022, 73–74).

This view does not represent reciprocal causation as introduced and discussed in this book, since both historically and in present-day biological debates, reciprocal causation has been advanced as a diachronic, temporally extended view of organism-environment reciprocity. In this sense, reciprocal causation should be cast within a Humean view of causation in which effects follow causes over time. An organism modifies its surroundings at time *t*, and at time t_{+1} certain elements of its reconfigured environment redound on the organism, changing it. Of course, the modification that an organism makes to its environment at time *t* is causally linked with how its embedding conditions had altered it at time t_{-1}. Reciprocal causation dynamics are thus extended processes, and the diachronic dimensions of this notion capture this facet.

This is not to say that there are no epistemic upshots to ideas such as (synchronic) causal non-separability as it applies to some biological systems. For instance, some authors have mobilized it to the organism-environment pairing itself and, in this case, "when the effect that the environment has on the organism is a consequence of the effect that the organism has on the environment, one cannot explain the effect of the environment on the organism independently of the effect of the organism on the environment" (Walsh 2022, 74). This is an important explanatory point that biologists should take heed of, and certainly synchronic nonseparability is not necessarily at odds with a diachronic view of reciprocal causation. However, I consider that we might not need to appeal to notions such as synchronic nonseparability when we already have a powerful epistemic and ontological construal of organism-environment reciprocity such as reciprocal causation: The very same explanatory upshots of causal nonseparability can be recovered by a framework of reciprocal causation.

To truly grasp the continuously reciprocal relationship between environments and organisms, it is crucial to consider their historical and diachronic context of interconnected loops. In cases where the impact of the environment on the organism arises as a result of the organism's previous influence on the environment, one cannot adequately explain the effect of the environment on the organism without also considering past effects of the organism on the environment. The asymmetry of causation is retained when thinking thorough the mindset afforded by reciprocal causation: In some moments of this complex dynamic, an organism, through its activities and actions (or the alteration of its features), is a cause of environmental variations, but at other times it integrates and copes with the effects of environmental changes qua causes of, for instance, phenotypic plasticity-induced phenotypes. In sum, according to this view, an organism and (elements of) its environment alternate in being both causes and effects in a sequence of interactions,[27] and we need to consider both relata when coming up with explanations or modeling particular aspects of the phenomena under study.

In contrast, ontological coconstitution, as a different way to construe organism-environment reciprocity, implies strong interdependence and some additional elements: As its proponents argue, it also suggests that the organism and its environment are intertwined, forming a unified and interreliant entity (i.e., an organism-environment unit) that cannot be, neither in principle nor in practice, meaningfully separated (see also section 4.3). This is the unit that purportedly changes as time passes (figure 5.1, panel II).

One potential way to spell this out in clearer metaphysical terms would be to propose that organism and environment both hold a constitution relationship to larger organism-environment systems (see also section 5.2.3). In other words, organism and environment are reciprocal parts that constitute a larger organism-environment whole. If we follow this interpretation, the notion of coconstitution would be something of a misnomer or shorthand for describing the reciprocal contributions of organism and environment: It is not that organism and environment coconstitute one another, but rather, that they both constitute a more encompassing, commingled organism-environment system. At first glance, one could think that this constitution relationship is transitive, irreflexive, and asymmetric: Parts in conjunction constitute the whole (i.e., organism plus environment constitute the organism-environment system), but the whole does not constitute any of its parts (i.e., an organism-environment system does not constitute either an organism or an environment). Simply put, organism and environment jointly constitute an organism-environment system, with organism and environment having close reciprocal part-part relations that yield a new unit. In this view, organism and environment are in principle conceptually distinguishable parts of the organism-environment unit, but ontically commingled. If we accept this vision, there are two possible routes to spell out these constitution relations: as instances of (1) material constitution or (2) diachronic constitution.

Material Constitution Defenders of standard metaphysical views of material constitution as a grounding relation would not find the above framing concerning the organism-environment dyad problematic. It could be granted that an organism, its environment, and the ensuing organism-environment system exist, but each has different *de re* properties. Material constitution theorists would claim that, even though an organism and an organism-environment system stand in a very intimate relationship, they are nonetheless distinct (and the same could be said about an organismal environment and the corresponding organism-environment system). That is to say, material constitution is not a relationship of identity but one of dependence (for discussion, see e.g., Wiggins 1968; Johnston 1992).

Material constitution then refers to a synchronous and one-to-one relation of determination between entities of different kinds that are both spatially and materially colocated. This relation can be effectively described by saying, X constitutes Y at a given time *t* if and only if two conditions of coincidence are met. First, spatial and temporal coincidence is required, indicating that X constitutes Y at time *t* only if both X and Y occupy the same spatial location at time *t*. Second, material coincidence is necessary, implying that X constitutes Y at time *t* only if X and Y share identical material components at that specific time (Wasserman 2004, 694). In other words, organism and environment would constitute an organism-environment unit at time *t* only if both organism and environment and the organism-environment unit occupy the same spatial location at time *t* and if organism, environment, and organism-environment unit share identical material components at that specific time.

However, Kirchhoff and Kiverstein (2024) have stressed a complication that arises when attempting to apply the concept of material constitution to open dynamical systems. These systems involve constituent elements that mutually constrain each other's dynamics, functioning as cohesive and integrated units. Considering that organism-environment systems fall into this category, it becomes challenging to impose the characteristics of synchronic material constitution on them. Attempting to determine whether the formation of organism-environment units adheres to synchronic constitution, where constituent elements constitute ensemble behavior and events at an instant *t*, is implausible. It is inappropriate to inquire whether the organism, environment, and organism-environment unit occupy the same spatial location or share identical material components at that specific time. This schema does not align with the formation of organism-environment units due to their temporal nature that ties them as outcomes of diachronic processes.

Philosophers such as Ylikoski (2013) have put forth arguments suggesting that material constitution, as it solely pertains to a synchronous time slice, fails to account for the behavior and coordinated activities of the constituent parts of a system. Consequently, asserting that a dynamical process possesses certain properties at a specific time point constitutes a significant abstraction (which, in some instances, might be explanatorily useful, of course). However, this abstraction should not be mistaken for how the properties of dynamical processes unfold and manifest themselves over time. Simply because a system—say, an organism-environment unit—can be described synchronically does not imply that it exhibits any of its distinctive characteristics within that snapshot. Just to give an example that is usually ignored in evolutionary work, the dynamic interaction between organisms and their environment over time dictates that statistical measures of covariance and regression, widespread in quantitative genetics, should also undergo changes during ontogeny. Relying on a single value at a fixed developmental time (e.g., some moment of adulthood) could potentially lead to a misleading representation of the system (see, e.g., Milocco and Salazar-Ciudad 2022).

Kirchhoff and Kiverstein (2024) caution against the fallacy of conflating claims regarding a dynamic system's behavior by concentrating on a synchronic instant, whose description may still be epistemologically useful within specific explanatory contexts. They emphasize the potential danger of mistaking epistemological tools employed for characterizing a dynamic system for ontological notions about the nature of the system itself. Those are some of the reasons why, they contend, a view of "diachronic constitution," which they ultimately take to be a causal relation, might fare better at capturing dynamic systems than material constitution (see also Kiverstein and Kirchhoff 2023).

Diachronic Constitution Let's explore *diachronic constitution* now. Kirchhoff and Kiverstein (2024) put forth the argument that the temporal dynamics observed in both microscopic and macroscopic behavior within dynamical systems force a conceptualization of cause-effect relations as a coevolving process. As a result, the relata involved in this relationship are not entirely distinct entities. Furthermore, they argue that the relevant dependence between the relata within dynamical systems can be best comprehended through the lens of symmetric dependence.

When explaining evolutionary dynamics, Walsh would seem to agree to interpret organism-environment reciprocity as diachronic constitution:

> Where systems of causes are dynamically, adaptive interacting, nonlinear, and cyclical, in the way that organism/environment interactions are, the dynamics of the system as a whole are "inseparable" . . . . Inseparability between the parts of a system entails that one cannot apportion the effects on the system into those that are due exclusively to one part and those due exclusively to another. The implication for adaptive evolution is that one cannot apportion evolutionary responses into those due exclusively to organisms and those due exclusively to the environment. Rather, the conditions to which adaptive evolution fits organisms are conjointly caused by an organism's environment and the organism's engagement with its environment. (Walsh 2022, 78)

Interestingly enough, for Kirchhoff, Kiverstein, and Walsh, diachronic constitution is an outcome of continuous reciprocal causation. Kirchoff and Kiverstein (2024) say that "diachronic constitution is a consequence of a common type of causation in the empirical sciences: continuous reciprocal causation" (p. 1), and thus, "diachronic constitution tracks continuous and reciprocal causal dependence in dynamical systems over time" (p. 23). Walsh would also agree that, developmentally speaking, reciprocal causation is a good description of organism-environment interactions (but perhaps not for explaining evolutionary dynamics in the sense outlined above; D. M. Walsh, personal communication). Hence, these authors grant the existence of reciprocal causation, which precedes and produces relations of diachronic constitution if sustained long enough, which ultimately yield inseparable organism-environment systems. Organism-environment systems are assembled in a post hoc fashion: They are defined by scientists by looking at a history of mutual interactions between organism and environment. If diachronic constitution obtains from sustained reciprocal causation, this might suggest that the latter is a more primitive—and perhaps a more focal—form of organism-environment reciprocity.[28]

As we have seen throughout this chapter, many authors, past and present, defend that ontological coconstitution is an appropriate construal of organism-environment reciprocity. However, there is an important point that most advocates of ontological coconstitution have obviated: the metaphysical status of the organismal environment qua relatum. And this bears weight on three important issues that put into the question the general adequacy of ontological coconstitution: (1) It incurs a category mistake; (2) it evinces a mismatch of corresponding dependences between organisms and environments; and (3) it is insensitive to the indexicality of organismal environments, and for this it confuses ontic indispensability with ontic inseparability. Allow me to discuss them in turn.

Category Mistake In the metaphysical exploration of chapter 3 (section 3.2), I came to the conclusion that an organismal environment encompasses a plurality of (ontologically heterogeneous) biotic and abiotic factors and conditions that cooccur in an extrinsically generated organism-directed, context-dependent class-as-many, linguistically masked in singular yet not conforming to this on ontic grounds. Any organismal environment is not a clearly delineated singular entity. If what I argued about environments is correct, then the position of ontological coconstitution is merging incommensurate relata. It is simply a category mistake to assert that organism and environment are interdependent parts or symmetrical components of an inseparable organism-environment unit. How could an organism be fused—or be on par—with a relational class? How could an aggregative abstractum, with radically different criteria of identification, identity, persistence, and countability, be united with a (more or less) concretum that falls under a sortal concept?[29] I think that, paradoxically, ontological coconstitution is a metaphysically misguided position of organism-environment reciprocity precisely because it does not take seriously what an environment stands for. Advocates of ontological coconstitution, from the times of Haldane to today, simply take the organismal environment for granted and never spell out how we should understand its ontology.

Mismatch of Dependences Second, the position of ontological coconstitution assumes that organism and environment hold the same type of dependence to one another. But the dependences between organism and environment do not arise in the same terrain. Consider again the principle of "reciprocal codetermination" that Lewontin and Levins (1997) advanced: "There is no organism without an environment, but there is no environment without an organism" (97). Their principle is entirely correct, but it holds for different reasons for each side of the binomial. As I mentioned in chapter 3 (section 3.1.1), organisms, as the condition of possibility for them being alive in the first place and sustaining their existence, are inherently dependent on their surroundings. For instance, organisms depend on their environments to acquire necessary resources, such as matter and energy, for their self-maintenance, long-term survival, and reproduction. Organisms simply cannot exist in isolation, without being embedded in particular environing conditions: They develop, partake in ecological and evolutionary processes, and exist within, through, and as a result of their environmental conditions. Environments are absolutely fundamental for organisms. Nevertheless, authors that endorse ontological coconstitution confuse *causal dependency* (i.e., an organism necessarily needs to be embedded in an environment, especially to harvest enough matter, energy, and resources to survive if the conditions are adequate) with *ontic indispensability* and *ontic inseparability*. Let's review these notions now.

The fact that any organism needs an environing context to be and keep being what it is does not mean that organism and environment are one and the same entity. In contrast, environments also depend on organisms, but for a very different reason: They require an organism to environ, to gain a reference. This is another way of looking at the relationality desideratum mentioned in chapter 3 (section 3.1.1). Environments come into being concomitantly as organisms do: The presence of a developing organism is a *structuring precondition* of manifold, ontologically heterogeneous circumstances that then can be said to constitute an environment. This is not causal dependency, but *ontic indispensability*. A glimpse of this can be found in the work of holist Kurt Goldstein:

> The environment of an organism is by no means something definite and static but is continuously forming commensurably with the development of the organism and its activity. One could say that the environment emerges from the world through the being or actualization of the organism. Stated in a less prejudiced manner, an organism can exist only if it succeeds in finding in the world an adequate environment—in shaping an environment. . . . An environment always presupposes a given organism. (Goldstein [1934] 1995, 85)

Indeed, an environment cannot be disentangled from its counter relatum because it is inherently relational (section 3.1.1). As I have said before, there are simply no environments if nothing is being environed. However, this metaphysical proposition should not, under any circumstances, be interpreted causally: Organisms do not cause their environments to be such (e.g., an organism does not cause the "environmentness" of an environment) nor create fully formed environments: They bring about important causal consequences to the make-up of their surrounding circumstances (e.g., through their niche construction activities) and structure which worldly components would have a potential impact on them through their behaviors, traits (e.g., particular sensory systems), and evolved capacities, but this not to say that organisms "carve out" or cause all the elements of the world that can affect them. Lewontin would say that the biology of an organism *determines* its effective environment (see above), and I agree with him, but this determination is not a causal one, but a *structuring* ontic one. The features and activities of organisms ontically determine what count as their environments (i.e., organisms are ontically indispensable for their environments), but they are not causally indispensable for the extrinsic relations that are formed between different abiotic and biotic components with that organism—these depend on and establish different causal relations.

In sum, the meaning of interdependence in the organism-environment relationship is not symmetrical. Defenders of ontological coconstitution do not take into account that there is a mismatch between the dependence that an organism has on its environment (causal dependency) and the relational dependence that an environment has on said organism (ontic indispensability).

Insensitivity to Indexicality At the same time, ontic indispensability—the type of dependence an environment has with respect to an organism—should not be confused with *ontic inseparability*. I believe the position of ontological coconstitution collapses both as if they entailed the same things. Environments (qua classes-as-many) at particular moments of a life cycle are ontically dependent on token organisms to count as such, but organisms are not inseparable from them: For once, they have multiple, dynamic boundaries with particular biotic and environmental components that mediate, filter, or shut them off (see section 4.6). Contrary to an organism, an environment, qua class, is never self-individuating. The bundle of heterogenous factors included as environmental components at a particular moment of the life cycle of an organism is relationally defined, as we saw before, but we are able to talk about continuities and breaks in environments because of indexicality. The abiotic and biotic members of the environmental class-as-many are shifting constantly, and some might be constant at different points in time. This represents a sharp contrast to how organisms establish their bounds as relata.

In general, the thesis of ontological coconstitution requires that what is embraced under the item of "environment," that which is inextricably interwoven with an organism, does not change dramatically from one point in time to the next. But this is not the norm for organisms. Ontological coconstitution abstracts away from the complexities of conditions that organisms encounter throughout their ontogeny. As I expounded in chapter 3, organismal environments display indexicality—namely, these are tied to the varied ontogenetic contexts of the same token organisms. There is a strong dependence of what an organism does or what developmental changes it is undergoing to what sort of surroundings environs it. As we have seen, the reference of an organismal environment is inescapably context-dependent, a changing bundle of heterogeneous, relevant factors without clear proximal limits. A position of ontological coconstitution would face a hard time making sense of diachronic radical shifts in abiotic and biotic conditions that organisms customarily confront throughout their life cycles.

Because of the phenomenon of indexicality, some organisms indeed *do come apart* from particular environments. Trematodes from our example in chapter 3 (section 3.2.3) are so separable from the particular environing conditions affecting them at a particular time of their life cycle (although always causally dependent on them) that they can move from the water of a lake to the interior of a mollusk and then back to water in a different stage—two completely distinct environments, if you will. But this phenomenon is not exclusive of organisms with complex life cycles: The same is true in many instances of relocation. When organisms physically move, they could then be affected by new environments. And yet the same is true with constitutive niche construction (even for sessile organisms): When organisms change their traits as a product of their normal development or plastic responses, they are altering which abiotic and biotic factors from their surrounding conditions are included as components of their environments.

In contrast to these three shortcomings faced by ontological coconstitution, the view of reciprocal causation seems to be a more promising candidate to clarify organism-environment reciprocity on ontological grounds. Foremost, it does not require that both relata are commensurate in the same stringent sense as ontological coconstitution. Causes and effects do not have to be the same kind of entities (e.g., abstract or concrete entities) in a continuous cycle of modifications.[30] With some conceptual refinement and qualified statements, reciprocal causation is perfectly compatible with the idea of an organismal environment as an indexical class-as-many. When it is said that an organism modifies its environment (e.g., through some kind of niche construction), it is meant that an organism modifies at least *one* of the components of the relational class that we call "its environment" at a particular moment of its life cycle. A niche construction event would never alter the totality of factors comprised under any environment at time *t*. The altered components could then cause changes in the organism. Likewise, a modification of an organism by an environmental change at a certain time would never be conveyed through the totality of the environment, but only through the participation of certain environmental factors. The changed developmental system would then be in a position to further alter the same or perhaps different components of the environment—and so on and so forth. The causal difference makers on the side of the environment in a reciprocal causation chain do not need to be the same factors (e.g., be they temperature values, chemical substances, the presence of predators) throughout a whole time series. And precisely because of this, a view of reciprocal causation can accommodate the indexicality of environments: The composition of members of the class "environment" might vary from one point in time to the next depending on the context of an organismal ontogeny, but this does not interrupt the diachronic history of interconnected causal loops between an organism and its fluctuating surroundings and allows them to be traced further into the future.

Ontologically speaking, we can make sense of reciprocal causation chains being established between a self-individuating, agential concretum (see chapter 6) and a relational, indexical (but causally efficacious) abstractum. Likewise, through the idea of reciprocal causation, we can still accept the environmental (existential) requirements of organisms (causal dependency) and the relational definition of environments (ontic indispensability of organisms as counter relata) that the "reciprocal codetermination" principle of Lewontin and Levins demands. I think that, overall, reciprocal causation is a stronger construal of organism-environment reciprocity than ontological coconstitution.

Walsh (2022) stands as one of the few scholars who has explicitly examined the philosophical nature of environments and has arrived at the conclusion, albeit through a different path than the one explored in chapter 3, that the environment is an abstraction.[31] I concur with this conclusion for organismal environments (section 3.3), but I recognize that environments-as-abstractions still hold ontological significance for biological science. By adopting a perspective of reciprocal causation, we can restore this significance and gain a deeper understanding of the interactions between organisms and their surroundings. In my position, I do not advocate for the traditional conceptualization of the environment as "a wholly external, autonomous, causally unified entity" (Walsh 2022, 3), which Walsh wants to undermine. As I have argued here, the autonomy assumption is violated because of the importance of organisms determining what counts as environment in the first place (see also Gerard and Maublanc 2023) and given their relentless niche construction activities. I maintain that the accurate allocation and disentanglement of causal contributions in prolonged interactions between organisms and their surrounding conditions remain attainable objectives in the investigation of complex developmental and evolutionary dynamics.[32] This is the case *precisely because* an organismal environment is not a causally unified entity. An environment, as a class-as-many of manifold biotic and abiotic factors, is a causally dispersed entity: At any point in time, it affects an organism through many different routes, in different scales, and with different consequences in development, ecology, and evolution. And because this is so, we can apportion specific causal contributions in causal reciprocity chains. We can inquire about what specific environmental factors produce what effects in organisms and what concrete elements in an environment were altered through organismal activities (or even how specific developmental changes affect the experience of particular environmental factors). Given that an organismal environment includes in its elements myriad different biotic and abiotic factors, it possible to break apart how these are causally affecting the physiology, development, and ecological interactions of organisms. Even more so, precisely because environments are not causally unified entities, the view of ontological coconstitution becomes suspect as well. Ontological coconstitution presumes the oneness of an environment on par with the oneness of an organism (qua reciprocal, constituting parts), but, as I have tried to show throughout this book, an organismal environment is simply not one singular entity.

In concluding this chapter, I will present additional arguments that underscore the superiority of reciprocal causation over ontological coconstitution as the most apt interpretation of organism-environment reciprocity.[33] This assertion is not only grounded in the ontological considerations that I have reviewed in this subsection, but it is also supported by epistemic justifications of considerable weight.

5.5.2 Epistemic and Heuristic Advantages of Reciprocal Causation

There are several domains related to biological practice in which reciprocal causation fares better than ontological coconstitution. These include methodological considerations, scientific explanation, modeling tractability, and predictive capability. Let's explore them in turn.

Methodological Considerations The recent arguments advocating for the inseparability of organisms and their environments introduce the possibility of anti-individualistic positions within the framework of NCT (see Baedke 2019). However, these views present methodological challenges to biological research, as it is crucial to clearly delineate organisms and their causal roles. Purportedly, NCT seeks to understand the organism as a distinct causal agent, capable of actively shaping its own environment and evolutionary trajectories. But views of ontological coconstitution lose the organism as a causal efficacious unit. For example, Pradeu (2010, 219) has argued that DST has this issue: "One loses the idea that it is the organism which evolves when one says that it is the organism-environment, or O-E, system which evolves, i.e., the [developmental system]." The same could be said of other views that champion ontological coconstitution: If organism and environment are mingled, there is simply no organism as a unit. We see this clearly in a critique that William McDougall raised against Haldane's view of undividable organism-environment wholes:

> It is true that many of these assertions of the unity or wholeness of the organism are followed by the assertion that "This unity extends over its relations to environment as well as over the mutual relations of its parts"; or that "the life of an organism must be regarded as an objective active unity which embraces its environment" . . . . Now Haldane, no matter how admirable his aims, cannot be allowed the privilege of eating his cake and keeping it intact. . . . I submit that, if each organism . . . is a unity and has wholeness, it cannot at the same time be (in perception, observation, thought, and space) inseparable from and indistinguishable from its environment; its unity or wholeness cannot "embrace" or "extend over" or include its environment. In short, Haldane uses the unity and wholeness of each individual organism . . . and then he turns round and, in asserting that the organism cannot be distinguished from its environment . . . , he virtually denies the unity and wholeness of the individual organism. (McDougall 1936, 424–425)[34]

Additionally, as highlighted in chapter 4 (section 4.3.1), the holistic stance of organism-environment coconstitution fails to address the problem of individuation of biological units; bring to mind again the critique that Needham waged against Haldane. Merely asserting that an organism and its environment form an inseparable unit does not provide a definitive solution for individuating such a system—namely, distinguishing it from other systems. Individuation becomes indispensable for discerning between the proximal and distal environments of a specific system. It involves determining the elements of the world encompassed within the system and the relevant spatiotemporal scales at play. Furthermore, individuation is crucial, among other tasks, for clarifying the units of physiological and ecological interactions, identifying which biological systems can legitimately be considered "causal agents," differentiating between individuals within a community or population, and delineating the components within multispecies collectives like holobionts (Baedke et al. 2021). In this regard, the holistic perspective of organism-environment inseparability endorsed within the position of ontological coconstitution fails to address the challenge of individuating biological units; instead, it sidesteps the issue altogether.

Postulating organism-environment systems with no bounds whatsoever is not experimentally tractable or translatable into biological interventions. So, in the methodological and empirical terrain, it seems that ontological coconstitution is not a fruitful construal. Even staunch holists like Haldane would grant that separating an organism from its environment is heuristically fruitful for biological research (see, e.g., Haldane 1917a, 107).

Scientific Explanation In contrast, within a purview of reciprocal causation, causal chains can be unpacked in different spatio-temporal scales, and scientists can identify difference-makers across both sides of the divide (i.e., in organismal activities and environmental components) and apportion causal contributions (something out of reach if one vouches for strong forms of ontological coconstitution). Relevant counterfactual dependencies between organisms and environments can be traced, which helps with designing experimental setups and selecting suitable variables to intervene on. In general, organism-environment reciprocal causation has proven to be an important ingredient in tackling developmental and evolutionary explananda (see, e.g., Baedke et al. 2020). Biologists, and even some anthropologists, are starting to appeal to reciprocal causation dynamics between organism and environment as meaningful explanantia in their work (e.g., Japyassú and Laland 2017; Kissel and Fuentes 2021; but see Hazelwood 2023).

Due to its abstract quality, the construal of reciprocal causation is so flexible that diverse types of niche construction can be accommodated with this perspective (for a detailed discussion, see Baedke et al. 2021). What's more, different kinds of interactions can be clarified by identifying characteristic causal patterns in sequences of organism-environment relations.

Modeling Scholars can also use the framework of reciprocal causation to capture and model the interactions between conspecifics or organisms from different species (see Baedke et al. 2021 and references therein). Recently, various mathematical models have emerged to address reciprocal interactions between organisms and environments, with some specifically focusing on niche construction (see, e.g., Torres et al. 2009). Drawing inspiration from Lewontin's early modeling ideas (Lewontin 1983), certain formalizations depict organism-environment interaction through systems of coupled differential equations, while others leverage causal graph theory (see, e.g., Krakauer et al. 2009; Otsuka; 2015; Tanaka et al. 2020).

Moreover, scholars have also argued that experienced environments can be modeled with a reciprocal causation view. To bridge the gap between conceptual frameworks of experiential niche construction and scientific practice, Baedke et al. (2021) proposed integrating experienced environments into a visual and conceptual model that is also able to accommodate several types of niche construction—from external/physical niche construction to constitutive and relational niche construction. There, they introduce the concept of the "experienced environment" (E_x) as a mediating interface between the organism and the physical environment. E_x encompasses the environmental cues (temperature, pressure, location, etc.) that can causally influence this interface and, consequently, the organism. The notion of E_x conveys several fundamental ideas. First, what constitutes a cue depends on the organism's sensory system and its ability to modulate its behavior in response to specific environmental factors. Second, experienced cues are transduced into biochemical and cellular processes, which regulate gene expression patterns or microbiome composition, among other processes, ultimately leading to metabolic, morphological, physiological or behavioral changes. Third, differences in E_x between organisms inhabiting the same physical surroundings imply that the environment is experienced differently by each organism (e.g., as favorable or unfavorable, stressful or nonstressful). Individual experiences are directly linked to the ecological performance of organisms within their environment, thus influencing their distribution and potentially shaping their evolutionary trajectories. Lastly, and perhaps most importantly, changes in E_x signal alterations in the organism's relationship to the physical environment, without modifying the intrinsic properties of the external environment itself—a case of mediational niche construction sensu Sultan (2015) and Chiu (2019). The integration of experienced and physical environments, one of the long-lasting challenges related to organism-environment reciprocity, can be broached—at least partially and in operationally profitable directions—in a view of reciprocal causation (see Baedke et al. 2021 for details).

Predictive Capability Beyond modeling vexing organism-environment interactions, reciprocal causation has proven to be a valuable framework for predicting organism-environment dynamics in empirical evolutionary scenarios. For instance, researchers have observed that when organisms mitigate environmental variation through their activities, through constructing elements or selecting suitable resources, the gradients of natural selection display reduced temporal and spatial variability. In fact, these gradients may even exhibit weaker selection than nonconstructed, abiotic sources (Clark et al. 2020). Building on this idea, Laland et al. (2017) argue that persistent cycles of niche construction processes introduce a consistent bias to selection by reliably generating and maintaining specific environmental states. Accordingly, this leads to predictions regarding the consequences of selection depending on its sources—constructed, nonconstructed, or a combination thereof. For example, it may result in reduced variance in field measurements of responses to natural selection. Laland et al. (2017, 5) predict that niche construction-driven selection will be weaker than, but will share important similarities with, artificial selection; however, it will show more regularity than natural selection stemming from external factors. This leads to distinct genetic and trait coevolution patterns compared to selection from autonomous environmental processes.

More recently, Milocco and Uller (2023) have taken steps forward to advance modeling approaches with predictive capabilities in order to gain insights into organism-environment interactions. Their proposal involves capturing and representing reciprocal causation between organisms and their environments through an input-output model of a coupled dynamical system. This model rings in a temporal connection between the inputs and outputs of the system, enabling predictions and the examination of the dynamics of a responsive organism in conjunction with its environment. Central to their framework is the distinction between the local environment, which is causally related to an organism and partakes in the organism-environment system (a useful abstraction in this case), and larger-scale environmental factors that constitute the external outputs of the system.

Milocco and Uller (2023) demonstrate how the framework can effectively explore phenotypic plasticity as a dynamic characteristic—a property that changes over time throughout an organism's development and lifespan. Importantly, their work highlights that modeling frameworks that accurately represent the diachronic nature of reciprocal organism-environment interactions should be sought, especially when these can be fitted with data even in the absence of detailed knowledge about the systems under study at the start of a research project.

In summary, reciprocal causation epistemically surpasses ontological coconstitution in various domains of biological practice. Its methodological advantages, ability to provide explanations, tractability in modeling, and predictive capabilities make it a valuable framework for understanding and studying the knotted reciprocity between organisms and their environments. On this, Buskell (2019, 270) says, "The positive epistemic value of reciprocal causation is linked to how well it brings complex webs of mutual influence into view. This might involve identifying new kinds of reciprocal causal links, providing new empirical tools, or showing how consensus practice fails to represent crucial causal features in evolutionary change." However, an important open challenge, Buskell recognizes, is for proponents of reciprocal causation to articulate "how exactly reciprocal causation facilitates or renders visible such mutual causal influence" (Buskell 2019, 270).

In this chapter, I examined the issue of organism-environment reciprocity. Through a historical overview, I explored three distinct accounts that have been discussed by past organicist and holistic authors: ontological coconstitution, mutual structural fitting, and reciprocal causation. As I also showed, these accounts experienced a decline in prominence during the latter half of the twentieth century. The study of organism-environment reciprocity was replaced by investigations on the reciprocity of other relata (e.g., gene-environment and population-environment), and the construal of organismal environments was superseded by population environments in ecology and evolutionary research. However, recent decades have witnessed a resurgence of interest in understanding the reciprocal relationship between organisms and their environments. This renewed attention, especially within NCT, suggests the importance of reflecting deeply on organism-environment reciprocity. Through my philosophical analysis,

I concluded that in light of ontological and epistemic considerations, reciprocal causation offers a more adequate framework than ontological coconstitution for understanding organism-environment reciprocity. By embracing this perspective, we can gain deeper insights into the nature of the organism-environment relationship and the set of interconnected issues it brings in its wake.

As we have seen, the asymmetrical nature of organism and environment qua relata—with distinct dependencies on each other—bears weight on which is the best construal of reciprocity. We now stand poised to dip into another foundational asymmetry of the organism-environment relationship: the agential status of organisms. In the next chapter, I will contend that only organisms, though deeply embedded and causally affected by their surroundings, are bounded loci of causation (i.e., agents) that perform goal-directed actions and exhibit intrinsic normativity. I will focus on the problem of organismal agency in order to argue that causal reciprocity should not be understood as causal symmetry. This will be the final piece before having all the historical and philosophical resources in hand to come back to the question of what kind of relationship is instantiated when organism and environment are juxtaposed as relata.

6 Breaking the Symmetry: Organismal Agency and Intrinsic Purposiveness

Living things are not completely at the mercy of their environment.
—Scottish biologist and philosopher Edward Stuart Russell (1945, 145)

6.1 Organisms as Agents? A Historiographic View

On December 22, 1945, with the throes of war slowly fading in the distance, Conrad Hal Waddington published a review of Edward Stuart Russell's recently published book, *The Directiveness of Organic Activities*, in the pages of *Nature*. There, Waddington asserted,

> DR. RUSSELL'S book is concerned to establish "the conclusion that directiveness and creativeness are fundamental characteristics of life . . ." Its purpose is primarily philosophical. . . . However, . . . instead of presenting a close train of argument, Dr. Russell indicates, by descriptions of actual examples, what he means by "directiveness". Look, he says, at this regenerating flatworm or developing egg; naïvely beheld, they cannot but seem to strive towards a well-recognized completeness, which is their goal. Consider again . . . the fact that a rabbit, which reacts to a loss of blood by rapidly making more, and to the transfusion of extra blood by getting rid of the excess, nevertheless does neither of these when the loss is rapidly followed by a transfusion; is it not clear, Dr. Russell asks, that it is the need of the organism, rather than any mere physico-chemical stimulus, which determines the animal's behaviour? (Waddington 1945, 731–732)

After discussing some empirical examples raised by Russell that he found unconvincing, Waddington expressed overt skepticism and went further: "The most unsatisfactory feature of Dr. Russell's book is that he makes no attempt to provide an explanation in other than self-consciously 'biological' terms, or even to discuss the explanations which most biologists nowadays advance" (Waddington 1945, 732). After his pioneering efforts to link genetics and embryology, Waddington, who began his career as an experimental embryologist aligned with the theoretical tenets of organicism (see section 5.2.5) that we reviewed in chapter 2 (section 2.4.1), was becoming increasingly dissatisfied with vocal calls to stress the autonomy of biology from the physicochemical sciences that did not take genetics and natural selection into account: "It may be granted that, whatever a 'free biology' may be, specifically biological laws may be quite useful tools of thought. . . . And biologists, still without reaching the domain of physics, can rely on a much subtler form of explanation, the appeal to natural selection and the gene" (Waddington 1945, 732). In fact, Waddington contended, when it comes to the organism as a whole, "the actual essence of the thing itself is to be found in the genes which control its synthetic activities, and which, being capable of maintaining their own specificity by self-reproduction, direct this synthesis always into the paths which are characteristic of that particular organism" (Waddington 1945, 732).

Waddington's perspective was not the only negative assessment of Russell's book. For instance, Gillette (1946, 181) wrote in *The Quarterly Review of Biology*, "In spite of his announced intention to avoid philosophical speculation and not to revive the old controversy between materialism and vitalism, in his enthusiasm for presenting directiveness as an outstanding feature of organic activities he makes many statements which leave him open to accusations of vitalism."

In a similar vein, the charge of smuggling vitalism back on board biological science was also made, on numerous occasions, against another book published that same year by another former organicist: *General Biology and Philosophy of the Organism*, written by the Canadian physiologist Ralph Stayner Lillie, then based at the University of Chicago (Lillie 1945). The emphasis on organismal purposiveness qua the distinctive mark of the "living" also took center stage in Lillie's book. Of this, the ichthyologist Carl L. Hubbs wrote in *The American Naturalist*, "We may be considering an important treatise on philosophy, but not . . . a particularly stimulating contribution to biology" (Hubbs 1946, 382). And the psychologist Milton Horowitz retorted that Lillie's book "solves no scientific problems and answers no factual questions. Indeed, the vitalistic point of view always creates its own difficulties" (Horowitz 1946, 322).

At first glance, these disparaging reviews of Russell's and Lillie's books, more examples of which could be cited, signal a grim horizon of the reception of ideas related to the purported special character of the goal-directed activities of organisms in the middle of the twentieth century. Hence, this might constitute yet another confirmation of the traditional historiography that some historians of biology built in the second half of the twentieth century, which suggests that the project to ground biology on the organism concept had been a scandalous, undeniable failure.

In a nutshell, the traditional historiography goes like this: In the interwar years, in contrast to successful mechanistic approaches, "forms of discredited or questionable biological views ranging from vitalism to an undefined purposiveness, or holism" (Carlson 1971, 152) circulated. Historians like Nils Roll-Hansen (1984) have gone in this direction and argued that, as exemplars of a larger speculative, holistic movement, E. S. Russell's and Woodger's attempts to undermine mechanistic thinking in biology were nothing but hand-waving verbiage with no actual impact on biological science (for a refutation of Roll-Hansen's historiographic portrayal of Woodger, see Nicholson and Gawne 2014). Along these lines, the fall of organicist and holistic positions, covered in chapter 5 (section 5.3), was amply justified and had it coming from the start.

In contrast, in this chapter, I want to argue that a generalized disparagement of organismal purposiveness and its cognate concept "agency" was not the generalized state of affairs of early-twentieth-century biological disciplines that, in one way or another, dealt with whole organisms. Even Waddington, who criticized Russell in 1945, would eventually grant some role to organismal agency in evolution in his later works: He framed this as the "exploitive system" (see section 5.2.5). I want to suggest that we need to move past the accounts of scholars like Roll-Hansen to understand that "organismal agency" was a scientific problem *in its own right,* deemed as such by practitioners in several fields of biological inquiry. Similarly, some authors have portrayed organicist and holist authors sharing a single, common perspective regarding organismal purposiveness (see, e.g., Blandino 1969, 163–165). However, on closer examination, it becomes evident that the viewpoints among early-twentieth-century biologists were, in fact, diverse and heterogeneous.

This historical arc matters because a good understanding of organismal agency and intrinsic purposiveness is, I contend, of paramount importance in clarifying the nature of the organism-environment relationship. While it can be fruitfully expounded through causal terms, especially applying the notion of reciprocal causation (as discussed in chapter 5), we need to recognize that this relationship is far from being symmetrical. Besides the diverging types of dependencies organism and environment hold to each other (causal dependency vs. ontic indispensability, covered in section 5.5.1), the manner in which organisms effect causal modifications to their surroundings differs starkly from how environments contribute their share to the causal tango of these relata. Likewise, an organism is a very different relatum compared to an environment (i.e., an indexical class-as-many, as explored in section 3.2.4). By putting these aspects into perspective, we gain valuable insights into the asymmetries that characterize the organism-environment pairing.

Throughout their developmental trajectories, in response to their environments, organisms partake in all kinds of interactions. However, they are not mere passive recipients of external forces. Instead, they show themselves as agents, as autonomous loci of causation. Organisms perform goal-directed actions and take an active role in shaping their relationship with their changing conditions of existence. By devoting time to explore the topics of organismal agency and intrinsic normativity, my aim is to further shed light on the causal reciprocity between organisms and their environment to show that it does not adhere to a state of perfect *causal symmetry*. Rather, it is the source of asymmetries and nuances that deserve thorough examination.

In the upcoming sections (6.2. and 6.3), I cover diverse perspectives on organismal purposiveness and agency within early-twentieth-century biology. An interesting observation is the varying prominence assigned to organismal environments in past accounts. In some instances, organismal environments were "backgrounded," seen as mere contextual settings for the exercise of agency (section 6.2). On the other hand, certain accounts "foregrounded" the significance of organismal environments, diagnosing their impact on agential activities (section 6.3). As with the case of organism-environment reciprocity (see section 5.3), views on organismal agency got sidelined in biology in the second half of the twentieth century, only to again become salient debated issues in recent years (sections 6.4 and 6.5). The standpoint that Waddington took with respect to Russell's book would prove to be prescient of the kind of arguments that evolutionary biologists would adduce to neutralize or dissolve the problem of organismal purposiveness (section 6.4).

As we will see, the theme of foregrounding-backgrounding the environment can also be extricated from present-day debates on organismal agency (section 6.5). In section 6.5, I will argue in detail that broaching organismal agency is a yet another necessary step toward reaching a better understanding of the organism-environment relationship. Organismal agency stands as one of the foundational asymmetries of this pairing. Due to their agential capacities, organisms distinctly define and mold the affordance landscapes available to them as bounded centers of causal influence across a diverse spectrum of actual and counterfactual scenarios. On the contrary, organismal environments are causally dispersed and lack the hallmark traits of agency, such as goal-directedness, intrinsic purposiveness, and normativity. The assortment of abiotic and biotic elements making up an organism's environment can indeed have important causal effects and consequences, but this is, I will assert, different from bona fide agency. While organisms and environments are engaged in protracted cycles of reciprocal causation, these interactions do not exhibit a state of perfect causal symmetry. When we talk about organisms and environments, we are invoking two profoundly different kinds of relata. Acknowledging this ontic asymmetry is decisive in comprehending the organism-environment relationship.

6.2 Charting Contrasting Stances on Organismal Purposiveness and Agency in Early-Twentieth-Century Biology

The observations that organisms have the ability to actively react to environmental changes, autonomously construct and maintain their organization and identity despite changes in material composition and form, self-regenerate, reproduce, find shelter and food, and much more, and seem to *do* all of these things in the world by themselves and for their own sake, have long puzzled philosophers and scientists.[1] How do we explain the apparent purposiveness of organismal activities? Do all organisms have agency and pursue goals of their own? Is the scope of agency restricted to goal-directed behaviors, or can it be also ascribed to certain developmental processes? What evolutionary consequences obtain, if any, from the agential activities of organisms?

Throughout the proximate and remote history of the life sciences, manifold answers to these thorny questions have been advanced. In the last decades, we have seen a surge of historical and philosophical scholarship addressing, among other topics, the bundle of problems related to the relevance of the posit of purposiveness as a constitutive property of organisms during the articulation of biology as a science of life in the nineteenth century and in preceding periods (see, e.g., Richards 2004; Cheung 2014; Riskin 2016; Nassar 2016; Gambarotto 2017; Zammito 2018; Steigerwald 2019; Jones 2023; see also section 2.2.1).

In particular, recovering the historiography of French historian Georges Gusdorf, Esposito (2024) has argued that the notion of "organismal agency" took shape during the nineteenth century, amid three central debates. The first debate focused on the dynamic correlation and, at times, tension between the "external" and "internal" factors that contribute to life. Here, the environment was seen as an external factor influencing organisms' existence and functioning (for some authors, overtaking the influence of internal developmental factors; see also sections 2.2.1 and 2.3.1). The second debate centered on the relation between organic "wholes" and "parts." In this regard, grasping how the various components that constitute an organism contributed to its overall functioning was a key aspect of fleshing out the bases of organismal agency. Lastly, Esposito (2024) contends that the dichotomy between "creativity" and "determinism," and especially the extent to which organisms exhibited self-determination and purposeful action, structured the discourse on organismal agency from early on. Throughout these debates, the notion of environment helped to set the stage of the problem space surrounding organismal agency.

Moving forward in time, historians have not systematically explored how organismal purposiveness and agency were discussed by experimental and theoretical biologists during the first decades of the twentieth century. Even though an increasing number of scholars have started to work on organicist and holistic perspectives of the interwar period and have called attention to the commonalities, frictions, and local, idiosyncratic configurations of different viewpoints and authors in the United States, Great Britain, and German-speaking countries (see section 2.4), a systematic exploration of how organismal purposiveness was framed by practicing scientists is still lacking.

The lack of systematic explorations of this topic is somewhat surprising because, if one browses the literature of the period, it is not hard to find explicit debates on organismal purposiveness. An instance of this is the 1923 debate at the Aristotelian Society held between Edward Stuart Russell, John Scott Haldane, and the psychologist Leslie McKenzie. There, the issue of organismal purposiveness turned out to be one of the central points of contention (Haldane et al. 1923). In another address that same year, Russell argued that "the individualized activity [of organisms] is not stereotyped or unchanging, but shows definite tendency or striving towards an end. Think, for instance, of a salmon ascending a stream, or of the growth and differentiation of a seedling plant. The activities of a living thing are coordinated to achieve some end related to its own development, persistence or reproduction" (Russell 1923, 123). For Russell, this meant that "the organism must be regarded then not as an arbitrary . . . unit, existing as such only 'for us,' as does a machine, but as a real unity existing in its own right and showing a measure of independence of its surroundings, an ability to go its own way" (Russell 1923, 124; see also section 4.2.2).

Russell and his colleagues are just a sampling of numerous early-twentieth-century biologists interested in these topics, belonging to organicist and holist trenches but crossing the mechanism-vitalism divide. Reconstructing how biologists reasoned about organismal purposiveness and agency in the first decades of the twentieth century and filling this historiographic gap is a problem of enormous difficulty. Many historiographic approaches could be proposed to address this issue, but here I focus on the general contrasting stances at play in different biological disciplines that scientists customarily assumed. This is especially important to understand how the organismal environment featured (or didn't) in their accounts. I argue that it is possible to identify (at least) seven distinct stances apropos organismal purposiveness and agency: (1) the neo-Aristotelian position, (2) the Drieschian answer, (3) purposiveness eliminativism, (4) the heuristic approach; (5) the holistic alternative to purposiveness, (6) organismal agency as a noetic principle, and (7) organismal purposiveness and agency as explananda for dynamic equilibrium-based research. The first four approaches confined the environment to the background (sections 6.2.1–6.2.4), while the last three foregrounded it to underscore the agential status of organisms (section 6.3).

After outlining these standpoints by focusing on representative defenders and covering the winding road of organismal agency in late-twentieth-century biological research (section. 6.4), I will discuss the contemporary controversy surrounding organismal agency and advance certain clarifications regarding this problem and its connection to teleological explanations (section 6.5). This chapter will end with a lengthy discussion of the importance of understanding agency through the lens afforded by the organism-environment relationship (section 6.6).

6.2.1 The Neo-Aristotelian Position

In a 1929 paper in *The American Naturalist*, zoologist Edwin Grant Conklin attempted to summarize the alleged big problems that the science of embryology was confronting in his time. Conklin identified three clusters of problems: the progressive differentiation of parts in development, the coordination and regulation of ontogenetic phenomena, and the problem of the teleology of development. On this last point, he said, "How can one explain the apparent teleology of development where the end seems to be in view from the beginning? . . . From beginning to end it appears that development is moving toward a goal" (Conklin 1929, 31). Conklin was convinced, as many of his embryologist peers of the time were, that accounting for, or not glossing over, the intrinsic purposiveness of development was a real problem for biologists: "To refuse to recognize the teleological aspect of development does not eliminate it; we can not explain it by merely explaining it away, for the problem remains and future generations will deal with it if we can not" (Conklin 1929, 32; see also Atkinson 1985). In another publication, Conklin aligned his views, both rhetorically and intellectually, with those of Aristotle:

> Aristotle, "the master of those who know," was great in every field of knowledge, but greatest of all in zoology. In his "De Partibus Animalium" he maintained that the essence of a living animal is found not in *what* it is, or *how* it acts, but *why* it is as it is and acts as it does. This question *why* is a hard one to answer . . . , but it is one that cannot be wholly disregarded, and, by facing it frankly, we may be saved from much shallow thinking, cocksureness and narrowmindedness. (Conklin 1944, 127; emphases in original)

Indeed, if one peruses the embryological literature of early-twentieth-century biology, one finds that Conklin was not the only scientist granting the constitutive purposiveness of developing organisms while associating these ideas with Aristotle (for an analysis, see Grene 1972). For example, in an article on the morphogenesis of muscoid flies, the Canadian entomologist William Robin Thompson—who I mentioned in the previous chapter as a defender of the idea of organismal environments in debates within population ecology (see section 5.3.3)—asserted,

> The vital movement, taken as a whole, is *essentially adaptive*, or, in other words, directed to the attainment of ends advantageous to the organism itself and to its maintenance in existence. This perpetual adjustment to changing conditions both within and without the organism has always been acknowledged to be the characteristic feature of vital action, admirably defined by Aristotle . . . in the statement that life is an "immanent movement." (W. R. Thompson 1929, 237; emphasis in original)

Yet another example is afforded by D'Arcy W. Thompson, who acted as translator of the biological writings of the Stagyrite and enshrined his legacy in the scientific study of life by proclaiming that he recognized the great problems of biology that are still ours today through his work at Lesbos Island (D. W. Thompson 1913, 15). Thompson, like Aristotle, held that form was fundamental in the study of life, and he contended that a recourse to final causes should be part of the toolkit of the biologist (Plochmann 1954).

Conklin and the two Thompsons are just representatives of a larger trend in early-twentieth-century biology: We can identify a definite neo-Aristotelian stance toward the problem of organismal purposiveness, wherein this is postulated as an intrinsic, salient property of what makes an organism what it is. Henderson, defender of organism-environment reciprocity and proponent of the idea of the fitness of the environment (see section 5.2.4), is another example of a neo-Aristotelian regarding organismal purposiveness. For Henderson (1917b, 74), grasping the "truly Aristotelian idea of internal teleology of the organism" is what makes someone a bone fide "organicist."

With the exception of Henderson, advocates of the neo-Aristotelian standpoint did not emphasize the environing conditions of an organism; they considered that these do not affect what an organism ultimately is. Conklin, in particular, highlighted that the environment is not the ultimate arbiter in evolution; what organisms do and how they react to it are the crucially important bits: "Assigning all evolution to externally caused mutations and to environmental selection neglects the fact that the organism is itself a living, acting and reacting system. *Life is not merely passive clay in the hands of environment, but is active in response to stimuli; it is not merely selected by the environment*" (Conklin 1934, 155; emphasis added).

For biologists that assumed the neo-Aristotelian vantage point, the idea of resorting to purposiveness and final causes only as heuristic devices was not convincing, and this had nothing to with the postulation of nonphysical forces acting on developmental trajectories (see Haldane et al. 1917, 459). In contrast to this, we reach the second stance with respect to organismal purposiveness in the landscape of early-twentieth-century biology.

6.2.2 The Drieschian Answer

Among biologists, historians, and philosophers, a common misassumption is that an ontological posit of organismal purposiveness entails, ipso facto, a descent into a metaphysical form of vitalism. This conflation is conceptually mistaken, and many early-twentieth-century biologists were aware of that. For instance, Conklin, also a staunch mechanist, claimed that purposiveness is a mark of organisms while maintaining that mechanistic explanations should be sought for all purposive phenomena (Conklin 1937, 232).

The German embryologist Hans Driesch, a bête noire for many biologists and philosophers, was also very clear on this point: "The main question of Vitalism is not whether the processes of life can properly be called purposive: it is rather the question if the purposiveness in those processes is the result of a special *constellation of factors* known already to the sciences of the inorganic, or if it is the result of an *autonomy* peculiar to the processes themselves" (Driesch 1914; emphases in original).

We all might be familiar with Driesch's (1892) trailblazing experiment separating sea urchin blastomeres and finding out that each one of them developed into a complete, although smaller, larva. These results contradicted Wilhelm Roux's earlier studies that suggested that inactivating a frog blastomere with a heated needle thwarted its development, and only a half-embryo was formed. In the experiments that were successful, Roux (1895) managed to raise embryos to the neural fold stage, obtaining both lateral and anterior half-embryos. This led him to a strong conclusion: Each individual blastomere possesses the remarkable ability of self-differentiation. In contrast, taking into account his empirical findings, including follow-up experiments with different cell-stage embryos (see also Driesch 1900), Driesch considered organisms to be *harmonious-equipotential systems* with the capacity to reproduce the whole if disturbed. Driesch called this plastic, self-regulatory capacity "equifinality" (for discussion, see Churchill 1969).

In contrast, a part that is not separated from the whole organism does not realize this potential.

What is responsible for the maintenance of this balance? Driesch asked. To begin advancing an answer, he considered the analogical correspondence between organisms and machines and thus the possible physicochemical options to explain this phenomenon, even the idea that some parts of a machine could contain the potential to become a new whole. However, all the possibilities he discussed were crossed out—all except one: In the form of a disjunctive proposition, he reached the idea that the factor *E* that explains purposiveness was nonphysical and nonchemical. In other words, he presented an *argument per exclusionem*: "*E* is either this, or that, or the other, and it was shown that it could not be any of all these except one . . . ; something new and elemental must always be introduced whenever what is known of other elemental facts is proved to be unable to explain the facts in a new field of investigation" (Driesch 1908, 142).

Driesch later renamed this factor *entelechy*, borrowing the term from Aristotle but recognizing that it meant something different under his guise (see also Ritter 1932). Historian Garland Allen (2008, 52) said that "rather than being the primary cause . . . of actualization, it was a regulator governing which of the various potentialities resident in a material system is to be realized and which is to be restrained. Entelechy was thus not the blueprint of an organism's organization, nor the creative agent that brings it about, but a kind of a mediator . . . that protects the tendency of the system from being disrupted by extraneous factors," such as an ablation or blastomere separation. For Driesch, the existence of an entelechy did not violate the physical laws of nature, as it was not a means of creating energy outside the laws of thermodynamics; it was simply a nonmaterial vital force that kept development on its route (for extensive discussion, see Chen 2023, 2024).[2] In Driesch's account, the environment played no instructive role whatsoever in ontogeny: It was sent to the background in his explanation of organismal purposiveness.

In the first decades of the twentieth century, Driesch's vitalism amassed some followers, but certainly it ignited a strong opposition as well: Biologists from many disciplines rejected his vitalistic standpoint (for an analysis, see Chen 2018, 2024) but, in doing so, vetoed any reference to organismal purposiveness altogether. Even though purposiveness and entelechy are conceptually distinct—for example, through the explanandum-explanans distinction, namely, that organismal purposiveness is what demands an explanation and the entelechy provides one possible answer—authors rejected both: They threw the baby out with the bathwater. So now we reach the third stance in this survey.

6.2.3 Purposiveness Eliminativism

In several fields of early-twentieth-century biology, the stance I call "purposiveness eliminativism" held sway. An illustrative example is the field of *Entwicklungsmechanik* (developmental mechanics). Roux, alongside some of his colleagues, claimed that all purposiveness in development is just superficial and apparent (*Scheinbar*) and thus misleading for embryology (Roux et al. 1912, 460). They maintained that biology can do perfectly well without agentic notions. Their only concession was that exclusively in instances when conscious intention is involved, the idiom of purposiveness is warranted, like in cases of human behavior.

In this perspective as well, the environment was backgrounded. Roux had made a distinction between the roles of "extrinsic" and "intrinsic" factors in development. In an experiment conducted in 1884, Roux explored the potential impact of gravity. His experimental setup involved placing frog eggs within a water-filled vessel that was subjected to rotation on a wheel. Roux's results hinted that development remained unaffected by this gravitational manipulation (Roux 1895, 256–276). Building on this finding, Roux went on to investigate the influence of other environmental factors such as light, temperature, and oxygen on amphibian development. To his surprise, he observed that none of these variables exerted any discernible effects. For this, Roux drew a distinction between nonspecific (environmental) conditions, which failed to impact development, and specific causes that could bring about developmental changes. He summarized his findings by stating that the development of a fertilized egg occurs as a consequence of the formative forces inherent within the egg itself. He characterized this process as the "self-differentiation of the egg," and he emphasized the autonomous nature of development (Hamburger 1997, 231). Russell provides insight into Roux's perspective on the role of the environment, stating that he conceptualized it as supplying "not the determining but the conditioning factors of organic activity. The main causes of form-production are to be sought inside, not outside, the organism" (Russell 1930, 100).

In general, purposiveness eliminativism was not exclusive of the *Entwicklungsmechanik* approach. An example of this rationale in a different field is embodied in the work of British plant morphologist William Henry Lang. He argued that the attitude of the morphologist should lead to rejection of the view that organisms exhibit intrinsic purposiveness: "The course of development to the adult condition can be looked upon as the manifestation of the properties of the specific substance under certain conditions. This decides our attitude as morphologists to the functions of the plant and to teleology. . . . Until purpose can be shown to be effective as a causal factor it is merely an unfortunate expression for the result attained" (Lang 1915, 784). Other botanists also outrightly rejected organismal intrinsic purposiveness and argued for its epistemic irrelevance: according to them, conceptualizing self-regulation processes occurring in a developing plant as "purposive" only "obscured" sequences of highly coordinated mechanistic, physiological events (e.g., Farmer 1903).

Another famous example of purposiveness eliminativism can be observed in the work of the American zoologist Herbert Spencer Jennings, who advocated for the "trial and error" principle to cast away organismal agency (see also Torrey 1907). Jennings (1905) argued that the seemingly goal-directed behavior displayed by so-called lower organisms such as bacteria and ciliates is not indicative of actual intrinsic purpose, but, rather, stems from individual regulations based on self-adjustment mechanisms. According to Jennings, organisms undergo specific internal processes, and interference with these processes leads to changes in behavior until one relieves the interference (Jennings 1905, 481–482). In this sense, he claimed, "It is clear that regulation taking place in this way does not require that the end or purpose of the action shall function in any way as part of its cause, as some vitalistic theories hold. *There is no objective evidence that a final aim is guiding the organism*" (Jennings 1905, 482; emphasis added; see also Jennings 1906).

Yet another influential example is afforded by physiologist Jacques Loeb and his tropism theory in the study of animal behavior. Loeb investigated so-called reflex-actions and construed them as reactions of isolated parts inside an organic body; in an analogous fashion, he conceptualized the reaction of whole organisms to be phenomena that can be reduced to physicochemical triggers (e.g., moths being inexorably drawn to a light). For Loeb (1919), a stumbling block for the scientific study of animal behavior was to grant that organisms pursue goals of their own; in particular, he claimed that a difficulty for the widespread acceptance of his tropism theory was "created by the fact that the Aristotelian viewpoint still prevails to some extent in biology—namely, that an animal moves only for a purpose, either to seek food or to seek its mate or to undertake something else connected with the preservation of the individual or the race. . . . *The analysis of animal conduct only becomes scientific in so far as it drops the question of purpose* and reduces the reactions of animals to quantitative laws" (Loeb 1919, 23; emphasis added). Beyond behavior, Loeb challenged the organicists' belief in the supposed inherent capacity of organisms to create a cohesive whole during development. He staunchly rejected the notion that the "adaptation of each part to the whole" was a fundamental characteristic of organisms (Loeb 1912, 23).

Needham (1930, 192) concurred with this general outlook and said that "[the neo-mechanistic position] knows teleology to be an unquantitative category, and banishes it from the laboratory to the domain of the philosophers, who are quite capable of dealing with it" (see also Heikertinger 1917).

If not as a constitutive element of what makes organisms what they are, did some early-twentieth-century biologists grant to purposiveness at least a heuristic role that could impact their research? This question leads us to our next stance.

6.2.4 The Heuristic Approach to Purposiveness

In the third edition of *The Cell in Development and Inheritance*, the American zoologist and geneticist Edmund Beecher Wilson recovered a heuristic role for purposiveness. With regards to segmentation patterns in animal embryos, Wilson pondered,

> We cannot comprehend the specific forms of cleavage without reference to the end-result of the formative process. . . . The teleological aspect of cleavage thus suggested has been recognized . . . most by [Frank] Lillie, who has urged that with this principle in mind "one can thus go over every detail of the cleavage, and knowing the fate of the cells, can explain all the irregularities and peculiarities displayed". . . . Such a conclusion need involve no mystical doctrine of teleology or of final causes. (Wilson 1925, 1005)

Wilson's remark alludes to a prevalent approach employed by embryologists of the time, such as C. O. Whitman, E. B. Wilson, and Frank Lillie, who frequently utilized teleological language in their investigations of cell lineages (see Churchill 1969 and references therein).

In his book *Biological Principles*, Woodger, a defender of intrinsic purposiveness as a constitutive feature of organisms, cited the case of Wilson as a representative example of a larger trend in interwar biology to treat purposiveness in a heuristic manner. Woodger (1929) attempted to clarify what exactly could be this heuristic role: "In regard to explanation a knowledge of the *outcome* of a biological process is often more illuminating than knowledge of what went before" (32; emphasis added). Along those lines, in fields like physiology, some scientists, including Walter B. Cannon, proposed that considering organismal teleology could lead to improved explanations for various regulatory physiological processes. By embracing the concept of organismal teleology, researchers believed that they could develop more comprehensive and accurate understandings of how a body's regulatory systems function (for analysis, see Ortega Lozano 2022).

Like the previously discussed perspectives in this section, the heuristic approach to organismal purposiveness and teleology paid lip service to the environment. The explanatory focus of these biologists did not encompass the environmental contexts in which organismal development unfolds.

The heuristic approach has a long philosophical pedigree and was most famously articulated by Immanuel Kant in his *Critique of the Power of Judgement*.[3] Besides noticing the extrinsic fit of "organized beings" (*organisierte Wesen*) within their surroundings in a way that could foreshadow design in a larger system of nature, Kant also emphasized the intrinsic fit between the parts of organisms that allows them to be functional wholes (see section 5.2.3), organized beings with generative and nutritive processes. In that sense, living organisms, for Kant, are perceived as natural purposes (*Naturzwecke*) "in which everything is a purpose and reciprocally also a means" (5:376; Kant [1790] 2000, 247). He explains this idea by the example of a tree (5:371). First, the tree always generates new trees of the same species. Second, "a tree also generates itself as an individual" through growth. And third, the various parts of the tree are mutually, reciprocally dependent on each other for preservation (Kant [1790] 2000, 243). Parts combine into a whole by being reciprocally the cause and effect of their form (5:374; Kant [1790] 2000, 245).

According to Kant, the acknowledgment of the teleological organization of organisms seems to transcend the boundaries that define acceptable scientific discourse. For him, scientific explanations should adhere to a mechanistic framework, such as the one present in Newtonian mechanics. However, he also acknowledged that biological phenomena, while partly compatible with mechanistic explanations, cannot be fully encompassed by them. As a result, a task of critical philosophy is to scrutinize the concept of purpose within nature and the proper and improper ways it is employed in scientific explanations (Körner 1955, 197–198), especially when explaining the existence of entities unrelated to human intentions (e.g., developing organisms).[4]

This led to the articulation of the widely known antinomy of the teleological power of judgment in Kant's work: The thesis "All generation of material things and their forms must be judged as possible in accordance with merely mechanical laws" is juxtaposed with the antithesis "Some products of material nature cannot be judged as possible according to merely mechanical laws" (5:387; Kant 2000, 259). Kant's chosen avenue to resolve this dilemma is that teleological judgment is valid only for cognitively limited beings like ourselves as a regulative ideal for inquiry, operating as a reflective judgment that does not unveil the ontological nature of the objects under consideration. Illetterati and Michelini (2008, 4–5) say that, for Kant, "teleological judgment referring to nature is legitimate only . . . as *reflecting* judgment, as *regulative* judgment, in those shapes that do not have any constitutive role for the way of being of the object taken into account. Teleological judgments . . . tell us nothing about the constitution and the way of being of the relevant objects; their only role consists in helping us to understand the way of being of that object and, consequently, to guide our research" (emphases in original; for discussion, see also Toepfer 2011c).

Kant distinguished judgment from understanding, the latter being the primary focus of the *Critique of Pure Reason*. According to Lindsay's (1919) reading of Kant, understanding pertains to the faculty of rules, while judgment is revealed through the application of rules to specific instances. The faculty of judgment seeks to unify the intricate details of the natural world. To achieve this, it employs two regulative principles: mechanism and teleology. However, reality cannot be explained simultaneously by both principles. Mechanism assumes that reality can be seen as a pattern or composition of recurring and interchangeable parts. On the other hand,

Kant suggests, teleology posits that the living world cannot be understood without recognizing purpose as an active force in change. Kant's resolution to the antinomy of the teleological power of judgment stated above is that both mechanism and teleology are merely *regulative principles*. They do not reveal the ultimate nature of reality, except for the fact that we can explain a significant portion of it by considering it as a machine (mechanism), or by attributing purpose to its operations (teleology in the living realm). Reality must be consistent with both these aspects, but we cannot assert more than that. We are forced to continue treating these principles as regulative uses of reason that guide our cognitive processes and enable us to acquire knowledge about puzzling entities like organisms, yet without asserting anything as objectively existing or true in this regard. In this sense, it is useful and illuminating to think of organisms, like works of art, as possessing purposiveness (Lindsay 1919, 82–84, 88). Nevertheless, according to Kant, organismal purposiveness is not just a projection that can be subjectively chosen or dismissed. Once biological entities become the subject of inquiry, there is no alternative for us, given our finitely rational make-up, but to see them as purposive for acquiring knowledge about them (Desmond and Huneman 2020, 57). We cannot do otherwise.

Kant's answer rippled in the articulation of biology as a scientific discipline in the nineteenth century. Until very recently, the mainstream historiographic view, as maintained by Lenoir (1989), was that, after Kant, appeals to regulative teleology, as opposed to granting constitutive purposiveness to organisms, shaped the incipient life sciences through methodological and theoretical approaches of a tradition that Lenoir dubbed "teleomechanism." Scholars have challenged the "Lenoir thesis" and have adduced evidence to contend that many authors in the fields of morphology and physiology, before and after Kant, sought to explain "purposiveness" as something constitutive of living organisms, and not as a regulative maxim of the reflective power of judgment with heuristic value for biological research (see, e.g., Richards 2000; Zammito 2012, 2018; Gambarotto 2017; for discussion, see Rupik 2024). More often than not, biologists that were heavily inspired by Kant's critical philosophy decided to part ways with him regarding the issue of organismal purposiveness, and this is particularly clear in the next stance I recount.

6.2.5 The Holistic Alternative to Purposiveness: The Environment in Tension

John Scott Haldane (see sections 4.3.1 and 5.2.3) is a prime example of a different position regarding organismal purposiveness in early-twentieth-century biology. In an early piece coauthored with his brother, the lawyer and philosopher Richard Burdon Haldane, they argued that, for the sake of building a better science and philosophy of science, "the mastery of the critical investigations of Kant and Hegel . . . will be absolutely essential. But such a discipline can form simply a part . . . of *preparatory culture*" (Haldane and Haldane 1893, 66; emphasis added). Indeed, when turning his attention to the study of organisms, J. S. Haldane was disappointed with Kant's treatment: "At the time of Kant and his immediate successors biology had hardly begun to be conscious of her strength. Living organisms seemed, as it were, to be at the best only dotted about here and there in the midst of a totally foreign physical universe. Kant assigned to them a very doubtful place in his philosophy" (Haldane 1921, 99). In short, Haldane thought that Kant drew his conclusions regarding the status of developing organisms too precipitously. He even proclaimed, "As regards life, . . . Kant was mistaken" (Haldane 1929, 65).

In the aforementioned piece, the Haldane brothers contended that "the distinguishing feature of vital activity is self-preservation . . . ; and this is just as true of the most complicated actions of the human body as of the movement of the amoeba towards a source of nourishment. . . . The fact is that every part of the organism must be conceived as actually or potentially acting on and being acted on by the other parts and by the environment, so as to form with them a self-conserving system" (Haldane and Haldane 1883, 54–55). As we saw in chapters 4 and 5, a leitmotif in Haldane's life work would be the emphasis on the importance of the environment in the self-maintenance of life. Embracing an ontological coconstitution perspective, Haldane viewed the organism and its environment as entangled, forming an interconnected whole that resists trouble-free separation.

Within this holistic position, Haldane reasoned that the concept of purposiveness loses its coherence. This is a position that he adopted early on in his career. "What is ideally implied in such words as 'function,' 'purpose,' 'means,' and 'end,'" the Haldane brothers argued, "is that we are looking at the organism, not as acted on by things outside it, but as in teleological connection with that which is different from, but not existent independently of it" (Haldane and Haldane 1883, 58). Later on, Haldane laid down his vantage point clearly: "It seems to me that as mere biologists *we have no need to make use of the concepts of either memory or purpose*. What we observe in all lives is simply their *tendency to maintain and reproduce themselves as co-ordinated wholes*" (Haldane 1931, 161; emphases added).

Other examples of authors that argued for replacing the concept of purposiveness with that of wholeness were the German botanist and philosopher Emil Ungerer and the Dutch ecologist Cornelis van der Klaauw. Ungerer (1919), in his studies on the regulation of plant activity, contended that if an organism preserves itself during development in the form of a whole, the processes that seem to condition the preservation of this wholeness can only be described as purposive in a mere descriptive, almost void sense. When thinking about organismal processes in a teleological sense, Ungerer (1919) maintained, it does not matter whether a given process can be subordinated to a purpose, but rather, whether it contributes to preservation of the wholeness of the organism in question. van der Klaauw (1935), following a similar line of thought, also disagreed with the Kantian interpretation of *Zweckmäßigkeit* and sought to address what he saw as a substantial drawback. In alignment with Ungerer, he proposed replacing the concept of purposiveness with the concept of *wholeness* (van der Klaauw 1935; see also Trienes 1992).

The holistic standpoint on purposiveness was widely debated in those years, and scholars have suggested that it replaced neo-Kantian positions after they peaked in significance in early-twentieth-century biophilosophical debates (Baedke et al. 2024). For example, at the Seventh International Congress of Philosophy held at Oxford in 1930, a whole session was devoted to the topic of organismal purposiveness, and the holistic stance dominated there. Biologists and philosophers, such as Haldane, Ungerer, Wildon Carr, Reinhold Friedrich Alfred Hoernlé, and others, agreed that self-maintenance could not be described as purposive. In the summary of this session, Hoernlé claimed,

> The antithesis of mechanism and purpose is out of date. . . . Evidences of this shift are: (a) The substitution (of the concept of purpose by that of the whole) . . . . Whole is a concept equally applicable to plant, animal, man. . . . (b) The biological protagonists . . . argued primarily on methodological, not on metaphysical, grounds. The crucial question was: What is the most profitable technique of investigation in biology? . . . *Biological processes must be holistically conceived, in order to be mechanistically studied.* (Hoernlé 1931, 44; emphasis in original)

The holistic alternative was sometimes paired with a stance of purposiveness eliminativism. A clear example of this can be found in the work of the German neurologist Kurt Goldstein. He explicitly agreed with Ungerer that the concept of wholeness should be given priority in the explanation of organisms and that all purpose-talk should be sidestepped in biology (Goldstein [1934] 1995, 324). In his holistic view, Goldstein granted to the environment a secondary role. He specifically quibbled with Uexküll's view of *Umwelt* (see also sections 4.3.3 and 5.2.5) and argued for the primacy of the organism:

> Each organism lives in a world that by no means contains only such stimuli as are adequate for it. It lives not merely in its "own environment" (milieu) but in a world in which all possible sorts of stimuli are present and act on it. . . . *An environment always presupposes a given organism. How could it then be determined by the environment? How could it achieve order only by the environment? Of course, as soon as it has an environment, it has order.* Order is only achievable if there is the possibility of obtaining an adequate environment. But the possibility alone is of no avail. *Environment first arises from the world only when there is an ordered organism.* Therefore, the order must be determined from somewhere else. . . . We are ultimately referred back to the organism itself. (Goldstein [1934] 1995, 85; emphases added)

The theory of *Umkonstruktion*, developed by the German eco-morphologist Hans Böker, provides another holistic perspective that supposedly challenged the traditional notion of organismal purposiveness. Böker's theory construed the organism as a cohesive entity in harmony with its parts and the surrounding environment. Böker posited that when the bio-morphological equilibrium is disrupted by environmental changes, the organism must strive to restore it to avoid perishing. The morphological perturbations triggered by environmental changes can lead to variations in interconnected aspects of the organism, setting off a series of changes that ultimately work together to restore the bio-morphological equilibrium (Böker 1935; for an analysis, see Fábregas-Tejeda et al. 2021).

While the preceding discussion may give the impression that Böker attributed a prominent role to the environment in evolution, he asserted that the active organism itself, as a whole, is the ultimate determinant of evolutionary processes. Böker emphasized that the production of variation is never directly dictated by specific environmental changes:

> [Reconstruction] depends on the intensity and speed of perturbating influences, and on the degree of the organisms' ability and readiness to react. Every single

> time a change of species through reconstruction is an active, a holistic event, never a passive mechanistic one. A. Meyer . . . describes this idea as follows: "Organisms are never only passive adaptations to the organic environment." . . . The actual reorganization of organisms is always an *active act of creation* of the organisms themselves that matches their inner world [*Innenwelt*]. Both Darwinism and Lamarckism have ultimately failed because of their conflation of active adaptive insertion [*Einpassung*] with passive adaption [*Anpassung*]. Both are theories of milieus that attribute to the environment active, creative forces in the process of species transformation. (Böker [1935] 2021, 120; emphasis in original)

Let us now turn our attention to the final two historical perspectives that I will delineate in this discussion. In contrast to what has been previously presented, they placed a significant emphasis on the role of the environment when considering organismal purposiveness and agency. First, to comprehend the postulation of organismal agency as a noetic principle in biological practice, we must delve into the writings of Edward Stuart Russell.

6.3 Foregrounding the Environment in Early-Twentieth-Century Positions on Organismal Purposiveness and Agency

6.3.1 Organismal Agency as Noetic Principle

In chapter 4 (section 4.2.2), I contextualized some of Russell's views regarding organism-environment separation. Adding to that reconstruction, I can say here that the emphasis on organismic purposiveness can already be seen in Russell's early works. For example, in a review of Ludwig Plate's (1908) *Selectionsprinzip und Probleme der Artbildung*, Russell said, "We have no space to do more than mention what is after all the main question as regards Darwinism, how far it is an explanation of that *purposiveness which is the essential characteristic of living things*" (Russell 1908, 178–179; emphasis added).

In another article, he argued that, in the study of evolution, one should not disregard the dynamic, active role of the organism: "We believe that it is one of the fundamental truths of biology that the action of the environment on the organism is never purely physical and direct. *The organism never remains passive in the presence of action*; it reacts to the influences of the surrounding environment. This reaction can lead to a permanent modification of the organism, but this modification is only an indirect effect of the action of the environment" (Russell 1914, 113; emphasis added; my translation). As he would later bitterly complain in *Form and Function*, "We tend . . . to lay the causes of form-change, of evolution, as far as possible outside the living organism" (Russell 1916, 307).

I think it is fair to say that Russell, at the time of publication of *Form and Function*, was a convinced defender of the neo-Aristotelian standpoint regarding organismal purposiveness. In fact, the book ends with a suggestive and vocal call to, once again, imbue biological research with Aristotle's insights: "We need to look at living things with new eyes and a truer sympathy. We shall then see them as active, living, passionate beings like ourselves, and we shall seek in our morphology to interpret as far as may be their form in terms of their activity. *This is what Aristotle tried to do*, and a succession of master-minds after him. We shall do well to get all the help from them we can." (Russell 1916, 364; emphasis added; see also Ritter 1933, 501).

Nevertheless, here, I want to argue that Russell changed his general stance in the next decades of his career and put forward a different view of organismal purposiveness and agency, one that, although sharing sympathies with the neo-Aristotelian view, has nuances of its own that can be extricated. Russell put forth a unique perspective in which organismal agency is conceived as a "noetic principle" for biological practice. Allow me to explain what this means.

Like many of his organicist counterparts, Russell embraced the tenet of organism-environment reciprocity and considered that the organism should be "regarded as an individual whole which is as much conditioned by its parts as one part is conditioned by the others. For an understanding of correlation a knowledge of functions, and of the functional relations of the organism to its environment, is clearly indispensable" (Russell 1916, 264). In his work, Russell argued that it was important to "regard the activities of a given individual from its own point of view, in relation to the world as perceived by it, not in relation to the world as perceived by us" (Russell 1924, 147). Although all organisms have particular experiences and different perceptual capacities, Russell contended that they share a common world due to evolution and thus are not isolated or unconnected bubbles, like Uexküll would have it. There is an underlying world that serves as the backdrop of very different perceived environments (Russell 1924, 85). Along these lines, Russell espoused a dynamical view of the organism-environment relationship that extended to an organism's goals. For instance, he claimed, "The objective purposiveness of organic activity . . . is self-regulatory, and adjustable to a wide range of circumstances" (Russell 1924, 57), and further that "to live successfully . . . a living thing must be able to follow and counter by appropriate response the changes in its environment which affect it, which have a vital meaning for it. Now this can be achieved only through perception. I use the word here . . . in a broad way [i.e., not related to conscious states], to cover all degrees of the receptive side of vital activity" (Russell 1924, 57).

In particular, Russell argued that the organism needs separation from the environment to have agency, or phrased differently, that individuality is a precondition for organismal agency (see also section 4.2.2). "The living thing can have," he said, "no more than the machine, an internal or self-existent purposiveness, unless it has at the same time real individuality and persistence" (Russell 1924, 16–17). We come to understand now that Russell sought to carve out organisms as subjects, as agents in the world that act on their own behalf: "*The living individual is then a subject, or better—if we lay emphasis on action rather than on presentation or representation—an agent*; and there is necessarily implied an object, or more generally, something sensed to which the individual responds. *Individuality has therefore as its necessary complement a sensed environment*" (Russell 1924, 59; emphases added). As we can see, the environment is foregrounded in Russell's views on purposiveness and agency: an organism dynamically responds to sensed elements in the fluctuating world in which it is embedded and it acts accordingly as a "fount of activity" (Russell 1924, x).

Russell's observations encompassed various examples that highlighted the agency of organisms. For instance, he studied starfishes, examining their response time when turned over, and he closely interacted with his fox terrier, Gina, playing ball with her and publishing studies on this behavior (Russell 1919, 1936). Likewise, he conceptualized some types of fish migration as instances of goal-directed activities and called attention to the importance of the individual environmental experiences of these organisms in their trajectories in the sea (Russell 1934, 41; 1937, 331). In these inquiries, Russell made explicit that organisms exhibit goal-directed behavior. They actively pursue their own purposes and respond to objects and elements in their sensed environments, with the valence of their reactions varying according to an organism's states and needs. But how, exactly, can Russell's standpoint be construed as advancing a noetic principle for biological practice?

Here, I must say that I am using the term "noetic" in epistemic contexts to refer to *scientific understanding,* and not as something related to the study of mind or the realm of the mental. For Russell, postulating organismal purposiveness and agency as ontological principles did epistemic work as well, especially for the scientific study of animal behavior.[5] He contrasted the prevalence of causal explanations in the physical sciences or in biological fields such as physiology with his subject matter: the activities and behaviors of organisms. In this regard, he asserted,

> In the physical sciences and to a large extent in physiology we are concerned primarily with causal explanations. . . . We do not get this sort of knowledge by using the direct or descriptive method. What we get may be better described as *an understanding of behaviour.* When you see a wasp, for example, standing on a gate post and busily chewing at the wood, you do not understand this action until you follow the wasp up and find out that it uses this material to construct its nest; you go on then to discover what the nest is used for—the care and upbringing of the young. In a word, *to understand the action of the wasp on the gate post you have to integrate this action into the whole directive cycle of activity.* . . . This is the sort of knowledge which we get by studying animals from the [agential] point of view, and it is extremely valuable knowledge too. *Without this knowledge we simply cannot make sense of behaviour;* even the most extreme believer in mechanism must use this sort of knowledge to make behaviour *intelligible* at all. (Russell 1934, 15; emphases added)

In other words, for Russell, the recognition of the goal-directed character of behavior is absolutely necessary for *understanding* it, for making it *intelligible* to scientific scrutiny. Russell's noetic standpoint was not confined to organismal behavior. He had the same estimation regarding, for instance, physiological processes. Russell asserted that physiological processes only "become intelligible" when scientists link them up "with one of the main biological ends which the organisms blindly pursues" (Russell 1945, 9), such as surviving or maintaining its organization.

This is a stance that Russell kept underlining throughout his career. For example, in his 1945 book, *The Directiveness of Organic Activities,* he stressed, "If we disregard ends and the directiveness of activities towards ends, we leave out what is distinctive in life, and we simply amass more and more data about the details of organic activities and the physico-chemical conditions of life, without connecting them up in a rational way with the functional life-cycle of the organism" (Russell 1945, 8). Once more, in Russell's perspective, acknowledging the purposive nature of organismal activities becomes an essential foundation for rendering them comprehensible under scientific investigations, a prerequisite for accurately grasping and interpreting them. Coming to terms with the goal-directedness of organisms lies at the heart of gaining an understanding of life for a science that decides to study it:

> If then it is true, and indeed the chief truth, about the living organism that its activities and those of its parts are directive towards living, reproducing and developing, we must, in our study of these activities, consider first and above all their biological significance or *function*, their relation to one or other of these biological ends. If we do not do so, but consider them separately, without relation to the life of the organism as a living, developing, reproducing whole, we shall never understand them, even though we succeed in working out their physico-chemical "mechanism" or mode of action. We shall acquire, and go on acquiring, a vast mass of unrelated facts of biochemistry and biophysics, but we shall never build up a real biology. (Russell 1945, 9; emphasis in original)

I consider that a way to spell out Russell's stance is to frame it as a form of transcendental argument. In employing this notion, I draw inspiration from contemporary philosophers of science such as Hasok Chang and Daniel S. Brooks, who have attempted to rehabilitate transcendental arguments in naturalist frameworks (Chang 2008; Brooks 2021). A transcendental argument in this context is one that accords to the following form: If we want to engage in a particular epistemic activity (e.g., understanding animal behavior and making it intelligible), *then* it is necessary to presume or postulate particular ontological principles—accordingly, organismal purposiveness and agency. An ontological principle, under this conception, is one that makes systematically intelligible at least one kind of epistemic activity. In this case, postulating organismal purposiveness as an ontological principle allows scientific understanding qua epistemic activity and allows for the design of experimental interventions and the interpretation of empirical results (see below).

Moreover, I think that Russell's stance can be said to be transcendental insofar as it delves into the *conditions of knowability* (see Van de Vijver et al. 2005) of the activities of other organisms (i.e., we need to presume their agentic status in order to begin to understand their doings), and from there, a biological science that seeks complementary mechanistic and causal explanations can then take off. In sum, for Russell, the presumption of agency and purposiveness is a *noetic principle* for practicing scientists dealing with developing, behaving organisms.

Russell's view certainly has some echoes of the Kantian interpretation of organismal teleology. For Kant, "seeing an organism as an agent is even a precondition (the transcendental ground) to being able to make a projection onto a natural system. For instance, if in some agential explanation, a repertoire of actions is projected onto a living organism, this presupposes seeing an organism as an agent. Assuming agency makes ascribing empirical methodology and even (behavioral) property to organisms possible. This is how the 'indispensability' implied by the Kantian option should be understood" (Desmond and Huneman 2020, 58). A key difference between Russell and Kant is that the former is interpreting agency ontically (i.e., as a constitutive feature of organisms), and this would be anathema in Kantian critical philosophy (but see the discussion in section 6.5.3).

The emphasis on the importance of granting intrinsic purposiveness to organisms in order to understand them can be glimpsed in Russell's early works. In *Form and Function*, he says, "Teleology in Kant's sense is and will always be a necessary postulate of biology. It does not supply an explanation of organic forms and activities, *but without it one cannot even begin to understand living things*" (Russell 1916, 35; emphasis added). Later on, his noetic stance was much more palpable in his empirical studies of fish behavior and in his valence theory on why only certain stimuli in an animal's (sensed) environment elicit behavioral responses depending on the context and needs of the organism (Russell 1935). For example, a token hermit crab (*Pagurus arrosor*) devours or keeps a commensal anemone in its shell if it is hungry, but not solely as a univocal reaction to the presentation of the environmental stimuli. Russell (1935) made sense of these observations through his "valence theory" and argued that a mechanical, deterministic stimulus-response view fails to apprehend the actual behavior of organisms because it disregards the internal states and needs of living agents that act on their own behalf. In the empirical terrain, Russell (1931) performed some detour experiments with sticklebacks. In this behavioral setting, the shortest, direct way to a visually attractive object is blocked for the test animal, although a roundabout way is left open. Russell tested sticklebacks to see if they could learn to find their way into a small jar filled with food, and then he evaluated if they could also learn to exit the jar. In his interpretation, the fish at first found the solution by chance, but after a few training trials the entrance-seeking behavior became purposive, with the clear goal of attaining the nourishing items in its sensed environment. After the behavior became purposive for the organisms, there was a measurable sudden decrease in the time it took them to get the food (Russell 1931). By presuming that organisms are agents, Russell could begin planning his experiments and make sense of the data he collected in particular behavioral settings. This is a concrete example of his noetic principle in action.

Let's turn now to the last stance I will characterize in this historical survey: organismal purposiveness and agency as *explananda* for dynamic equilibrium-based research.

6.3.2 Organismal Purposiveness and Agency as Explananda for Dynamic Equilibria-Related Research

As we will see, many scientists shared this standpoint, but the main proponent of this view that I want to highlight here was the Canadian-born organicist Ralph Stayner Lillie. In 1901, he received his PhD from the University of Chicago with a work on the excretory organs of *Arenicola cristata* (Maienschein 1988). Throughout his career, Lillie transitioned between various academic positions at different universities in the United States. He was eventually appointed as professor of general physiology at the University of Chicago in 1924, where he stayed until his retirement (Gunderman 2005, 1473; for a general biographical sketch, see Gerard 1952). Lillie was a prolific author: he published around 125 articles, ranging from embryological research in marine invertebrates and general physiology to philosophical pieces in professional philosophy journals.

Lillie began his scientific career conducting experimental research on the effects of solutions of inorganic salts on the ciliated epithelium of the larvae of lugworms and polychaetes (see, e.g., Lillie 1902). He also performed experiments on the environmental induction of parthenogenesis in starfish eggs by manipulating temperature variables (Lillie 1908). Early on, the main scientific question that interested him was the problem of what at the time was called "irritability," namely the excitatory capacity of developing organisms to react to environmental changes. How are these ontogenetic and physiological reactions articulated by the organism as a whole? To try to answer this question, Lillie started studying the effects of anesthetics on marine invertebrates as a good contrast case because anesthesia, he reasoned, "is a physiological condition in which the normal responsiveness of the living system . . . is temporarily decreased or abolished" (Lillie 1916, 311). What must happen for the organism to stop reacting as a whole?

As he said in a 1920 paper, "One of the many remarkable peculiarities of the living system . . . is the regular and rapid transmission of metabolic and functional influence between its different parts . . . and it is a familiar fact that slight changes in the surroundings may, by acting as 'stimuli,' instantly alter the whole behavior of the organism" (Lillie 1920a, 165). In agreement with Child (see section 4.2.1), Lillie contended that the answer was to be found in the structural properties of "protoplasm," and that "its fundamental biological significance is apparent, since it is the chief means of controlling and coordinating the various separate functions and activities of the organism" (Lillie 1920a, 165).

In general, Lillie tended to emphasize the active character of organisms, which are not only passively acted on by the environment, but also react to it:

> All living systems . . . react to changes occurring in their immediate environment, or to changes in their relations to the environment, by exhibiting characteristic alterations in their own special activity. . . . Changes in the *relation* of the organism to the environment, although the environment itself remains unchanged, may also furnish the conditions for stimulation and response; thus an animal moving from place to place continually encounters new conditions to which it reacts. (Lillie 1924, 167; emphasis in original)

The counterpart of active reaction is that organisms "have the power of self-maintenance; i.e., of maintaining their identity and a certain constancy of structure, chemical composition, and activity in spite of continual changes in their surroundings and in their own living substance" (Lillie 1923, 25). To Lillie, this highlights a fundamental challenge for biologists: the question of how to explain the self-maintenance of organization. This issue raises a second crucial problem, one that, according to Lillie, distinguishes biology from other sciences: the nature of organization itself. He posits that understanding organization provides biology with a unique and irreducible subject of study. In his first monograph, Lillie vouched for an emergentist view of organization:

> *When the materials and energies of the surrounding world unite to constitute the organism, new qualities and modes of activity inevitably come into existence; these special properties of living beings form the subject-matter of the biological sciences*. . . . For this reason the physical and chemical characterization of the constituents, reactions, and processes whose combination or synthesis produces life is not in itself sufficient; the biological interest centers in the conditions and special mode of this combination, and in the nature of the resulting unity. *The problem of the nature of vital organization remains the fundamental one for biology*. (Lillie 1923, x; emphases added)

After having laid out these two problems, we can now move to how Lillie tried to tackle the goal of continuous self-maintenance as an explanandum for physiology. In a 1915 article, he addressed the problem of organismic purposiveness head on, foregrounding the organismal environment in his ruminations:

> Apparently, *the general "purpose" of most animal actions is to take some advantage of conditions existing in the environment, or to modify the relations between the individual and the environment in some way favorable to the species*. It is these externally directed actions which form the greater part of what is known as "animal behavior," and they represent an important . . . *means by which the animal adapts itself to its environment*. . . . This is why they impress us as "purposive." The "teleological" characteristic of living beings appears most conspicuously in this aspect of their life. But from the physiological point of view it is necessary to reach some purely objective or physicochemical definition of the term "purposive" as applied to such actions. (Lillie 1915, 589; emphases added)

Lillie wanted to know if biologists could establish that purposive behavior is a plausible or perhaps essential attribute of material systems characterized by the unique composition and organization of living organisms and their distinctive interactions with their surrounding conditions. To tackle this issue, he said, "My procedure and methods of reasoning will be those of objective natural science purely; and purposive actions . . . will be considered simply as events in external nature, disregarding their possible conscious or psychic accompaniment" (Lillie 1915, 590). Lillie argued that purposive actions obtain in complex physicochemical systems like organisms that (1) exhibit metabolism and (2) preserve an equilibrium with a changing environment.

His argument depended on a physiological notion of "adaptation" (see section 2.3.2 for background). Lillie defined an adaptation in a very general way as a "species conserving characteristic." He went on: "An organism is said to be adapted to its environment when it exhibits physical properties, structural characters and activities of such a kind that its development, growth, and continued existence in that environment are ensured. . . . The question then becomes: what is the nature of the interaction that makes such continued existence possible?" (Lillie 1915, 590–591). Lillie put forth two primary assertions regarding organisms and their environments. First, he posited that all organisms undergo a distinctive conversion of matter and energy sourced from their environment into their own organization. Second, organisms engage in activities that aim to preserve and sustain their own existence, such as securing food and resources. From a biochemical perspective, these processes manifest as constructive metabolic processes, while energy utilized by the organism in its activities is supplied through what Lillie referred to as "destructive processes," primarily oxidations. To achieve self-maintenance in a moment of the life cycle, Lillie claimed that an organism needed

> Primarily a balance between gain and loss of material and energy—a metabolic equilibrium. The physiological processes within the organism proceed automatically . . . so long as the supply of food, oxygen, and other necessary material from without is maintained; the external activities or "behavior" of the organism, thus, have reference chiefly to . . . maintaining the necessary supplies. . . . *Under these conditions there is equilibrium—balance of constructive and destructive processes—and life continues normally.* Those external activities which directly or indirectly have the effect of maintaining this equilibrium are the ones which are characterized as adaptive. *Adaptation, as a condition characterizing the relations between organism and environment, is thus defined [as] the maintenance of the organic equilibrium*; adaptive features of external structure or behavior are those which contribute to this end. (Lillie 1915, 592; emphases added)

Adaptation, for Lillie, could be spelled out in physiological terms: It refers to a metabolic equilibrium between an organism and its environment. With this in mind, we can see how Lillie rounded up his argument. He expounded that purposive actions, which aim to ensure the self-maintenance of an organism, have the consequence of securing an energy supply that can be utilized for future requirements. In this sense, any action that facilitates the continued existence of an organism can be deemed as "purposive" (Lillie 1915, 602). Lillie contended that all instances of adaptation fundamentally represent equilibria of varying complexity from a physicochemical perspective (Lillie 1915, 606). Purposive actions, in this view, arise through the establishment of a dynamic equilibrium between the organism and its ever-changing environment.

Over the following years, Lillie would continue defending this standpoint, especially in his philosophical publications. He connected the two problems that I mentioned earlier (i.e., the nature of biological organization and the self-maintenance of said organization) by arguing that the persistence and individuality of an organism are due to "an automatically controlled balance of diverse activities" (Lillie 1920, 483)—in other words, that these are the outcomes of dynamic equilibria. And thus, for him, the nature of vital organization can be explained as an "integration of processes," which are equilibrated "as to secure persistence and unity to the whole living system in its environment," and for that, "life occupies its unique position among the phenomena of nature" (Lillie 1920, 486–487).

His general outlook grounded on the constitutive purposiveness of organismal activities and the dynamic equilibrium between the organism and its environment would persist for some years.[6] For his idea of adaptation as dynamic equilibrium between organism and environment, Lillie was heavily inspired by the work of the German physiologist Paul Jensen (1907), who was convinced that organismal purposiveness and agential behaviors could be understood within the bounds of physiological science and by recourse to dynamic equilibria. Indeed, Jensen and Lillie were not isolated defenders of this standpoint in early-twentieth-century biology.

The idea of organisms undergoing "compensatory regulation," following changes in their environments, was common currency in early-twentieth-century biology, especially in physiology (e.g., Zeleny 1905, 3; Pike and Scott 1915). In general, there were many defenders of a physiological understanding of adaptation in those years (e.g., Mathews 1913; Perry 1917; for an overview, see Schanck 1954, chapter IV). The organicist Francis Sumner also embraced this view of adaptation and argued that organismal purposive activities, from developmental phenomena such as regeneration to behavioral activities, can be explained by *"the reattainment of a condition of equilibrium which has been overthrown"* (Sumner 1919a, 356–358; emphasis in original; see also Sumner 1919b). Additionally, in another 1919 paper in *The American Naturalist,* written almost ten years before Walter B. Cannon introduced the term "homeostasis" (see the discussion on organic regulation in section 5.2.4), Hooker (1919) argued that Le Chatelier's principle of chemical equilibrium was enough to account for the purposiveness of organisms. This principle states that if a system in chemical equilibrium is subjected to a constraint that shifts the equilibrium (e.g., a change of temperature or concentration), a reaction takes place that opposes the constraint, i.e., one that partially annuls its effects and restores the equilibrium. Hooker claimed,

> Inasmuch as the reaction of a system is directed according to the theorem of Le Chatelier, *every system in equilibrium is teleological*. The means that produce the reaction are directed to a definite end, to overcome the constraint, and the reaction might be said to take place in order that the system may be preserved. *This is evidently the source of the "purposefulness,"* that has occasioned endless biological discussion. *The living organism, however, is teleological only to the same extent as the ice-water-vapor system.* (Hooker 1919, 509; emphases added)

In the next years, biologists recognized that living systems, even when "resting," can never fully be in a state of "equilibrium," in which the total flow of energy is zero. In 1930, Archibal Hill pointed out that biological systems are, rather, in a "steady state" (Hill 1930; see also Hill 1928). The Austrian theoretical biologist Ludwig von Bertalanffy, famous for his organicist commitments, contended that a true equilibrium can only occur in closed systems, while in organisms, paradigmatic open systems, the increase in entropy that stems from their activities is compensated for by an influx of free energy from the environment, a state of stationary nonequilibrium (*Fliessgleichgewitch*), different from the traditional thermodynamic equilibrium (for an overview, see Bertalanffy 1953). Bertalanffy and others, like Burton (1939), applied the steady state view to organismal processes such as embryonic development and behavior. Bertalanffy argued that the steady state dynamics of an organism solved the riddle of intrinsic purposiveness that had long puzzled Driesch:

> Organic processes are explicable by means of two assumptions: (1) that the internal forces of living systems are directed towards states of equilibrium, and (2) that this direction holds for the system as a whole. This self-regulation of the organism is wonderful enough in its details, but, in principle, every connected system of the inorganic world in which the groupings of internal forces is directed towards equilibrium, behaves in the same way (Principle of Le Chatelier . . .) In this way Driesch's paradox is elucidated. . . . *We can . . . suppose that the "purposefulness" and "striving towards a goal" of organic processes is nothing else than the outcome of communicating systems of causally determined processes, the inner dynamical conditions of which tend towards equilibrium.* (Bertalanffy 1933, 103–104; emphasis added)

Until today, this stance has been commonly adduced to try to explain organismal purposiveness (see, e.g., section 6.5.4).

In this section I recounted how, in the early decades of the twentieth century, biologists from diverse disciplinary trenches addressed the problem of organismal purposiveness. Charting a taxonomy of positions is a useful first step when facing the theoretical and practice-associated heterogeneity of different biological disciplines, but of course, subsequent detailed historical analyses are still needed. According to my reconstruction, it is possible to identify (at least) seven stances on organismal purposiveness and agency in early-twentieth-century biology: (1) the neo-Aristotelian position, (2) the Drieschian answer, (3) purposiveness eliminativism, (4) the heuristic approach, (5) the holistic alternative to purposiveness, (6) organismal agency as noetic principle, and (7) organismal purposiveness and agency as explananda for dynamic equilibrium-based research. The first five considered the organismal environment as part of the background of agency, and the last two foregrounded it as an important component for articulating their accounts regarding agential processes and behaviors.

In what follows, I will cover the winding road of organismal purposiveness and agency in twentieth-century biological research (section 6.4), and then I will turn to the contemporary debate in biology and the philosophy of biology on these issues (section 6.5) to flesh out the asymmetries of the organism-environment relationship.

6.4 The Winding Road of Organismal Agency in Biological Research

In chapter 5 (section 5.3), I offered a narrative of the gradual marginalization suffered by the Organicist Movement within biological science. In particular, the decline of perspectives that reciprocally coupled organisms and their environments in the domains of ecology and evolution took center stage. In particular, as I recounted, the tides of postwar biological research shifted the focus toward alternative relata, such as genes and populations, and their respective environments. This shift, symptomatic of the broader sidelining of organicist and holistic viewpoints, had far-reaching consequences, resulting in the gradual disappearance of the organism as an active agent in the processes of development and, more notably, evolution (for analyses, see Walsh 2015; Baedke and Fábregas-Tejeda 2023; Cortés-García and Etxeberria Agiriano 2023).

As molecular biology and the physicochemical investigation of cellular components gained scientific prominence, discussions surrounding organismal teleology underwent a shift that, according to Sloan (2012), led to its diminished significance within the discipline. Moreover, from the 1960s onward, genes assumed the mantle of sole agents of evolution, often in ways that turned out to be empirically and conceptually unwarranted. According to the gene-centric rationale, the activities of genetic replicators were deemed causally necessary for the production, differential survival, and reproductive success of entire organisms (Walsh 2017, 243; see also Riskin 2020). An example of this reductionist stance can be found in Monod's influential book *Chance and Necessity*, where he argued that the purposiveness observed in organisms could be fully explained by invariant molecular mechanisms transmitted across generations (Monod 1971; for discussion, see Walsh 2017).

The receptiveness toward organismal agency was heterogenous among evolutionary biologists around those decades. For instance, the British ethologist William Homan Thorpe (1945) argued that the agential behaviors of organisms not only affected their ontogeny but have downstream causal effects in the speciation patterns of populations (see also Thorpe 1978). For example, he studied how genetic changes could follow and make an acquired behavioral preference hereditary, a phenomenon in line with what Julian Huxley and George Gaylord Simpson later called the "Baldwin effect." In fact, historian of science Gregory Radick (2017) has recently contended that Thorpe's thinking infiltrated into the Modern Synthesis. Thorpe's empirical work, as publicized by Julian Huxley in his 1942 book, *Evolution: The Modern Synthesis*, served to popularize and revitalize "the fortunes of what became one of the mainstays of agential science [in evolutionary research], the Baldwin effect" (Radick 2017, 35). Therefore, Radick claims, the Modern Synthesis was not completely inimical to animal agency.

While it is true that certain aspects of organicist thinking were incorporated into evolutionary theorizations under the umbrella of the Modern Synthesis, such as the recognition of the important role of organismal behavior in cases of the Baldwin effect—although as a peripheral factor that did not shake the central tenets of the synthetic theory of evolution, as held by Simpson (1953)—the broader range of phenomena associated with organismal agency went largely overlooked. One contributing factor to this marginalization was that organismal (intrinsic) teleology "transmogrified" into teleonomy, as aptly described by Krieger (1998; for a detailed account of this transmogrification, see Nahas 2024b).

Colin Pittendrigh introduced the term "teleonomy" as a way to study processes in biology that appear to be goal-directed, such as adaptation, without invoking the controversial concept of teleology (Pittendrigh 1958). In turn, Ernst Mayr redefined teleonomy to refer specifically to "systems operating on the basis of a program, a code of information" (Mayr 1961, 1505). For Mayr, this meant that a "*teleonomic process or behavior is one which owes its goal-directedness to the operation of a program*" (Mayr 1985, 140; emphasis in original). Mayr believed that all apparent goal-directed processes in development, including behaviors exhibited by organisms, are ultimately controlled by the genetic information encoded in DNA (see also Mayr 1964). Mayr advanced his understanding of teleonomy at a time when biologists were widely influenced by ideas coming from cybernetics and information theory (see Mayr 1985, 134, 142, 144; for a historical appraisal of the extent of these influences, see Nahas 2024a, 2024b). Drawing an analogy between organisms and computers, Mayr proclaimed, "The purposive action of an individual, insofar as it is based on the properties of its genetic code, therefore is no more nor less purposive than the actions of a computer that has been programmed to respond appropriately to various inputs" (Mayr 1961, 1504). However, artifactual views of organismal purposiveness faced criticism from some authors who were associated with the Organicist Movement (e.g., Bertalanffy 1951; see Nahas 2024b). Despite these criticisms, Mayr's interpretation of teleonomy and its connection to genetic programs became widely adopted among biologists. In the 1990s, Mayr confidently asserted that the idea of a genetic program shaped by natural selection "is now such a standard evolutionary concept, so heuristic in its applications, that it is unlikely that objections . . . will result in its abandonment" (Mayr 1993, 93).

Furthermore, prominent evolutionary biologists such as Williams argued for shoehorning all agential processes under the scope of teleonomy, considering it to be a valid and meaningful explanation (see Williams 1966, 258–269). Afterward, the problem of organismal teleology did not entirely go away, and a scientific and philosophical debate emerged in the 1970s regarding the proper status of teleological and teleonomic explanations in evolutionary biology (e.g., Ayala 1970; Wimsatt 1972). Despite this debate, the field favored the teleonomic perspective of organic purposiveness. If organisms *seem* to be agents to us, it is because genetic programs that encode purposive-like traits have been selected for in evolutionary time. In the words of Mayr, "Each particular program is the result of natural selection, constantly adjusted by the selective value of the achieved endpoint" (Mayr 1961, 141). If not vitalism, as Dobzhansky and his colleagues emphasized, the only conceivable alternative was "to regard internal [organismic] teleology as *a product of evolution by natural selection*" (Dobzhansky et al. 1977, 96; emphasis added). In this perspective, organismal agency was deemed the outcome of evolution by natural selection and not an evolutionary cause in itself. Dobzhansky was especially insistent on the epistemic inoperability of organismal, agential views of evolution:

> Classical theories of evolution fall into two broad groups. Some assume that evolutionary changes are autogenetic, i.e., directed somehow from within the organism. Others look for environmental agencies that bring forth evolutionary changes. . . . Autogenetic theories have so far proved sterile as guides in scientific inquiry. Ascribing arbitrary powers to imaginary forces with fancy names like "perfecting urge," "combining ability," "telefinalism," etc., does not go beyond circular reasoning. (Dobzhanksy 1950a, 210)

Regarding organismal purposiveness as a façade, as an ordinary result of the workings of natural selection, creates a form of (intrinsic) purposiveness eliminativism, as discussed in section 6.2.3. This viewpoint has been widely embraced by evolutionary biologists from the latter half of the twentieth century up to the present day. One additional issue that arose in evolutionary biology was the conflation of discussions regarding finality in evolution, such as divine design, an overarching *telos* in the living world, or orthogenetic trends, with the restricted issue of organismal purposiveness and specifically of organismal agential, goal-directed processes. This confusion led many evolutionary biologists to erroneously assume that, using Okasha's (2018) useful terminology, "agential thinking 1" (the question of teleology in nature) and "agential thinking 2" (the examination of evolved entities, such as organisms, as agents pursuing intrinsic goals) were essentially one and the same problem. This lack of careful analysis resulted in any discussion of organismal agency being equated with (cosmic) teleology in its broadest sense. Hence, it is not surprising that a considerable number of evolutionary biologists continue to be hesitant about considering the idea that organisms are causal difference-makers in their environments as they pursue intrinsic goals, such as surviving or reproducing, and this might have consequences as to how evolution unfolds.

Before covering the contemporary debate on organismal agency and the stances that have emerged in the philosophy of biology to conceptualize it—with distinct weights assigned to the organism-environment relationship and the ontic-epistemic placement of the environment qua relatum—I will briefly elucidate which type of teleology is at stake in these discussions of organisms and their ecological and evolutionary roles.

6.4.1 Disentangling Types of Teleology in Biological Explanations

Prima facie, what groups distinct kinds of teleology under the same general umbrella or justifies keeping the same broad linguistic descriptor is that, when used with explanatory functions, they give an account of something by appealing to *final causes* or by reference to an *end* or *goal*.[7] Many biologists and philosophers have tried to disentangle different senses of teleology to identify which of them are relevant for biological endeavors (see, e.g., Mayr 1998). Here, following Toepfer (2004), I argue that we need to differentiate at least four kinds of teleology that have been influential in the history of philosophy and biology (see also Toepfer 2011, 798):

Cosmic Teleology This corresponds to the view of a subtending *telos* in the cosmos that directs or controls the general state of affairs in a particular direction or toward the attainment of certain purpose (for discussion, see Henderson 1917a). With regard to living beings, this could translate into steered evolutionary trajectories in accordance with a plan or changes in the inorganic or organic world directed toward the fulfillment of a specific goal (e.g., the continued survival of mankind). The view of cosmic teleology ties in with the historical pedigree of Christian Wolff's school of metaphysics, which espouses teleology as a field of study that uncovers the underlying purpose and design of the world through the examination of natural phenomena. According to this perspective, teleology involves the pursuit of insights into divine intelligence by analyzing the structures and functions of natural products (for an analysis, see van den Berg 2013).

Intentional Teleology In this view, an agent anticipates a particular goal, and this intention is relevant for triggering and controlling the performance of an action (e.g., the planned construction of a house or heating some *Glühwein* to endure winter). For many scholars working in the philosophy of action, human action is primitively characterized in terms of intention, and this, in turn, is purportedly explained by a causal chain (e.g., of bodily movements) that ultimately can be traced back to an agent's mental states that give reasons for acting (for a recent overview, see Paul 2020). When these discussions are extrapolated to nonhuman organisms, Allen and Bekoff (1995, 15) speak of "teleomentalists"—namely, those scholars that "regard the teleology of psychological intentions, goals, and purposes as the primary model for understanding teleology in biology."

External Teleology According to this viewpoint, an entity is useful or purposeful for a different entity. For some authors, the external character involved is twofold: "(a) *the agent* whose goal is being achieved is external to the object that is being explained teleologically, and (b) *the value* aimed at is the agent's value, not the object's" (Lennox 1992, 325; emphases in original). This type of analysis is usually cast over machines and their constituent parts, or other human-made artifacts. For instance, a watch is said to be purposive insofar as it was designed and assembled for the attainment of a particular human goal (i.e., telling time), and in that same line, the purpose of, say, its battery is to keep it ticking (Goudge 1961, 192–193). The design stance of external teleology has also been commonly applied to organisms and their traits, especially when these are seen as shaped and optimized in evolution by bouts of a blind, aimless process: natural selection acting on random genetic variation (see Dawkins 1986; for a critical analysis, see Lewens 2004). As I mentioned above, when taken to an extreme, the position that ties external teleology with an adaptationist, optimality rationale of organismal traits paves the way for another instance of the stance that I have categorized as organismal (intrinsic) purposiveness eliminativism (section 6.2.3).

Internal Teleology In contrast, for this view, sometimes also referred to as "immanent teleology," the proper locus of teleological adscriptions is not a part or a suite of traits of an organism, but rather, the organism (or another evolved entity) as a whole that possesses goals of its own (Lennox 1992, 326). The intrinsic purposiveness of organisms—in contrast to the extrinsic purposiveness of machines, always set by an external designer (see Nicholson 2013)—means that organisms, through their activities, pursue their own goals, such as surviving and maintaining their organization, reproducing, and overcoming challenges throughout their life cycles—for instance, by undergoing compensatory morphological or physiological changes in response to changes in their environments (Walsh 2008; see also Baedke 2025).

After outlining these different types of teleology, it is important to stress that cosmic teleology plays no role in today's scientific debate surrounding organismal agency. Likewise, intentional teleology, with the onerous self-conscious, mentalistic, and rational undertones discussed within the philosophy of action literature, is rarely postulated in the biological sciences.[8] A complicated issue has been the relationship between external and internal teleology. In the history of biology, external and internal teleology have been advanced in various forms in biological explanations (see also Nahas 2024b). Distinguishing them is crucial.

Some authors have contended that Darwin's evolutionary theory provided grounds for integrating external teleology reasoning into biological thought (e.g., through the use of legitimate functional language), while others have suggested that he replaced teleological thinking through natural selection or banished it altogether from biology (see, e.g., Lennox 1993; Ghiselin 1994). For instance, Brandon (1981) has argued that explanations of adaptations are indeed (external) teleological explanations. In opposition to this reading, many scholars have expressed their dissent with linking Darwin's thinking with external teleology: they contend that the Darwinian position in the *Origin of Species* is that natural selection does not *cause* the adaptive "design" of organisms, but *results from* the ordinary activities of organisms in their struggle for existence (see, e.g., Walsh 2010; Bradley 2022). In this interpretation, Darwin exemplified the connection between the predominantly adaptive characteristics of organisms, the presence of variation, the possibility and reliability of inheritance, and the exponential growth of populations outlined by Malthus. In light of this, he postulated that natural selection inevitably emerges as a result of these interconnected factors (Grene 1961, 30). In other words, in this view, natural selection—and thus adaptive evolution—is the outcome of the purposive activities of developing organisms.[9]

Regardless of which side is correct in this dispute about natural selection, it is a fact that Darwin's theory has been mostly discussed through an external teleology lens (see Huneman 2017)—which of course can dispense of a strong, rationalist intentional teleology. In this view, an organism is understood as a collection of functional, fitness-enhancing traits that are considered to be "tuned" or optimized by natural selection. Instances of purposive behavior, for instance, are said to be accounted for through the "selected effects" account of function.[10] On this, Ruse (2000) asserted, "Organisms are adapted, hence they are teleological, and (for the Darwinian) this teleology can be explained through, and only through, natural selection" (223).

Against this undertow, a number of authors have stressed that we should not forget about (and better conceptually integrate) internal teleology. Moreover, scholars have maintained that internal teleology is more basic than external teleology (e.g., Russell 1916, 232). For example, Bertalanffy argued that "selectionism does in fact not explain organic wholeness, but rather presupposes wholeness in organisms' functions of life. Only because organisms are 'whole-maintaining' and 'persisting' [*dauerfähige*] beings, they can struggle for existence with one another" (Bertalanffy quoted in Fábregas-Tejeda and Baedke 2023b, 254). However, the integration of internal teleology into biological frameworks was—and it still is today—much more contested.

As we saw in section 6.2.3, some defenders of purposive eliminativism approved the validity of intentional teleology, only granted to humans as fully cognizing subjects; nonetheless, they denied that internal teleology can be ascribed to nonhuman organisms. This stance is common among present-day scholars. In contrast, defenders of the neo-Aristotelian position (section 6.2.1), the Drieschian answer (section 6.2.2), the heuristic approach (section 6.3.4), the noetic principle (section 6.3.1), and dynamic equilibria views (section 6.3.3) endorsed, in one way or another—for example, as a guide for research and explanation or as an ontological posit regarding a constitutive feature of organisms qua living beings that deserves to be explained or that does epistemic work in itself—internal teleology. In this sense, the historical debate on organismal agency has been articulated around ideas of intrinsic purposiveness, which we also see resurfacing in ongoing discussions (see below).

While the study of organismal agency experienced a decline in the latter half of the twentieth century, it is now beginning to regain momentum (see section 6.5). However, in early-twentieth-century biology, the investigation of organismal agency went beyond being merely a protest science. It represented a collective and heterogeneous inquiry driven by the shared general assumption that organisms are active agents in the world, demanding novel scientific ideas and approaches to figure out their distinctive characteristics. In what follows, I cover contemporary discussions of internal teleology and organismal agency in biology and the philosophy of biology and focus on how it matters for understanding the asymmetrical nature of the organism-environment relationship.

6.5 The Contemporary Debate on Organismal Agency: Foregrounding the Environment

In recent times, we bear witness to the renaissance of organismal purposiveness and agency as focal targets within controversies in biology and the philosophy of biology (see, e.g., Walsh 2015; Sultan et al. 2022; Corning et al. 2023; Fábregas-Tejeda et al. 2024 and chapters therein). This resurgence shows the enduring relevance of these concepts within the life sciences. In particular, current debates on the agential status of organisms in ecology and developmental and evolutionary biology reopen a bundle of old questions regarding the ontological and epistemological dimensions of agency. Is agency a capacity that belongs to organisms, or is it a heuristic tool for scientists to temporarily deal with the intricacies of organismal development while mechanistic research takes off? Can we dispense with it for explaining biological phenomena, or is it an inescapable outcome of our rational makeup without which we cannot fully grasp the properties of living beings? Furthermore, what is the structure of teleological explanations, and what kind of relations should they trace when whole organisms are considered explanatory relata?

A research area in which organismal agency is being amply discussed with respect to organism-environment couplings is niche construction theory (NCT).

6.5.1 Organismal Agency and Niche Construction

As we saw in chapter 5 (sections 5.4.1 and 5.4.2), niche construction covers the processes through which organisms modify their (physical or experienced) environments, thereby altering their conditions of development and the selective pressures acting on themselves, conspecifics, or other organisms with which they share ecological relations. Under the broadest interpretation, proponents of NCT admit that *all* organisms are, to a certain extent, niche constructors. Specifically, "obligate niche construction," as referred to within NCT, represents a sine qua non aspect of life. Living organisms, as open systems in a state far from thermodynamic equilibrium, are compelled to extract matter and energy from their surroundings to sustain their internal order and dynamic stability (Odling-Smee et al. 2003, 167–179).

For proponents of NCT, this aspect countenances the agential status of all organisms, extant and extinct:

> For reasons . . . that derive from consideration of the properties of life and the principles of thermodynamics . . . , we view agency to be an essential and inescapable aspect of nature. . . . Organisms are self-building, self-regulating, highly integrated, functioning, and (crucially) "purposive" wholes, which through wholly natural processes exert a distinctive influence and a degree of control over their own activities, outputs, and local environments. Indeed, organisms must have these properties in order to be alive. . . . (Laland et al. 2019, 131–132)

Undoubtedly, NCT, particularly when combined with empirical data and conceptual resources coming from studies of phenotypic plasticity, has presented scholars with fresh insights and motivation to reconsider organisms as agents in development, ecology, and evolution (for discussion, see Affifi 2020; Sultan 2015; Aaby and Desmond 2021). Turner (2016) has argued that organismal agency marks "the crux" where NCT and traditional approaches in evolutionary biology depart. According to Turner, agency emerges from organisms' homeostasis. In order to survive and uphold order within their internal milieu and immediate surroundings, organisms must actively interact with their environment and effect purposeful changes in specific directions (Turner 2016, 209–211). The "extended physiology" (Turner 2002) and active engagement of organisms with their environment is what allows them to achieve the states of equilibrium that keep them buzzing (Turner 2016).

All the same, Baedke has argued that this broad thermodynamic understanding of niche construction is not enough to resolve the problem of organismal agency:

> A consistent theory of agential causation that would strengthen especially the status of niche construction as a theory is missing. Attempts to provide such a framework . . . draw on classical understandings of purposefulness of organisms through thermodynamics. . . . However, it remains unclear how this framework can incorporate those cases of purposeful behavior of organisms that . . . [are]

> able to flesh out the different causal roles the agent is performing in changing its environment (by modifying it) and by changing its relation to its environment (by experiencing it). There is an important analytical difference between the two that any theory of agency needs to incorporate, as both cases can have very different evolutionary effects. (Baedke 2021, 83)

Aaby and Desmond (2021) have made a step in that direction and put forth a distinction between *agential* and *contributional* niche construction, making explicit their differential reliance on organismal goal-directedness. Agential niche construction arises from purposeful behaviors directed toward specific goals, while contributional niche construction lacks this goal-directed element. In cases of agential niche construction, a teleological explanation becomes essential in understanding why organisms actively engage in niche construction. On the other hand, in cases of contributional niche construction (e.g., the release of detritus or products of secondary metabolism into the environment), the accompaniment of a teleological explanation does not affect the organism's active role as a contributor that alters certain selection pressures: It would be epistemically inadequate to provide one.

But what exactly is a teleological explanation when organismal activities are involved? We turn our attention to this issue next.

6.5.2 Teleological Explanations

Godfrey-Smith (2009b) has suggested the label "agential thinking" to frame a particular way of understanding evolution.[11] According to this viewpoint, "we think of evolution in terms of a contest between entities with agendas, goals, and strategies" (Godfrey-Smith 2009b, 5). However, Godfrey-Smith issued a cautionary note against excessive reliance on agential thinking: "I see the agential view of evolution as something of a trap. It has real heuristic power in some contexts, but also has a strong tendency to steer us wrongly. . . . And once we start thinking in terms of little agents with agendas—even in an avowedly metaphorical spirit—it can be hard to stop" (Godfrey-Smith 2009b, 5). This skepticism toward agential thinking has been the prevailing attitude in the philosophy of biology for several decades, although recently there has been a gradual swing toward alternative, more welcoming perspectives.

Teleological explanations in biology have faced numerous criticisms, both in the past and in current controversies (for overviews, see Mayr 1998;

Allen and Neal 2020). Among them are objections related to vitalism, the incompatibility of teleology with naturalism and mechanistic explanations, the issue of excessive or unwarranted mentalistic assumptions (e.g., attributing actions to minds where there are none, as in the case of some agential construals of natural selection), the empirical challenges in testing teleological claims, and the theoretical incoherence and metaphysical complexities associated with the concept of *backward causation* (i.e., future goals causing some past states of affairs).

Many authors have tried to resolve this unease related to biological teleology. According to Walsh (2008, 113), teleology is "a mode of explanation in which the presence, occurrence, or nature of some phenomenon is explained by appeal to the goal or end to which it contributes." In light of this, we need clarity regarding how teleological explanations, which rely on the notion of "goals," actually explain. In the context of biological sciences, two key questions need to be addressed in order to explore this matter further: (1) What are the relevant relata involved in teleological explanations within developmental, ecological, and evolutionary contexts? and (2) What types of dependencies are traced by teleological explanations? For instance, are these dependencies causal or modal in nature? Answering these questions will shed light on the nature and scope of teleological explanations in the life sciences.

With regards to question 1, it matters which relata are in fact conceptualized through an agential lens. In the contemporary debate over biological agency, the position that advocates for "pan-agentialism" (for discussion, see Desmond and Huneman 2022), in which all living systems or even nonliving entities, from molecules to ecosystems, are regarded as agents with actual intrinsic goals, is relatively scarce. Usually, biologists and philosophers grant an agentic status to either organisms, cells, or genes.[12] Okasha (2018, chapter 2) has argued that there is only a restricted sense in which genes can be considered bona fide agents in evolution. This pertains to gene-selectionism and, more particularly, to the view of "selfish genes," which postulates that these have the (metaphorical) "ultimate goal" of outcompeting other alleles in terms of their representation in biological populations, and to achieve this they co-opt organisms as vehicles of self-replication. Okasha claims that "to say that genes 'want' to further their own replication, or encode strategies that further that goal, is ambiguous.

Is this the logical point that the spread of any gene is necessarily at the expense of its alleles, or the empirical point that some genes spread via mechanisms that harm their host organism? Only in the latter case are we compelled to treat the gene rather than the organism as the agent, in order to apply agential thinking" (Okasha 2018, 46). Conceptualizing genes as agents has been epistemically and explanatorily fruitful for evolutionary biologists; then again, becoming aware of the limitations of the gene's-eye view (e.g., being clear on the explananda it can tackle and what falls outside its explanatory purview) is an important consideration for making the most of this persuasive stance (see Ågren 2021).

In particular, Okasha (2018) has argued that the merit of the genes-as-agents view is that "it enables us to make sense of intra-genomic conflict and its phenotypic consequences, and thus to extend the adaptationist paradigm to a range of phenomena that would otherwise be baffling" (50). Phenomena such as sperm competition, sex-ratio distortions, transposable elements, and so on, have resisted explanation under the traditional adaptationist logic, and the gene's-eye view allows biologists to make sense of them. But it's one thing to grant this bestowal of intelligibility and do research on these topics; it's quite another to conclude from that assessment that all the features of organisms, including their intrinsic purposiveness and agency, are the direct consequence of selfish gene action and gene selection.

Epistemic upshots notwithstanding, the ontological problem is that genes, as fragments of molecules that yield a (somewhat) coherent set of potentially overlapping functional products that contextually depend on inter-, intra-, and extra-cellular factors, do not exhibit intrinsic purposiveness. Attributing internal teleology to a molecule would lead to what organicist E. S. Russell called the "fallacy of reduplication." For him, this epitomizes "a curious inversion of the problem" in which "the functional manifestations of the whole [i.e., the organism] are assumed to be due to an internal mechanism, or entelechy, or agent, which as it were doubles the role" (Russell 1930, 154). In another passage, he says that it is "unjustifiable to . . . ascribe to [lower units] the powers and capabilities which we know only as belonging to the organism as a whole" (Russell 1930, 49).

Wilson (2005) has argued that a biological agent "is an individual entity that is a locus of causation or action. It is a source of differential action, a thing from which and through which causes operate. . . . The notion of an agent is linked, but not identical, to that of a cause. Agents are individuals, and causes often are not" (6–7). He goes on to draw attention to the importance of having individuated agents: "Crucial to being an agent . . . is having a boundary, such that there are things that fall on either side of that boundary" (Wilson 2005, 7). In chapter 4, I argued that my proposal of shifting boundaries can account for organism-environment separation. Likewise, many other authors have convincingly argued that organisms can be described under sortal concepts and are a special kind of biological individual (see Okasha forthcoming; Prieto 2024). For agency, Wilson demands a criterion that organisms indeed fulfill: Agents need to be separated from their surroundings and qualify as individuals. Likewise, organisms are loci of causation and action: As we have discussed, they can, for instance, modify elements of their surroundings in pursuance of intrinsic goals such as maintaining themselves or reproducing (Nuño de la Rosa 2023). Organisms-as-agents, acting on their environments and producing manifold outcomes in the world, are valid relata in teleological explanations.

Regarding question 2, above, teleological explanations should not be mistaken as explanations based on unactualized future contexts, hinging on the spurious notion of token backward causation. The normative requirement in teleological explanations lies not in the cause-and-effect relationship, but in the connection between the goals of an agent and the means employed to achieve those goals. This idea, referred to as *hypothetical necessity* by Aristotle, entails explaining a phenomenon teleologically when it arises from goal-directed activity and is hypothetically necessary for achieving the anticipated outcome (for discussion, see Walsh 2008). Within this framework, Walsh (2015) and Fulda (2017) consider means as the elements within an agent's repertoire encompassing behavioral and developmental capacities of organisms that contribute to the attainment of their goals. These goals can range from evading predator attacks to adapting morphological or physiological features in response to changing environmental conditions. Importantly, the relationship between goals and means, characterized by hypothetical necessity, is *modal* rather than causal in nature:

> Hypothetical necessity is not a causal relation—goals don't cause their means, they hypothetically necessitate them—but it is a natural one nevertheless. Hypothetical necessity entails that, without the action in question, the goal would not have occurred and, with it, the goal . . . occurs reliably. Its dual is the relation of

> conducing. Whereas ends hypothetically necessitate their means, means conduce to their ends. Conducing is not the same as causing; *m* conduces to *e* only if *e* is a goal and, under the circumstances, *m* would reliably cause *e* across a range of counterfactual conditions. (Walsh 2018, 173)

Building from Woodward's (2003) interventionist view, Walsh has argued that teleological explanations are explanatorily autonomous: They point to regularities and counterfactual dependencies between means and goals that cannot be reduced to standard causal-mechanistic explanations (Walsh 2008, 2015, 2018). In this view, teleological (modal) and mechanistic (causal) explanations are complementary to each other, as they provide access to different sets of counterfactual dependencies in biological phenomena. In that sense, for example, it is possible to show that the goal-directedness of a system-as-a-whole has explanatory consequences that cannot be solely pinned down to the interactions of its parts (Walsh 2006, 2008). In a similar vein, Jacobs (1986) contended that agency represents an attribute of an organism that is important in elucidating its constituent parts and ongoing processes both synchronically and diachronically. In that sense, many philosophers concur that agency has (top-down) explanatory value for organismal processes. Moreover, Uller (2023) has argued that the concept of organismal agency possesses epistemic import by its ability to shape the structure of evolutionary investigation by adhering to specific criteria of explanatory adequacy. Agential explanatory frameworks recognize the presence of nonselective causal influences on adaptive evolution, thereby prompting inquiries into how bridges between developmental and selective explanations can be built (Uller 2023).

As highlighted by Nahas and Sachs (2023), the dispute about the integration of organismal agency into biological science is related to the question of whether adopting an organismal approach that untangles internal teleology would enhance our understanding and explanation of puzzling biological phenomena (e.g., of certain developmental and evolutionary processes). However, as they also point out, this inquiry represents a distinctive metatheoretical commitment or motivation of particular communities of biologists and philosophers of biology, separate from the philosophical quandaries concerning the consequences of naturalizing teleology for our conceptions of nature, mind, and our own place in the natural world (see, e.g., Gambarotto and Nahas 2023). The debate over the naturalization of teleology is a polymorphous one, as it encompasses diverse scientific and philosophical objectives, each with its own unique significance, implications, and success criteria. For this reason, in what follows, I will not cover the entire spectrum of positions related to organismal agency in science and philosophy, but only those that have been advanced in attempts to improve the epistemic edifice of biological science while underscoring the centrality of the organism-environment relationship.[13] In these, we will see some resonances and continuities with former positions in early-twentieth-century biology (see section 6.2).

Besides the purposiveness eliminativism stance that began to solidify in the second half of the twentieth century (recounted in section 6.4), which is still widespread in evolutionary biology,[14] there are three main approaches in extant biology and the philosophy of biology to deal with organismal internal teleology that are explicitly related to the organism-environment relationship: (1) the Kantian projective approach (section 6.5.3), (2) the autonomy school (section 6.5.4), and (3) the so-called ecological approach (section 6.5.5).[15]

Desmond and Huneman (2020), proponents of the Kantian approach, argue that defenders of both approaches 2 and 3 can be classified under a "neo-Aristotelian position" regarding organismal purposiveness (see also Okasha 2024, 6), as presented in section 6.2.1. For them, the neo-Aristotelian position entails recognizing the epistemic indispensability of organismal agency and adopting an ontic view of agential activities, which involves expanding biological ontology to embrace organismal purposes. However, emphasizing the ontic realism of organismal purposiveness as a criterion to count as a bona fide neo-Aristotelian is too broad, as it overlooks the differences among various authors. For instance, all the diversity of positions described in sections 6.2 and 6.3, except purposiveness eliminativism and the heuristic approach, would collapse into neo-Aristotelianism if our sole classification criterion is presumed ontic realness with regard to organismal purposiveness. Some philosophers, like Walsh (2021) and Nahas (2024b), indeed find value in the biological works of Aristotle and see some parallels between the Stagirite's ideas and the contemporary debate on organismal agency (e.g., between Aristotle's concept of *bios* and the active role that organisms enact in the evolutionary process), but their ecological standpoints are quite distinct from other positions in the contemporary debate (e.g., the stance advanced by the autonomy school). For these reasons, my tripartite categorization for contemporary discussions is preferable. In what follows, I cover the three positions successively and chart their particularities with respect to how they frame the organism-environment relationship.

6.5.3 The Kantian Approach in the Contemporary Philosophy of Biology

In recent years, there has been a renewed interest in rehabilitating Kantian perspectives concerning the purposiveness and agency of organisms (see Jones 2023, chapter 5). As outlined in section 6.2.4, while early-twentieth-century biologists found inspiration in Kant's ideas, they more often than not deviated from his projective viewpoint and embraced an ontic interpretation of the seemingly purposive behavior and development of organisms. According to these biologists, such characteristics are to be considered ontological features that make organisms what they are, and biological science cannot do so without facing the challenge of enlightening them. In contrast, some contemporary philosophers have again raised alarm bells about attributing ontological status to organismal purposiveness and agency.

In this vein, Desmond and Huneman (2020) have suggested that it is worth reexamining the question of whether agency is a genuinely "real" capacity of living beings. Throughout history, attributing agency to organisms has often been regarded as metaphorical or a construct of human imaginings, perhaps reflecting our tendency to anthropomorphize phenomena that we have a hard time explaining otherwise. Should we embrace an ontic perspective or not when it comes to dealing with prima facie goal-directed processes and behaviors in organisms (Desmond and Huneman 2020, 34)?

In a nonontic view of agency, agential explanations take, for instance, behaving organisms *as if* they were agents with goals. We can adopt an *intentional stance* due to its epistemic practicality, as through this lens we can describe entities "whose behaviour can be predicted by the method of attributing belief, desires and rational acumen" (Dennett 1987, 49; for a Kantian reading of the intentional stance, see Ratcliffe 2001). In this sense, organisms are regarded as agents possessing goals and actively engaging with their environments as an explanatory fiat without matching ontological substrates. The rationale for this position is that scientists can gain certain insights on organismal behavior or development (see section 6.2.4) if they presuppose goal-directedness dynamics in organisms as integrated wholes, rather than viewing them as mere assemblages of individual parts that pull in different directions.

The perspective proposed by Desmond and Huneman, drawing from Kantian ideas (see section 6.2.4), offers a nonrealist interpretation of agential explanations while recognizing the epistemic indispensability of agency. For them, the attribution of agency to organisms does not entail a literal understanding akin to attributing quantifiable physical properties. However, it also surpasses being a mere heuristic or predictive tool: It is an inescapable consequence of our own limited rational make-up. Given our status as cognitively constrained beings, perceiving agency in organisms becomes an unavoidable demand of reason (Desmond and Huneman 2020, 59).

In biology, Desmond and Huneman (2022) maintain, agency refers to how scientists study organisms behaving in a seemingly goal-directed manner in response to changes in their environments. Construing behaviors as "goal-directed" has explanatory value in elucidating the patterns of behavioral robustness displayed by organisms. A goal, in their view, represents the (epistemically intelligible and scientifically detectable) objective pursued by an organism, irrespective of disruptions, eventualities, or alterations in its environment. Organismal agency, therefore, encompasses three aspects: (1) a set of goals that can help interpret patterns of interrelationships amid (2) a vast array of potential environmental conditions and (3) a diverse range of potential behaviors.

For Desmond and Huneman, the notion of a goal is deployed to account for how environmental states and behaviors are informationally integrated. Therefore, Desmond and Huneman (2022) contend that measures of information integration correspond to (operationalized) measures of agency. They further claim that agency manifests as a specific configuration of interconnections between environmental conditions (inputs) and corresponding behaviors (outputs). Along these lines, they argue that the concept of agency "imposes a structure on observed behavior, and if the observed behavioral patterns do not exhibit this general structure, there is simply no need to describe them as 'agential'" (Desmond and Huneman 2022, 23). In consonance with their Kantian commitments, they view agency primarily as an explanatory construct that grants intelligibility by foregrounding behaving organisms in an environmental setting without saying anything *de re* about those organisms.

A rehabilitated inspiration on Kant's work has also percolated in the positions I recount below. In this regard, Gambarotto and Nahas have reasoned that

> Kant's legacy for biology reflects two fundamental attitudes toward teleology today. On the one hand, the canonical Kant is seen to have saved the concept of teleology from dogmatic metaphysics by turning it into a regulative principle, and thereby making it safe for science. One the other hand, a minority of philosophers and theorists finds in the third Critique an undertow that takes intrinsic purposiveness as the inherent feature of self-organizing systems, and as a new causal principle within nature. (Gambarotto and Nahas 2022, 54)

Indeed, Gambarotto and Nahas have shown that, in recent decades, scholars with varying and sometimes conflicting research agendas often make reference to Kant as a "notable precursor" (Gambarotto and Nahas 2022, 48), just as we saw was the case for organicist and holistic biologists in the early twentieth century. They concur with Weber and Varela (2002) that the root cause of these divergent interpretations might be the "unstable middle position" between mechanism and natural teleology set out in Kant's third Critique. For this reason, Gambarotto and Nahas (2022) call for the recognition of the existence of two distinctive Kantian legacies within biology and the philosophy of biology. The first legacy aligns with the heuristic approach discussed earlier (see section 6.2.4 and this subsection). The second legacy discloses a naturalist approach to organic teleology that responds to Kant's apportioning of teleology to a mere regulative principle as it overlooks the importance of his distinction between "extrinsic" and "intrinsic" purposiveness. Followers of this naturalist approach to teleology expand what was professedly *in nuce* in Kant's writing to assert that the organization of organisms is inherently teleological, oriented toward its self-maintenance. The focus on organization as the foothold for organismal agency and how the environment comes into play in this viewpoint will be explored in the next subsection.

6.5.4 The Autonomy School and Organismal Teleology

Since the second half of the twentieth century, some authors have put forth the case that the sciences that study dissipative systems and far-for-equilibrium thermodynamics allow the fleshing out of Kant's notion of self-organization. In this vein, for instance, Kauffman and Clayton (2006) have submitted that there are certain minimal physical conditions that are necessary for applying genuine teleological language to a biological system: autocatalytic reproduction, work cycles, boundaries for reproducing individuals, self-propagating work and constraint construction, and choice and action that have evolved to respond to salient environmental elements (for discussion, see also Moss and Newman 2015). What all these conditions share is that they relate to how organismal organization is determined and enacted. In fact, Kauffman and Clayton (2006) claim that "ultimately we will only understand biological agency when we have developed a theory of the organization of biological processes" (501).

Defenders of the autonomy school in the philosophy of biology, which brings together the insights and decades-long work of many scholars, deem Kant to be the most important forerunner in their quest to explain organismal organization. Not infrequently, they make strong historiographic assertions such as this: "The origins of the theory of autonomy can be traced back to Kant" (Virenque and Mossio 2023, 11; but see Wolfe 2024 for a fair critique). Furthermore, they cite Bernard's views on regulation (see sections 5.2.3 and 5.2.4) as an additional inspiration for developing their views on organismal teleology. Mossio and Bich claim, "In Bernard's view, biological organisms are teleological not only because of their capability of compensating for external variations by means of internal modifications, but also because the conservation of the internal *milieu* serves the main intrinsic goal of maintaining the specific internal conditions for the organism to exist" (Mossio and Bich 2017, 1095).

More precisely, the autonomy school can be said to be expanding and refining the theory of autopoiesis by Humberto Maturana and Francisco Varela (briefly covered in section 4.5.3; for discussion, see Bich and Arnellos 2012). Indeed, a focus on autonomy, as part of a larger autopoietic framework, was recognized early on by Maturana: "The fundamental feature that characterizes living systems is autonomy, and any account of their organization as systems that can exist as individual unities must show what autonomy is as a phenomenon proper to them, and how it arises in their operation as such unities" (Maturana 1975, 313). In line with these ideas, Maturana and Varela (1972) put forth the claim that autonomy within living systems emerges from the process of cyclical self-production known as *autopoiesis* (see also Varela 1979). According to this viewpoint, a living system can only be accurately described as a network of interconnected processes that continually and recursively generate and manifest as a tangible entity within physical space. These processes are responsible for the production of the system's components, and the system itself is referred to as having autopoietic organization that separates it from its environment. Every occurrence within an autopoietic system is subordinate to the fulfillment of its autopoiesis, and failure to align with this results in the system's breakdown. Maturana (1975) argues that the specific autopoietic states adopted by an organism are determined by its structure, which is the product of its evolutionary and ontogenetic interactions with the environment and which shapes its coupling with the surrounding medium while autopoiesis is being realized.

Autopoiesis theory is building on the larger forgotten tradition that I covered in section 6.3.2: the suite of authors that saw organisms' intrinsic purposiveness as an explanandum for dynamic equilibrium-related research. When mapped against the landscape of early-twentieth-century positions, the theory of autopoiesis and the ensuing autonomy school (see below) are not radically new approaches to the problem of organismal teleology and agency.[16] They are tied in with this tradition that foregrounded the environment in the existential dynamics of an organism. This is clear when we see to which author they trace back the idea of "organizational closure."

Mossio and Bich (2017) have stressed that the Swiss biologist and psychologist Jean Piaget introduced the theoretical concept of organizational closure, which he considered complementary to the steady-state views of thermodynamic openness previously emphasized by Bertalanffy (see section 6.3.2). Piaget stressed, "The living organization is an *equilibrated system* (even if one avoids the term and substitutes Bertalanffy's 'stable states in an open system')" (Piaget [1967] 1971, 36). The idea is that organizational closure is not foreclosed by thermodynamic openness: "The opening . . . is the system of exchanges with the environment, but this in no way excludes a closure, in the sense of a cyclic rather than a lineal order. This cyclic closure and the opening of the exchanges are, therefore, not on the same plane, and they can be reconciled" (Piaget [1967] 1971, 155). The central idea was to forge a connection between the notion of a stable flow of matter and energy exchanges between a system and its environment and the concept of a circular internal order that facilitated the reconstitution of the system's components. Piaget's formulation of closure envisioned the dynamics of organisms as a manifestation of self-determination, achieved through a network of interconnected processes and components that sustained the unity of the living system (Mossio and Bich 2017, 1098). Mossio and Bich (2017) agree with Piaget's characterization of the "interaction between the organism and the environment in terms of (individual) adaptation, defined as the assimilation of external influences which involves an internal self-regulation (accommodation)" (1098). In this way, we can see how the autonomists have unwittingly recovered the physiological notion of adaptation that was common currency in discussions of dynamic equilibria in early-twentieth-century biology (see section 6.3.2).

Within the autonomy school, authors propose a perspective wherein biological organization is buttressed by a teleological causal regime. Central to their argument is the linkage between organization and teleology through the concept of *self-determination*. For Mossio and Bich (2017), self-determination entails a circular relation (e.g., part-part reciprocity; see also section 5.2.3) between causes and effects, wherein organismal organization generates effects that, in turn, contribute to the maintenance of the organization. They link this view of circular self-determination to intrinsic purposiveness and internal teleology:

> By relying on this circularity, we argue that the conditions of existence on which the organisation exerts a causal influence can be interpreted as the goal . . . of biological organisation: because of the dependence between its own existence and the effects of its activity, biological organisation is legitimately and meaningfully teleological. However, teleology is interpreted here in a specific sense, precisely because the final cause of the organisation is identified with its own conditions of existence. (Mossio and Bich 2017, 1090)

In this view, reciprocal causation (circularity) is necessary but not sufficient for intrinsic teleology: not all circular causal regimes embody self-determination; rather, self-determination in the context of biological systems is specifically manifested as *self-constraint*. This self-constraint takes the form of organizational closure (first introduced in section 4.5.3), in which a network of mutually dependent constitutive constraints operates to sustain the system (Mossio and Bich 2017, 1106–1110).

In the purview of the autonomy school, biological organization can be grasped as a causal regime where a collection of structures, serving as constraints on the dynamic flow of energy and matter and thus the far-from-thermodynamic equilibrium, create a reciprocal interdependence—that is, organizational closure. In particular, for defenders of this bio-philosophical framework, what sets organisms apart from other natural systems is that they exhibit *closure of constraints*. This entails the emergence of a second-order emergent causal regime where the individual constituents, functioning as constraints themselves, launch a reciprocal interdependence through which the system maintains its integrity collectively through self-determinations (for a detailed exposition of closure of constraints, see Mossio et al. 2013). In that wise, "self-constraint implies that the circular organisation specifies its own dynamics: that is why only self-constraint involves self-determination, and why biological organisation (by realising closure) is teleological" (Mossio and Bich 2017, 1091).

Organisms, as thermodynamically open systems that harvest matter and energy from their environments, exhibit intrinsic teleological dynamics as their own actions and activities contribute to found and preserve their own conditions of existence and viability. For the autonomists, organisms achieve this by the deployment of "regulation or control mechanisms" that make it possible to effectively manage the consequences of a disturbance, adjusting its processes to respond effectively to specific variations in both internal and external environmental (existence-altering) conditions (Bich et al. 2016; on the idea of "control mechanisms," see Bich and Bechtel 2022).

Moreover, the concept of organizational closure, which as we saw, paves the way for the characterization of organisms as autonomous systems, foregrounds the environment (for a counterview, see Etxeberria forthcoming): Organisms are not self-contained or isolated entities but constantly engage in interactions with their surroundings. These interactions, however, are not arbitrary but are regulated by the functional components and subsystems that operate within and enact organizational closure. As a result, the interactive capacities of organisms with their environments count as purposive. Within autonomy theory, this purposeful capacity is referred to as *agency*. Agency thus represents the dynamic and interactive dimension of organismal activity that emerges from organizational closure (see Moreno and Mossio 2015, 89–109). It encompasses the various functional capacities of a living being that are dedicated to purposefully regulating its relationship with the environment. Moreno (2018) states that agency "requires that the processes on which it exerts a causal influence belong to the external environment, which, in turn, means that these processes have not already been constrained by the autonomous system" (291).

In this perspective, agency is in continuity with an organism's ability to maintain its organizational integrity and adapt to the challenges presented by its external surroundings (i.e., adaptivity, see section 3.1.2). Besides the control mechanisms that facilitate integration, which enable organisms to exhibit versatility and adaptivity in their actions when confronting heterogeneous environmental circumstances, autonomists contend that there is an additional perceptual dimension to adaptive agency that is of great importance. To qualify as an adaptive agent, an entity must possess the capability to *sense* its environment actively, "measuring different conditions, triggering correlative effector actions, and monitoring its own constitutive processes so as to avoid or prevent dysfunctional situations" (Moreno 2018, 296).

For the autonomy school, then, agency aids in grasping the nature of autonomous living systems and their continuous engagement with their environments. The status of "adaptive agent" arises when a system acquires the capacity to dynamically adjust itself in the face of internal or external disturbances and influence its interactive abilities with the surrounding environment. As argued by Moreno and Mossio, such an adaptive agent, enacting a state of organizational closure, not only achieves autonomy but also achieves the status of a *living system*. This implies that the ability of a system to adapt, sustain its organization, and actively engage with its environment bestows upon it the property of being *alive* (see Moreno and Mossio 2015, chapter 4). In this view, there is a tight coupling between agency and life.

Let's move on now to the final position in the contemporary debate on organismal agency that foregrounds the role of the environment.

6.5.5 The Ecological Account of Organismal Agency

In recent years, philosophers Denis Walsh and Fermín Fulda have championed the ecological account of organismal agency. Unlike the autonomy school, the ecological approach does not concentrate on the specifics of organizational regimes within organisms. Instead, it accentuates agency as a *dynamic characteristic* of an organism, enabling it to prioritize and tailor its goal-directed actions in connection to the opportunities and possibilities presented by its environment. In this vein, for Walsh, organismal agency "consists in a capacity of the system to pursue goals, to respond to the conditions of its environment and its internal constitution in ways that promote the attainment, and maintenance of its goal states. Agency is observable in the sense that what we see when we observe an agent is its dynamics, the way that the agent negotiates its situation" (Walsh 2015,

210). For instance, the ecological approach casts phenotypic plasticity as a purposive ability displayed by organisms to ensure their survival. Through this capacity, organisms exhibit adaptive responses to both internal and external disruptions, enabling them to maintain their viability in a dynamic and ever-changing environment by mobilizing suitable phenotypic traits to cope with and surmount biotic and abiotic challenges (Walsh 2021; see also Walsh and Sultan 2024).

According to this perspective, agency cannot be detached from the manner in which an organism is dynamically embedded in its world. This integration shapes agency as an emergent property that arises from the organism's interactions within its ecological context (Fulda 2023). In contrast to the autonomy school, the ecological account stresses that agency, much like other dynamic aspects of complex systems, transcends the lower-scale causal processes that instantiate it (e.g., regimes of self-constraining and organizational closure). Instead, it finds its definition in the broader context of large-scale ecological regularities, which shape its adaptive dynamics during ontogeny (Fulda 2023, 4). For this view, agency qua capacity and observable, gross dynamical feature of an organism should not be confused with the suborganismal causal realizers of, say, phenotypic plasticity or robust, environment-sensitive goal-directed behaviors. Proponents of the ecological account would counter that adopting an "inside-view" of agency and staying there (i.e., peering into the inner workings of a behaving, developing organism and not leaving that level of analysis) would be like mistaking the forest for the trees.

When its evolutionary consequences are highlighted, the ecological account, often referred to as "situated adaptationism," is presented as an alternative conception of adaptive evolution. Instead of adhering to the traditional view that sought to dethrone Paley's natural theology, but that still talks of organic "design" and evolutionary responses seeking "solutions" to particular environmental challenges already set out by preexisting "conditions of existence," situated adaptationism frames adaptive evolution from the individual organism's vantage point. It envisions the organism as an active participant, "situated" within its environing conditions and in dynamic engagement with them (Walsh 2012, 90–91).

For Walsh, there are three interrelated commitments that ground the traditional adaptationist logic in evolutionary science: (1) suborganicism (i.e., an overt emphasis on organismal parts at the expense of the organism as a whole), (2) autonomy of organism and environment (i.e., granting the causal and ontic autonomy of these relata), and (c) explanatory externalism (also discussed in section 5.3.4). In the conventional adaptationist interpretation, the environment plays a marginal role in explaining development and a definite role in explaining evolution (see also section 5.3.1):

> [The] environment provides resources and background conditions for the inheritance and development of form. It is a facilitator, but [genetic] replicators are the difference makers. . . . The environment . . . plays a wholly separate role in evolution. The exigencies of the organisms' environment determine which organisms survive and reproduce. By this process, the environment selects, winnows, and changes biological form. The adaptationist program recognizes a division of causal labor between replicators and environment. Put crudely, replicators cause inheritance and development, while the environment causes adaptive change. (Walsh 2012, 93)

In this sense, environment is synonymous with "population selective environment" (see section 5.3.3) and only rarely recovers the specificities of what in this book has been called the organismal environment. "In this view," Lewontin famously voiced his discontent, "the organism is the object of evolutionary forces, the passive nexus of independent external and internal forces, one generating 'problems' at random with respect to the organism [i.e., the environment], the other generating 'solutions' at random with respect to the environment [i.e., gene mutations]" (Lewontin 200, 47). Walsh retorts to this view: "The organism is not simply an interface between the inner realm of replicators and the autonomous free-standing environment. The folly of the adaptationist program . . . is precisely the demotion of the organism, and its detachment from the environment. In contrast, Lewontin argues, *the organism-environment relation is crucially involved in adaptive evolution*. To understand the process of adaptive evolution, we must understand this relation" (Walsh 2021, 94; emphasis added). And, reciprocally, to fully understand the organism-environment relationship, I contend in this chapter, one needs to take seriously organismal agency.

Walsh has put forth the argument that Modern Synthesis–inspired evolutionary theories and models, rife with abstractions on what constitutes ensembles of gene types and populations, have inaccurately portrayed the metaphysics of evolution. Rather than regarding evolution chiefly as a molecular process, he advocates for a perspective that recognizes its ecological character: organisms, and not genes, are the main players of evolution, an insight stemming from Darwin that biologists seemed to have forgotten as the twentieth century unfolded (Walsh 2015).

As a replacement for the traditional adaptationist view, Walsh and colleagues' ecological account builds on organismal agency—and natural selection as a consequence—as a thoroughly ecological phenomenon (for discussion, see Aaby 2021b). As Walsh (2018, 182) puts it, agency "is the process by which a goal-directed system marshals the resources of its adaptive repertoire in response to the affordances it both experiences and makes." Here, Walsh is drawing from the conceptual vocabulary of Gibsonian ecological psychology in which "affordance" refers to the available opportunities for or obstacles to action that certain environmental components offer to an organism given its traits and capacities (for an in-depth exploration of affordances, see Heras-Escribano 2019). By bringing in the affordance concept, Walsh underscores an important element of organism-environment mutuality: "What the agent can do depends on what the environment affords. But, conversely, what the environment affords depends very largely on what the agent can do" (Walsh 2012, 98). For instance, a steep rocky slope affords very different things to an inexpert climber such as myself—a sedentary, bookish *Homo sapiens*—compared to a token bighorn sheep (*Ovis canadensis*), which tend to be agile climbers in their native settings. This has led some authors to argue that affordances are relational, emergent properties of the whole organism-environment system that do not inhere in either environmental factors (as dispositional properties) or the organism (see, e.g., Stoffregen 2003).[17]

Endorsing Gibson's outlook implies considering that an agent and its environment form an inseparable, dynamic, and interdependent system. Environments are relationally defined with respect to an organism: An environment holds significance *for* a particular agent, with its prominent features shaped by the agent's ability to halt or take action. Concurrently, the agent's capacity to act is influenced by the distinctive components of the environment. Perception, in this context, becomes a dual process: reacting to the environment while actively shaping and defining it—constituting it (Walsh 2012, 98). This is again reminiscent to what Lewontin and Levins' called "reciprocal codetermination" between organism and environment (see sections 3.1.1 and section 5.5.1).

Gibson (1979, 8) said that "it is often neglected that the words animal and environment make an inseparable pair. Each term implies the other. No animal could exist without an environment surrounding it. Equally, although not so obviously, an environment implies an animal (or at least an organism) to be surrounded . . . the environment is ambient for a living object in a different way from the way that a set of objects is ambient for a physical object." Building on this perspective, Walsh has claimed that the environment embodies the physical world perceived and engaged with *by* an organism. The organismal environment at any time point, in connection to an organism's traits and capacities, yields a structured collection of affordances—an affordance landscape. This landscape operates in a state of constant flux, with alterations rippling throughout the system in the organism's struggle for existence (Walsh 2012, 99).

In the evolutionary domain, for Walsh, structures considered adaptations have emerged as direct responses to an affordance. Therefore, we cannot conceive of an adaptation as a solution devised by a developing system to address (purportedly organism-independent) environmental challenges; we need to conceptualize said environment as an "affordance landscape." The adaptive predicaments encountered by an organism are not only always influenced by the external setting it inhabits but are also shaped by its own way of life and which phenotypic traits it already possesses (Walsh 2012, 100).

As we saw in the previous chapter, an organism has various means to shape its physical and/or experienced environment. It can bring about alterations in the physical parameters of its surroundings, adjust its sensitivity to manifold conditions, or even undergo transformations by changing its location or form. Fulda (2023) has called the developmental capacity of organisms to constitute environmental conditions into affordances without this involving altered physical parameters "ecological transduction," which can be either active or passive. In line with the ecological account of agency, the plastic phenotypic repertoire of organisms is tied to evolutionary dynamics:

> Evolution is adaptive because organisms are adaptive. Organisms are adaptive precisely because their developmental capacities constitute a complex interactive, "commingled" system of organism and environment. Plasticity introduces a distinctive perspective on the process of evolution. . . . Organisms respond to the affordance landscape by implementing compensatory changes that conduce

> to the fulfillment of their goals of survival and reproduction. Often enough these changes result in adaptive phenotypic novelties, and often enough they are (or become) sufficiently intergenerationally stable to count as *evolutionary* novelties. Each change of form or behavior ushers in a new landscape of affordances. (Walsh 2012, 107; emphasis in original)

Walsh reasons that if organisms indeed construct their affordance landscapes—and thus are partly responsible for structuring their conditions of existence and what counts as such (see also Gerard and Maublanc 2023)—then the environment to which they adapt in the evolutionary process cannot be treated as a self-standing ontological *given*, as something specifiable independently of the organism itself that irrevocably causes adaptive evolution (Walsh 2012, 108). In this view, organismal agency is of the utmost explanatory importance for evolution because it stands for the gross dynamical property of a goal-directed system to bias its repertoire in response to changing affordances (Fulda 2017, 2023). Organisms actively construct and modify their affordance landscapes through adaptive alterations to their own structure and functionality. Some of these changes, passed down through generations, exhibit remarkable stability over time, ensuring the continuity of adaptive traits and behaviors within populations. For the defenders of the ecological account, framing things through this lens demands an agential, organism-centered view of evolution (Walsh and Rupik 2023).

If biology is to embrace organismal agency (e.g., to tackle certain evolutionary explananda; for discussion, see Uller 2023), and in particular the ecological account of agency, new robust theoretical frameworks need to be developed and worked out. Agent theories, in contrast to what Walsh (2018) has branded "object theories," require a unique and interconnected set of concepts—among others, "affordance," "goal," and "repertoire" (see also Fulda 2017). Fulda has argued that a system is said to have a goal (G) if it possesses a repertoire that empowers it to achieve and sustain G effectively under a broad range of both actual and counterfactual conditions (C). This capacity is realized by the system's ability to appropriately respond to the affordances offered by C and by actively deploying its repertoire to accomplish G. Goals, in this context, refer to stable end states that goal-directed systems consistently and reliably tend to attain and maintain. Such systems exhibit robust and adaptive behaviors, either persisting in their trajectory despite disturbances or adjusting their trajectory to accommodate perturbations (Fulda 2023, 5). This robust, flexible adaptive dynamics of organismal goal-directed behaviors is identified by its unique modal profile, one that accords to hypothetical necessity under an interventionist framework, an issue that was fleshed out in section 6.5.2.

For the ecological approach, the degree of agency exhibited by an organism can vary significantly in phylogeny based on the versatility of its developmental and behavioral repertoire (Fulda 2017, 70). This variability gives rise to a continuum of agency, ranging from the simplest adaptive agents, like bacteria, to the most sophisticated cognitive agents, such as humans, which can act on, be influenced by, or shape a wide range of affordances. In that sense, for the ecological account, full-fledged cognition is not at all a prerequisite for agency, which comes in degrees and has evolved through the emergence of new organismal goals, novel environmental conditions, and varying ontogenetic repertoires (Fulda 2017, 86).

In sum, the ecological approach characterizes organismal agency as the "capacity of the system to pursue goals, to respond to the conditions of its environment and its internal constitution in ways that promote the attainment, and maintenance, of its goal state. . . . [Agency] consists in the capacity of a system to cope with its setting, to attain its goal by responding to its affordances *as* affordances" (Walsh 2015, 2010; emphasis in original). The environment is foregrounded as contributing to a landscape of affordances, shaped by and shaping the actions of organisms, with protracted evolutionary consequences.

As discussed in this section, contemporary perspectives on organismal agency consistently foreground the environment as a crucial factor in their analyses. It seems, then, that agency unfolds within the dynamic relationship of organisms and their environments. The final section of this chapter will combine the insights garnered on organismal agency to delve into the asymmetries that characterize the organism-environment relationship.

6.6 Symmetry Broken in the Organism-Environment Relationship: Agency and Normativity

The preceding discussions underscore a critical observation: Irrespective of which standpoint in contemporary philosophy of biology proves to be the correct or more promising one—something that I do not intend to arbitrate in this book—addressing organismal agency necessitates integrating considerations of organismal environments. Neglecting the significance of the environment, as was common in some early-twentieth-century approaches (see section 6.2), is no longer a viable approach if we seek to gain insights into how organisms autonomously interact with and act in their environing worlds pursuing intrinsic goals, such as surviving, maintaining their organization, or reproducing.

In our historical survey, we could attest to how some early-twentieth-century positions either backgrounded or foregrounded the environment. However, in the contemporary debate, we have come to realize that this is not a fixed dichotomy. Instead, we observe a diverse range of perspectives that foreground the environment in different ways. For instance, the ecological approach exhibits a neo-Aristotelian inclination (as was discussed in section 6.5.2), but it foregrounds the environment as contributory to an organism's affordance landscape to spell out agency, which sharply contrasts with former neo-Aristotelian stances of organicist biologists that tended to pay lip service to the environment. Similarly, in the past, proponents of the heuristic approach tended to background the environment altogether. Still, when examining the recent (Kantian) heuristic approach proposed by Desmond and Huneman, we find that the organismal environment is once again foregrounded and given an important epistemic role.

This reveals that the relative "position" of the environment in accounts of organismal agency (i.e., in the background or in the foreground) is not unmovably determined by a particular stance; rather, it is fluid and changeable. The saliency of environments can shift, and their importance can be stressed or relegated to the background, contingent upon the specific theoretical perspectives adopted by scholars. A modest but important conclusion of the historical analysis in this chapter is that environments bear differential weight in past and present accounts of organismal agency.

At the same time, and flipping the epistemic coin, the very issue of organismal agency reveals something important about the organism-environment relationship. As has been shown throughout this book, we need to pay attention to the particularities and singularities of each relatum. Organism and environment form an interdependent binomial (see section 5.5.1) and are engaged in a tight bond over ontogenetic time, but they are not equivalent relata on ontological grounds: They are strikingly different as members of this relationship of central importance to biological science.

And precisely because of this, their contributions to this association are not entirely symmetrical.

Even though organism and environment can be said to be entwined in a relation of reciprocal causation, as we concluded in the previous chapter, the way they enact their role as causal modifiers and the type of causal consequences they bring to the other relatum are unalike because organismal agency comes into the picture. In this sense, reciprocal causation should not be understood as perfect causal symmetry. This is something that I will explore in section 6.6.1. Finally, I will touch on additional asymmetries that exist in the link between organisms and their environments: only the former exhibit intrinsic purposiveness and normativity, and this is something that sets organisms apart from their environing worlds (section 6.6.2). The organism-environment relationship, we must conclude, is an asymmetric one.

6.6.1 Reciprocal Causation Is Not Perfect Causal Symmetry: Organisms as Agents

What asymmetries within the organism-environment relationship has this book uncovered so far? In chapter 3, my metaphysical assessment led us to the idea of environments as classes-as-many, consistently indexed to individual organisms throughout their life cycles. However, it was also suggested that an organism's actions and developmental changes are crucial to determine what specific surroundings environ it. I contended that, due to indexicality, the reference of an organismal environment must be taken to be context-dependent. Environments ontically depend on organisms (section 5.5.1), and claims such as "the environment of organism X is Y or comprises Us and Ts" are never absolute, requiring an organismal ontogeny for uncovering the extension of biotic and abiotic factors.

In contrast to organisms that have shifting boundaries throughout their ontogenies (section 4.6), environments, as extrinsically defined entities (section 3.2.2), lack self-individuation. The assortment of heterogeneous factors included as environmental components at a particular moment in an organism's life cycle is established relationally (always with respect to the counter relatum), enabling some discussions about continuities and fluctuations due to indexicality. This stands in stark contrast to how organisms self-establish their boundaries as distinct entities. Along those lines,

Toepfer (2012) has argued that, based on the causal reciprocity of its parts (see also section 5.2.3), "an organism can be represented as a closed causal system that is functionally separated from the stream of causal events in its environment" (114). Indeed, organisms are self-individuating entities (causally closed but thermodynamically open), and organism-environment separation is something that is theoretically possible to delineate, even for multicellular developing organisms (although doing away with the idea of a single, all-encompassing boundary, as argued in section 4.6). This by no means implies that organisms are causally independent of their environments (sections 3.1.1 and 5.5.1): organisms causally depend on them at all times. Again, we can highlight that environments ontically depend on organisms, while organisms causally depend on environments (e.g., for harvesting matter and energy). The mutual dependence between these relata is not on the same footing.

Adding to this set of differences, here, I point to additional important asymmetries of the organism-environment pairing (for a different perspective, see Etxeberria forthcoming): In this relationship, only an organism is a bounded locus of causation (i.e., an agent) that performs goal-directed actions and exhibits intrinsic normativity. And this is an important sense in which reciprocal causation should not be conflated with *perfect causal symmetry*. Of course, there is general causal symmetry in the very abstract sense that both organism and environment, when interacting as relata, act as causes and experience effects, and partake in bona fide causal relations in contrast to noncausal relations (e.g., noncausal dependences; for analysis, see Woodward 2010). Nevertheless, beyond this broad, abstract characterization, philosophers would want more fine-grained distinctions of what exactly organisms and environments qua reciprocal causal relata are. It is important to distinguish between causal relations that stem from organisms and those than stem from environments, even though both participate in reciprocal causation loops that diachronically extend over ontogenetic time and might bear some evolutionary consequences, as we saw in chapter 5.

For instance, the way certain dispersed components of environments affect the development and functioning of organisms is different from the way organisms modify the surroundings in which they are embedded in pursuance of their own goals, such as self-maintenance or reproduction. Organisms modify their environments as discrete causal sources: Their actions are targeted to particular environmental components (for instance, in cases of agential niche construction) or can have additive effects when considered in conjunction with those of conspecifics (in cases of contributional niche construction, e.g., populations of organisms passively releasing a particular kind of detritus to their surroundings). In contrast, environments are causally dispersed and fragmented: What has causal effects on organisms are actually different abiotic and biotic components acting on particular developmental or physiological processes in different moments of their life cycles. Given the ontological heterogeneity of its components, its dynamicity, and its ontic dependence on the context of organismal ontogenies (e.g., indexicality), the organismal environment is not a unified cause and should not be presumed to be such. This is one of the cautionary corollaries that we reached at the end of chapter 3 (section 3.3.2). At any instant, the totality of an organismal environment is not pushing in a single direction or channeling a univocal developmental trajectory. Even if we adopt Walsh's view of the affordance landscape, this thesis of boundedness versus scattering would hold: Organisms shape and act on particular affordances that they help to constitute through their repertoires at a given time as bounded units, and those particular affordances alter or shape organisms in discrete but meaningful ways. The entirety of an organism's surroundings is never presented as a single-world affordance; there are always *multiple affordances* offering opportunities for action to a particular organism. With respect to the affordances a developing organism encounters at a particular moment of its life cycle, there is continuously a one-to-many space of possibilities for the organism to act on or a many-to-one space to react to.

Moreover, in any reciprocal causation loop, the change exercised by an organism on its environment will not be identical to the change exerted by the environment at a particular time or during the next causal iteration. Although we typically represent the causal entanglement between organisms and environments graphically with identical, bidirectional arrows, the content represented by each arrow differs significantly. Organisms are differentiable causal sources of environmental changes compared to how environments are causes of organismal changes.

Philosophers of explanation have highlighted certain concepts that are used to mark distinctions among causal relationships. In terms of explanatory goodness, we could highlight (1) proportionality and (2) invariance (or nonsensitivity).[18]

Proportionality. When addressing causal explanations, philosophers emphasize the importance of citing causes that are proportionate to their effects. This entails maintaining the same degree of precision in the explanans, explanandum, and contrast classes. If an explanation includes causes that are excessively detailed or overly permissive, it may lack proportionality with its effect (for discussion, see Yablo 1992; Blanchard 2020). A proportional explanation, on the other hand, conveys accurate information on the appropriate level of analysis, only citing causes that are sufficient and necessary for the effect. In this context, the organismal environment, considered a whole from an epistemic perspective (something many biologists customarily assume; see section 3.3), may represent a cause with low proportionality. The environment encompasses numerous biotic and abiotic components that might not be causally linked to a specific phenomenon of interest, yet they are still included in the overarching causal explanans scientists call "organismal environment." For instance, biologists may claim that "the environment induced the emergence of a plastic phenotype Y," when a more proportional causal statement would be "environmental component X, and not environmental components Z or T, induced the emergence of plastic phenotype Y." In contrast, organismal activities are proportional causes when referring to particular environmental modifications. The level of causal analysis is bounded to a token organism behaving or developing as a whole and bringing about particular changes to certain elements of its environing conditions. For instance, citing the behavior of an organism as a cause of an environmental modification is not an excessively detailed or overly permissive cause: it is a proportional one. Behavior happens at the level of an organism embedded in and interacting with its environment (see Trestman 2012).[19]

Invariance. An explanatory relationship with plenty of details exhibits less invariance—or greater sensitivity, if you prefer that phrasing—when confronted with changes in the values of explanans, explanandum variables, or background conditions. This means that it remains valid only under a limited set of these alterations. In contrast, a more invariant explanatory relationship enables inferences to a wider range of counterfactual scenarios where variables or background conditions adopt new values without disrupting the causal relationship. Such invariant regularities, some philosophers of causation claim, are epistemically preferable, as they provide answers to a broader array of "what-if-things-had-been-different" questions (Woodward 2003). With respect to invariance, we have seen that the ecological approach to organismal agency centers around defining an organism as having a goal G when it possesses a repertoire that vests it to effectively achieve and sustain G across a wide range of actual and counterfactual conditions C (for further discussion, see Walsh 2012). This capability is exemplified by an organism's adept response to the affordances presented by C and its active utilization of its developmental and behavioral repertoire to accomplish G. Here, goals represent stable end states that goal-directed organisms consistently and reliably tend to attain, in spite of perturbations, through their plastic, yet robust, adaptive behaviors. Related to this point, some philosophers have contended that all organisms can said to be agents in a certain sense because they all exhibit adaptive responses to environmental stimuli (Okasha 2024).

As argued by authors that vouch for the ecological approach, such as Walsh and Fulda, the relation between the goals of an agent and its means is noncausal, as it accords to a (modal) hypothetical necessity dependence (as expounded in section 6.5.2), but I maintain that the relation between an organismal activity—which can be agential or not—and its ensuing consequences in environing conditions exemplifies a causal relationship. If we focus on a specific instance of goal-directed behavior, wherein an organism shapes its affordance landscape by altering it in a particular direction or with certain intensity (e.g., producing changes in a physical environmental parameter), we can observe that this process remains invariant or is less sensitive across a broad range of actual and counterfactual scenarios. For instance, even when presented with variations in other background environmental parameters, diverse starting conditions for the behavior, or even contingent obstacles encountered, it would likely result in a consistent outcome, where the organism flexibly regulates its behavior to attain its goal, especially when modifying the environment is an integral part of achieving that goal. As Russell (1945) maintained, what is distinctive of the goal-directed activities of organisms is "the active persistence of directive activity towards its goal, the use of alternative means towards the same end, the achievement of results in the face of difficulties" (144). Compared to how organisms exert environmental modifications with a high degree of causal invariance, environments qua causes do not exhibit robustness and flexibility across counterfactual scenarios: They are less invariant or more sensitive to changes in background conditions or to the actions and reactions of organisms.

This asymmetry between organisms and environments as reciprocal causes with respect to proportionality and causal invariance is another consequence of what kind of relatum an organism is with respect to its environment: an agent. As we will see in the next pages, an organismal environment in toto lacks all the qualifications of agency as well.

6.6.2 Agency, Intrinsic Purposiveness, and Normativity of Organisms and Environments

In recent discussions, Okasha (2024) has put forth the argument that there are two primary motivations for employing the concept of agency when examining organisms. The first motivation is centered on shedding light on which unique characteristics organisms possess in comparison to other biological entities (e.g., other kinds of biological individuals). The second motivation centers on the heuristic or epistemic value of regarding organisms *as if* they were agents or agent-like beings, as this perspective aids in comprehending their evolved features and behaviors. I think that the first motivation can be expanded to argue that, by considering organisms as agents, we can better articulate what sets them apart from their respective environments. How are organisms different from their environments? This is a question that becomes feasible to answer through the lens of agency and normativity.

Moreno (2018) maintains that, first and foremost, an agent emerges as the originating force behind a collection of actions. This entails the presence of a relatively stable system that stands apart from its environing conditions, exhibiting a discernible "individuality." Furthermore, the agent's interactions with the environment must encompass not only a causal engagement but also a distinctive causal asymmetry, wherein the agent generates actions that bring about modifications in the environment. Not all organismal activities are goal-directed, but in those that are, each action has a functional character as it represents a directed process, aimed at attaining specific objectives. For this reason, an action is susceptible to failure if these objectives are not achieved. Hence, it has a *normative* dimension. To fulfill these conditions, an agent must possess the capacity to elicit targeted responses in relevant features within its environment (Moreno

2018). For Moreno, the actions of an agent encompass a constellation of processes enacted on its environment, which are instigated by the agent as an integrated whole in pursuance of specific goals. This is another way of framing the idea that was mentioned in the previous section: An organism, in contrast to a dispersed environment, is a bounded locus of causation.

In that respect, actions are performed by whole organisms, and not by isolated organismal parts. Burge (2009) has argued that "the relevant notion of action is grounded in *functioning, coordinated behavior* by the *whole organism*, issuing from the individual's *central behavioral capacities*, not purely from sub-systems" (260; emphases in original). He has tried to make this characterization clearer by means of examples:

> A spider pursues, jumps on, bites, and eats its prey, approaches and inseminates its mate, navigates past an obstacle, or runs across a web. These actions are distinguished from processes occurring only within the spider. The spider ingests; its stomach digests. Only sub-systems operate in the circulation of fluids, and production of protein, semen, or wastes. . . . An amoeba's ingesting its food is action. Digesting its food is not. A paramecium's swimming forward or backward is action. The plasmolysis that causes shrinking of the paramecium in highly concentrated solutions is not. The crawling of a tick toward a heat source is active and attributable to the whole organism. Protein transfer through its membranes is not. (Burge 2009, 256)

In the paramecium case, for instance, the organism "is not just being moved around by its environment. . . . A significant contribution to the movement comes from within" (Burge 2009, 260). In this sense, for Burge, behavior is imputable to individuals (i.e., whole organisms) and not to their subsystems or their activities. This is one of the reasons why it is warranted to say that an organism, qua performer of goal-directed actions, is a bounded locus of causation.

The panorama looks very different if we focus on organismal environments. An environment has manifold causal consequences, but not intrinsic purposiveness, nor agency. Environmental components cannot be normatively evaluated. For instance, the wind bends a plant if it carries enough force, but there are no grounds to claim that it had the goal to bend said token plant. Wind blowing just happens, and it might have important causal consequences on organisms, but that's it. A change in pH might alter the physiology of a marine organism, but ions cannot be assumed to be normatively required to exert a causal influence on an organism: ions just interact, due to their concentration and Brownian motion, with the receptors and ion channels they encounter on the surfaces of cells.

Here, a distinction that is commonplace in the philosophy of action might help us better understand the difference between organisms and organismal environments: that between "happenings" and "doings," in other words, what an agent *does* and what merely *happens to it* (e.g., which bodily movements it makes and which occur without it making them). For instance, to illustrate the distinction between happenings and doings, Frankfurt (1978) advances the following example: "Consider the difference between what goes on when a spider moves its legs in making its way along the ground, and what goes on when its legs move in similar patterns and with similar effect because they are manipulated by a boy who has managed to tie strings to them. In the first case the movements are not simply purposive, as the spider's digestive processes doubtless are. They are also attributable to the spider, who makes them. In the second case the same movements occur but they are not made by the spider, to whom they merely happen" (162). Organisms *do* things in the world—and also some things just happen to them, as not all organismal processes and activities are agential. In contrast, an organismal environment is never the subject of a "doing"; its manifold processes unfold in the domain of happenings.

Could the same thing be alleged about organisms (e.g., conspecifics, predators, prey) as participants in an organismal environment? Clearly, all those individual organisms can plausibly be thought of as agents, pursuing intrinsic goals and being founts of their own doings. But the normativity on which they act is their own; it stems from themselves,[20] and cannot be ascribed to the environment(s) in which they partake—after all, a given organism could be part of the relational, bijective environments of as many other organisms as it is interacting with at a particular time. The environment of organism X, composed among other elements of, say, conspecific C and predator P, has no intrinsic goals and has no means to attain a uniform outcome on organism X. The agency and intrinsic purposiveness of conspecific C and predator P do not transfer to the environment of organism X. It would be a mistake to judge an organismal environment, in all its ontological heterogeneity, as subservient to overarching goals. There is no supreme "unity of purpose" (see Okasha 2018, chapter 2) for an environment in the way that could plausibly be articulated (in some instances) regarding organismal parts jointly contributing to the goals of an organism as a whole (e.g., surviving). Organismal environments are relational classes-as-many of heterogenous causal elements and processes: They simply do not have the ontic status of agents with individuality, goals, and repertoires.

For these reasons, environments also do not possess intrinsic normativity, which refers to a way to appraise if an agent has succeeded or not in attaining its goals. Normativity in the biological context relates to natural norms, i.e., standards or levels of possible performance that are adequate to fulfill a purpose (Burge 2009, 269). For the ecological approach of Walsh and Fulda, affordances have normative implications, but these, I think, still fall on the side of the organism capitalizing on the open opportunities for action. Fulda explains himself on this point:

> For an organism to experience its ecological setting as an affordances landscape is for the elements of that setting to have value or significance for the attainment of the organism's goals. The fact that glucose affords food to the bacterium implies a *ceteris paribus* that getting it is good for pursuing its lifestyle. Its ecological value in turn imposes a requirement on the bacterium to implement the appropriate item from its repertoire, namely, swimming in the direction of the glucose, the nourishment. (Fulda 2017, 85)

So, although affordances are normatively laden, this is not a normativity that can be predicated of the environment-as-a-whole. Affordance landscapes have value or present success criteria only with respect to the organismal agent that is embedded in them. The normativity they exhibit, if anything, is nevertheless reliant on the normativity of an organism. Once again, the ontic primacy of organisms emerges in this discussion (see also section 5.5.1).

In this chapter, I recounted the importance of organismal agency and intrinsic purposiveness in early-twentieth-century biology. I showed that debates revolving around organismal purposiveness and agency spanned multiple disciplines and research traditions. I presented a taxonomy of seven distinct theoretical positions related to agency and intrinsic purposiveness: (1) the neo-Aristotelian position, (2) the Drieschian answer, (3) purposiveness eliminativism, (4) the heuristic approach, (5) the holistic alternative to purposiveness, (6) organismal purposiveness as a noetic principle, and (7) purposiveness as an explanandum for dynamic equilibrium-based research. These positions either backgrounded or foregrounded the organismal environment. I then recounted how organismal agency was marginalized from the agenda of biological science in the second half of the twentieth century and how it is back at the center of important debates in biology and the philosophy of biology. I advanced certain conceptual clarifications regarding the problem of organismal agency and teleological explanations and detailed three main positions that are competing in the present-day debate: the Kantian account, autonomy theory, and the ecological approach. In contrast to what was the case in early twentieth century, all of these positions foreground the environment as an important component of their ruminations on organismal goal-directedness and purposiveness.

I also argued that the issue of organismal agency is eo ipso fundamental to understanding the organism-environment relationship: It reveals important asymmetries between organisms and environments when considered as mutual relata. Organisms qua agents define and shape their affordance landscapes and act as bounded loci of causation across a diverse range of actual and counterfactual scenarios (i.e., when their activities are mobilized in causal explanations, a high degree of invariance is found), while organismal environments are much more sensitive and less proportional in causal terms and, in addition, lack the staple marks of agency (e.g., goal-directedness, intrinsic purposiveness, normativity). The collection of environmental components that comprise a single organismal environment can only be said to have causal consequences on the development, physiology, behavior, or ecological interactions of an organism, but not bona fide agency. Even though organisms and environments partake in reciprocal causation loops throughout ontogenetic time, these interactions do not entail perfect causal symmetry.

From the convergence of all of these insights, we seem to be confronting again one of the central philosophical conclusions of this entire book: The organism-environment relationship, against some expectations, is an asymmetrical relationship. Organismal agency is one of the foundational asymmetries behind this discernment. One of the central relationships in all of the edifice of biology is an asymmetrical one, and we need to come to terms with this in order to better understand it.

In the final chapter of this work, I harness all the insights garnered through the lens of its integrated history and philosophy of science (&HPS) approach. This synthesis is undertaken to grapple with the central inquiries that conform the cornerstones of this book: What kind of relationship is instantiated when organism and environment are considered relata? How has this relationship been construed throughout the proximal history of biology, and what roles does it play within biological science?

7 The Organism-Environment Relationship Redux

> An apparently simple idea—that organisms interact with environments—came to have complicated and lasting consequences.
>
> —American philosopher and historian Trevor Pearce (2014b, 14)

This book has peered into how the organism-environment relationship came into biology and the different ways that it has been theorized throughout the twentieth century, while raising epistemological and ontological questions about the nature of this scientific pairing. One may wonder about the rationale behind this specific emphasis on the relationship itself, as opposed to dedicating exclusive focus to either organisms or environments in isolation. Indeed, these are analytical possibilities, and some scholars have opted to highlight one of the constituent members of this trident over the others as its most important conceptual spike. As Patrick Geddes and J. Arthur Thomson declared in colorful prose,

> On the one hand there is the Environment in its action upon the organism; and on the other the Organism in its reaction to and action upon the environment; the dynamic relation, in its twofold aspect, is called Function. . . . To some the fundamental fact is the living Organism—a creative agent, a striving will, a changeful Proteus, selecting its environment, adjusting itself to it, self-differentiating and self-adaptive. . . . To others it has always seemed that the emphasis should be laid on Function. . . . To others, again, what counts for most is the Environment. This wakes the organism to action, feeds it or starves it, gives it new experiences or imprisons it within the old. Environment prompts the organism to self-expression, yet moulds it and prunes it, punctuates its life, and finally puts in the full stop, of death. (Geddes and Thomson 1911, 184, 188)

The focus on the relationship connects the two poles, organism and environment, which I do not think could be treated separately altogether. I hope that by now the rationale behind my choice is clear for the reader and has proven to be a productive one. In this concluding chapter, I synthesize and interlace the historical perspectives and philosophical propositions expounded throughout this work. I present my answers to the fundamental questions and the problem space outlined in chapter 1, including explorations into the nature of organisms and environments qua relata, the plausibility of their separation and their limits, reciprocity, and the directionality and conformance to symmetry of this central pairing. I offer viewpoints on the organism-environment relationship on both its ontological character and its epistemic and heuristic importance within biological science, embracing the *de re* and *de dicto* dimensions outlined in chapter 1.

The presentation of the main conclusions, intercalated with outlook perspectives, is structured as follows: first, a revisitation of the general historical and historiographic implications arising from my reappraisal of the organism-environment relationship (section 7.1); second, a review of the *de re* and the *de dicto* ramifications of the organism-environment pairing for biological science at large (section 7.2). If any readers have made it this far, now is a good time to thank them for their patience and perseverance—I hope it has paid off.

7.1 Reconsidering the History and Historiography of a Central Dyad in Biological Science

"Organism" and "environment" are two concepts with profound historicity. This book has shown that we must not take their existence for granted, assuming they have been perennial fixtures in biologists' theoretical and empirical toolkit. Rather, their formation is a response to fundamental inquiries into the scientific examination of life and some effective means to engage with it. Both concepts bear the marks of their histories: On the one hand, the organism concept signals a move from a general principle of corporeal order to become a generic designation of a living being with an organized body capable of establishing internal and external relations with itself and its surrounding conditions (section 2.2.1); on the other hand, "environment" comes from a long trajectory that involved the Newtonian view of medium reinterpreted through the *milieu* conception and then translated into English to signify an aggregation, under a single compound entity, of myriad biotic and abiotic factors that affect the physiology, development, and evolution of an environed entity. This was a move that displaced concepts and ideas that theorized the surroundings of an organism in plural forms (e.g., Lamarckian *milieux*, Cuvierian conditions of existence, Darwinian conditions of life, circumstances). As Trevor Pearce has argued, the introduction of the environment concept came not only with a new linguistic descriptor but also with particular metaphysical consequences for the theories and practices of biologists ever since (see, one more time, Pearce 2010, 2014b).

A lesson from this &HPS book is that looking into the historical record allows us to spell out the ontological and epistemic assumptions of concepts, which in turn, can inform philosophical analyses and assessments. The introduction of the environment concept had profound metaphysical implications for how scientists conceptualized, construed, and explained the relation between developing organisms and their ontologically heterogeneous, ever-changing surroundings. These are consequences that I explored at length in chapter 3 and that rippled throughout the problems concerning organism-environment separation, the most appropriate construal of reciprocity, and attributions of agency in chapters 4–6.

Chapter 2 offered a general exploration of how biologists began speaking of "organism" and "environment" qua singular entities and studied the interconnectedness of both as a prerequisite for acquiring scientific knowledge about living processes. In particular, I provided an overview of the historical trajectories of the organism and environment concepts leading up to the late nineteenth century, when their juxtaposition gained prominence (section 2.2). The idea of apposing organisms and environments was a move that contributed to opening the (theoretical and experimental) epistemic space of biology as a new, integrated science of life (section 2.2.1). In my historical narrative, the role and tremendous sway of Herbert Spencer as a popularizer of the term "environment" and the idea of "organism-environment interaction" was also underscored (section 2.2.1), vindicating what Pearce had shown in his own scholarly work. Moreover, I recounted Spencer's answer to the question of what kind of relationship is instantiated when organism and environment are considered relata: For him, this is a relationship of "correspondence." Organisms do not simply suffer alterations but change in the direction of adjustment to the vagaries that affect them; they dynamically respond to match their ever-changing environments. Behind this simple, pervasive idea, Spencer thought, lies the principle of the correspondence between organism and environment. In other words, the organism-environment relationship truly grapples with what life is all about. To survive, developing organisms have to internally track whatever is going on in their environing worlds that is taking a toll on them and continuously adjust to it by generating appropriate secondary alterations. This view has been incredibly influential and can find present-day echoes in discussions concerning organism-environment interactions in diverse domains of biology and philosophy (see, e.g., sections 6.3.2 and 6.5.4).

Subsequently, I conducted a brief exploration of how the organism-environment pairing had already evolved into what I called a "framing device" for inquiries across various biological disciplines during the early twentieth century, from discussions about biochemistry, natural selection, or homology to debates about the environment as a "molding" cause in development and evolution (section 2.3.1). Around those years, evolutionary discussions regarding the organism-environment relationship persisted, largely echoing the perspectives and debates bequeathed by the preceding century, but also, new approaches arose and biologists increasingly manipulated the environments of developing organisms in their experiments to gain new insights into morphogenesis and reproduction, while also testing for the transgenerational effects of environmental action (section 2.3.2). In chapter 5 (section 5.2.1), I also succinctly covered the origins of the Haeckelian conception of ecology, which put organism-environment interactions front and center of its scientific agenda, and this is another example of how important and portable this dyad was for biological science at large (see section 2.3), which in part accounts for its epistemic success and rapid dissemination.

As the broad-spectrum case study of this book to rummage into how early-twentieth-century biologists reasoned about the organism-environment relationship, I introduced the Organicist Movement of the interwar period, characterized by an international community of scholars united in their collective pursuit of advancing a robust epistemic foundation for biological science rooted in the organism concept. The main theoretical predicaments behind this movement were the debate about the onto-epistemological foundations of biology qua science and the enduring mechanism-vitalism dispute (section 2.4.1). For organicists, the organism, as a central ontological entity that transcends the mere summation of its constituent parts, occupies a central position in the formulation of biological explanations. Specifically, the multilevel organization of an organism, which progressively emerges during its developmental trajectory in interaction with environmental conditions, must serve as the cornerstone on which both explanatory frameworks and methodological approaches in biology are constructed (sections 2.4.1 and 2.4.2).

Ecologist Jacob Gruber once made the observation that in "post-Darwinian biology," whenever "organism-environment relationships were examined, *they represented discrete inquiries*, relatively unsystematized and without reference to any particular body of theory" (Gruber 1954, 416; emphasis added). For Gruber, this panorama seemed to contrast starkly with the groundbreaking insights that Darwin's *Origin* had previously brightened regarding the relation between organisms and their surrounding conditions. While Gruber (1954) is correct in underscoring the importance of the bond linking organisms and their (organic) conditions of life in Darwin's work (as I briefly covered in section 2.2.2), his assessment of the state of post-Darwinian biology does not hold to nuanced scrutiny. As I have shown in this book, from the second half of the nineteenth century onward, especially in the early twentieth century through the impulse granted by organicist and holist authors, the organism-environment relationship was examined and systematized in light of organism-centered theories and frameworks. It is worth noting that by the time Gruber advanced his perspective, the epistemic and explanatory import of this relationship had already begun to recede in biology, as outlined in chapter 5 (section 5.3). This context may have contributed to Gruber's historiographic oversight. Nevertheless, this book has demonstrated that, at least in the first decades of the twentieth century, organicist and holistic scholars, while offering varied viewpoints, articulated sophisticated positions along distinct axes of the organism-environment relationship. These axes encompassed considerations such as (1) the separation of these relata and what enacts it or collapses it in ontogeny (chapter 4); (2) different hallmarks of reciprocity, spanning from ontological coconstitution to mutual structural fitting and reciprocal causation (chapter 5); and (3) reflections surrounding organismal purposiveness and agency, alongside the place of the environment within these disquisitions about the activities of organisms (chapter 6).

Through chapter 4, I supplemented the historiography of theorizations on the organism-environment relationship in the form of a taxonomy of three distinct stances on the separation of these relata that have thus far eluded systematic classification: single boundary seekers, boundary skeptics, and boundary renegotiators (sections 4.2–4.4). The insights gained from this revisitation of early-twentieth-century positions (e.g., of single boundary seekers such as Charles Manning Child and Edward Stuart Russell, discussed in section 4.2, or the holistic or dialectical inseparability positions of authors such as John Scott Haldane, Marcel Prenant, and Jakob von Uexküll, covered in section 4.3) and their contemporary counterparts, plus the analysis of boundary pragmatism as a fourth new standpoint that emerged in recent times (section 4.5.2), were put to the service of philosophical reflection to advance my own position on the topic, that of "shifting boundaries defenders" (see section 4.6; see also below). Future historical studies would necessitate an expansion of the twentieth-century pool of authors examined to assess the robustness of these categories or determine if additional conceptual refinements are needed. Along these lines, I think intriguing paths for investigation exist for historians and philosophers alike. These paths involve examining whether these theoretical positions exerted any tangible influence or constraints on the empirical (e.g., experimental) practices of those who espoused them and whether there were discipline-specific nuances that could not be recovered within this large-scale, abstract historical reconstruction.

In the quest to comprehend this relationship, the concept of "reciprocity" emerged as a focal point of inquiry in the history of biology. Foremost among the objectives of chapter 5 was to systematize the answers that have arisen by an appeal to reciprocity. To this end, a historiographic/philosophical categorization was advanced, offering distinct formulations articulated by some of the members of the Organicist Movement (section 5.2). Within this categorization, three primary construals of reciprocity were identified and fleshed out. (1) *Ontological coconstitution*: This holistic perspective, famously developed in the works of physiologist John Scott Haldane, postulates that organisms and their environments are inextricably interwoven, forming a compound living system that defies every attempt at separation (sections 5.2.3 and 4.3.1). (2) *Mutual structural fitting*: Advocated by the biochemist Lawrence Henderson and his followers, this perspective contends that there exists a reciprocal dimension to the concept of fitness, as both organisms and their environments are structurally suited to one another (section 5.2.4). And (3) *reciprocal causation*: This conceptualization suggests that organisms and environments, as distinct systems, engage in alternating causal relations with each other (i.e., one of them is a cause at time t and an effect at time t_{+1}, while the other is framed as the opposite, effect at time t and cause at time t_{+1}), resulting in temporally extended loops of interactions with both developmental and evolutionary repercussions. In contrast to the rejection of causal thinking in the ontological coconstitution perspective, many biologists in the early twentieth century asserted that organism-environment reciprocity could be effectively articulated in terms of reciprocal causation. To illustrate this standpoint, I focused on the cases of Adolf Meyer-Abich and Conrad Hal Waddington (section 5.2.5), two examples of early-twentieth-century biologists that considered that organism-environment reciprocity could be effectively articulated in terms of reciprocal causation.

A strong suit of this &HPS analysis was to show that there are important theoretical challenges that scientists face when studying and conceptualizing organism-environment reciprocity (section 5.3.4) that are reoccurring in contemporary debates in niche construction theory (NCT) and allied fields (section 5.4.2); the past can be a useful guide for the present. These challenges involve the blurring of meaningful boundaries between organism and environment and the integration between perspectives that highlighted individualized experienced environments with physical views of the environment.

Similar to the need for a more comprehensive examination of the reach of vantage points concerning organism-environment separation, a more extensive historical investigation of the topic of organism-environment reciprocity should be conducted within the works of the authors of the Organicist Movement and beyond. Early-twentieth-century biology was replete with discussions of organism-environment couplings and interactions, leaving open the possibilities that alternative perspectives on how to conceptualize reciprocity might have been proposed. Subsequent research could also scrutinize the history of various philosophical viewpoints concerning the framing of organism-environment reciprocity and assess their potential influence on biology (or perhaps, their potential to enrich it). For instance, the pragmatist conception of "transaction" does not aim to predefine the ontological boundaries between organism and environment in any a priori fashion but, rather, seeks to encompass the active, dynamic negotiation that transpires as scientists study these relata through several practices and uncover their bi-directional exchanges (see Dewey and Bentley 1946a, 1946b, 1949).

In chapter 6, I covered the problem of organismal agency and intrinsic purposiveness and how organicist and holist biologists approached it within their larger assumptions on the organism-environment relationship. This exposition showed that discussions on organismal purposiveness and agency were not completely confined to the periphery of biological science during the first decades of the twentieth century, as traditional historiography has claimed; instead, they permeated various disciplines and research traditions (sections 6.1–6.3). I presented a systematic taxonomy comprising seven distinct theoretical positions defended in early-twentieth-century biology to apprehend or dispense with organismal agency and intrinsic purposiveness: (1) the neo-Aristotelian position, (2) the Drieschian answer, (3) purposiveness eliminativism, (4) the heuristic approach, (5) the holistic alternative to purposiveness, (6) organismal purposiveness as noetic principle, and (7) purposiveness as an explanandum for dynamic equilibrium-based research.

These positions, in their diverse manifestations, either foregrounded or backgrounded the role of the organismal environment. This meant that, in certain cases, organismal environments merely served as contextual settings for the exercise of agency and, in contrast, some accounts recognized their sizable impact on agential activities. This variability in the relative importance of the organismal environment illustrates yet another instance of the organism-environment relationship serving as a biological construct in constant flux, undergoing renegotiations and reevaluations. This fluidity was characteristic not only of early-twentieth-century biology in a wider sense but also held particular relevance within the context of the Organicist Movement. In section 6.5, I showed how this theme resonates with present-day debates on organismal agency, which tend to foreground the environment. In this sense, the prominence and significance attributed to environments can fluctuate, and their role can be emphasized or diminished, depending on, among other things, the particular theoretical frameworks and perspectives embraced by researchers. A modest yet nontrivial conclusion drawn from this historical analysis is that environments assume varying degrees of importance in both historical and contemporary accounts of organismal agency.

A salient historiographic insight that emerges from the discussion at hand is that the conventional tripartite characterization of meta-theoretical commitments when approaching the scientific study of living beings—mechanism, organicism, and vitalism—often utilized by historians to analyze the early decades of the twentieth century, proves to be insufficient in comprehensively showcasing the diverse palette of stances pertaining to organismal purposiveness and agency. Certainly, there remain several open questions awaiting further exploration. We need to ascertain the prevalence of each of the identified stances and meticulously reconstruct nuanced subvarieties while clarifying potential instances of overlap. Moreover, for a thorough and systematic exploration of the topics of organismal purposiveness and agency, it would be important to map the network of authors involved, delineating their connections, influences, and social arenas. Another open thread pertains to the continuities and breaks of these discussions with earlier debates from the nineteenth century and their resonance in the context of late-twentieth-century developments (briefly touched on in section 6.2).

In the course of my historiographic analysis, it became evident that the distinct views of organism-environment reciprocity, and the stress on organismal agency, once occupied a more prominent position within early-twentieth-century research (sections 5.3, 6.2, and 6.3). Yet, as we move further in the historical trajectory of the twentieth century, we observe a noticeable decline in their favor during the postwar years. As part of the broader decline of organicist and holistic perspectives in biology, influenced by a variety of factors including reduced institutional support, changing research priorities among their proponents, and the ascendancy of molecular approaches (see section 5.3), the conceptual bedrocks of "organismal environments" were considerably simplified, and the theoretical significance of exploring organism-environment reciprocity was marginalized (section 5.3.1). This diminishing weight attributed to organism-environment reciprocity was paired with successful investigations into the reciprocity of other biological relata, such as gene-environment and population-environment (section 5.3.2). Moreover, the conceptualization of organismal environments was overshadowed by "population environments" in ecology and evolutionary research (section 5.3.3). For instance, I showed that there was a debate among population geneticists and population ecologists in the 1950s and 1960s regarding the most suitable construal of "environment." While figures like Theodosius Dobzhansky favored a pluralistic perspective allowing both concepts to coexist, others staunchly advocated for one over the other, resulting in the dominance of the construal of the population environment during the latter half of the twentieth century (section 5.3.3). Returning to Gruber's assessment, the historical evidence I brought to the fore throughout this book suggests that a shift in conceptualizations of the organism-environment relationship occurred not between Darwinian and post-Darwinian biology but, rather, between interwar biology and postwar biology.

In line with this, in chapter 6 I delineated the progressive marginalization of organismal agency during the latter half of the twentieth century. This period entailed a discernible turn wherein organismal agency and intrinsic purposiveness receded from their places as both problems to tackle and explanatory resources to employ in scientific inquiry (e.g., when investigating organism-environment interactions). A good case in point of this shift is observed in the group of evolutionary biologists who sought to explain away these concepts by resorting to natural selection–based explanations that capitalized on external teleology (section 6.4).

In a more general vein, what remains to be done for enriching the historiography of how the organism-environment relationship was construed within the Organicist Movement, as mentioned in chapter 1 (section 1.21), is to develop more detailed case studies of *situated understandings* of how this relationship was mobilized in scientific practice. These historical investigations would revolve around, or indeed beckon, contextually grounded reconstructions of how scientists understood the interrelationship between organisms and environments in their day-to-day activities. Within the purview of this book, however, the primary emphasis lay in a distinct project: the exploration of diachronic alterations in *broad understandings* pertaining to the organism-environment relationship within the sphere of biological science at large. I believe that my historical analyses can serve as groundwork for ensuing discussions and inquiries into how concrete scientists, in concrete socio-material backgrounds, approached the organism-environment pairing in their concrete investigations. Adding the influence of social arenas and larger cultural-political factors to the development and negotiation of biological ideas about organisms and environments would be welcomed by historians of science that aim to go beyond conceptual history and fairly abstract reconstructions. In general, detailed narratives on the study and conceptualization of the organism-environment relationship during the interwar period that strike a balance between internalist and externalist elements of scientific praxis are still open to be undertaken by historians, and this is the direction that I think holds more promise.

At the onset of the twenty-first century, Laubichler (2000) lamented that "nobody has yet attempted to write the 'history of the organism' similar to the many 'histories of the gene' that are produced by a small cottage industry at an increasing rate" (301–302). While some scholars have made valuable contributions in this direction, a systematic history of the organism remains to be written. Nevertheless, I hope that when such a historian or group of historians take on this task, the insights presented in this work will be taken up to shed some light on the evolving understanding of the organism-environment relationship within twentieth-century biology.

7.2 The Organism-Environment Relationship: Philosophical Conclusions

In the contemporary landscape of biological sciences, there exists a growing emphasis on organism-environment interactions, spanning developmental, ecological, and evolutionary domains. This recognition underscores the escalating challenges associated with defining precise boundaries between organisms and their environments. Simultaneously, intense debates have reemerged regarding the agentic status of organisms within these complex relations. Given the current state of scientific research and debates, historical and philosophical investigations into the nature of the organism-environment relationship have assumed a heightened relevance. As aptly diagnosed by Pearce (see also section 1.1.1),

> Today, organism-environment talk is more common than ever before. Variation and plasticity are once again major topics in the biological sciences . . . , and philosophers are increasingly attending to the fact that organisms modify their biological and social environments. . . . Thus looking back at the history of the notion of organism-environment interaction, we also look forward—to a century in which we continue to build with old tools made new. (Pearce 2014b, 28)

Beyond standard historical practice, embracing an &HPS approach offers a valuable route through which historical insights can be coupled with philosophical reflections to enrich and inform ongoing contemporary debates.

Let me restate the problem space I outlined in chapter 1 that has been broached in this book: Can we separate an organism from its environment? If organism and environment are said to stand in a relation of reciprocity, what is entailed by this claim? What is the most suitable construal of organism-environment reciprocity for biological science? Is the relationship between organism and environment wholly symmetrical, or are there some particular characteristics of these relata that break the symmetry of the pairing? And, in the background of all these important questions, what kind of relationship is instantiated when organism and environment are considered relata?

From the outset of this book, a central contention has been that in order to unravel the ontological and epistemological foundations of the organism-environment relationship, a necessary step is to gain a good understanding of the nature of an organismal environment as relatum. As articulated in chapter 2 (section 2.2.2), achieving this understanding demands exploring the specific metaphysical implications that the environment concept ushered into the theories and methodologies of biologists. In this vein, chapter 3 provided a philosophical reelaboration and expansion of the historical thesis advanced by Trevor Pearce in his pioneering work. This serves as an exemplar of the symbiotic relationship between philosophical and historical inquiry under an &HPS approach, wherein philosophical scholarship can also complement historical reconstruction (for the opposite direction, see the previous section).

In section 3.1, I posited that the biological conceptualization of organismal environments diverges from other cognate notions such as the "external world," "immediate surroundings," or "physical media." As part of embracing this conceptual singularity, biologists have delineated three essential desiderata for characterizations of organismal environments: (1) *Causal effectiveness* alludes to the partition or segregation of a subset of worldly components and processes that exert influence on the developmental trajectory of a specific organism; (2) *relationality* signifies that an environment needs a particular organism (i.e., its counter environed relatum) to gain reference; and (3) *dynamicity* underscores the fluctuating composition of diverse biotic and abiotic elements within a given organismal environment as a function of time.

Subsequently, I argued that a metaphysical lens is necessary in order to gain a better understanding of organismal environments (section 3.1). In section 3.2, I harnessed tools from analytic metaphysics to explicate the ontic status of an organismal environment. My contention was that the biotic and abiotic constituents circumscribed within a specific environment lack a shared intrinsic ontological measure or property. Instead, they hold individual extrinsic causal connections with a token developing organism (section 3.2.1). This gave rise to the question of whether it is metaphysically tenable to regard the surrounding context of a developing organism as a unitary entity. I maintained that we should be cautious and suggested that we could think of an organismal environment as an ontological construct best apprehended as a "class-as-many," as opposed to a traditional sortal or concretum. A class-as-many considers all the separate members of a class at once and nonetheless is not itself an (actual, concrete) entity; the collection denotes a concept (i.e., an organismal environment) but not an entity whose extension is something different than the structured sum of the individual components. I buttressed this viewpoint by proffering arguments that emphasized why an organismal environment diverges from sortal concepts. These arguments drew on criteria of identification, identity, persistence, and countability and on how hard it is to make sense of them when applied to organismal environments (section 3.2.2). I argued that there are sufficient reasons to contend that an organismal environment might be metaphysically viewed as a class-as-many that partitions the elements of the world in which an organism finds itself embedded and developing, with the criterion of membership for the changing set of biotic and abiotic environmental components (dynamicity) of having an actual or potential causal relation (extrinsic connection, effectiveness) with the token organism (relationality) in a moment of its life cycle (indexicality). In this sense, this view is able to recover the three desiderata that scientists demand from the concept of organismal environment while also being sensible to the challenge of indexicality (section 3.2.3). This last complexity arises because the composition of an organismal environment is contingent upon ontogenetic contexts and the activities undertaken by a token organism throughout its life cycle. Such considerations complicate straightforward attributions regarding the extensional reach of an organismal environment (section 3.2.2).

In terms of work to be done, the extension of the concept of "organism" should also be clarified. This task requires meticulous metaphysical analysis to complement the assessments made regarding organismal environments (for discussion, see Nuño de la Rosa 2010; Austin 2016; Wolfe 2014a). Within the scope of this book, the specific question of which biological individuals should be classified as organisms, as opposed to different types of entities (for instance, whether a siphonophore should be considered an organism or a colony and whether its constituent zooids should be deemed organisms or organismal components), was not explicitly addressed. Scholars such as Guido I. Prieto (2024a, 2024b) are actively engaged in this ontological project of demarcation, which falls beyond the immediate scope of this book. However, I consider that the insights I have advanced concerning organismal environments can be readily applied to entities such as siphonophores or zooids, considering them as organisms. These entities would still represent the counter relata within the context-sensitive category of effective abiotic and biotic factors that exert influence over their development, physiology, ecological interactions, and evolutionary trajectories. Therefore, the absence of a detailed delineation of the extension and scope of the organism concept does not impede the overarching historico-philosophical project undertaken within these pages. Similarly, the abstract, projectable qualities of the organism-environment relationship (*vide infra*) remain applicable even if the units under consideration extend beyond the conventional textbook exemplars of multicellular organisms (e.g., animals, plants, and fungi) or unicellular bacteria and archaea.

After clarifying the ontic status of organismal environments (for epistemic discussions, see below), the next piece of the puzzle to understand the organism-environment relationship was to ask if it is possible to separate an organism from its environment. After all, as Sandeman said, "It is obvious that we cannot speak of a relation where we have not two sides which already act upon one another" (Sandeman 1886, 151). Bona fide organism-environment separation seems to be a condition of possibility for there being an organism-environment relationship in the first place. Chapter 4 tried to address this issue by asking if there is something, call it a *boundary*, that instantiates the separation between these relata during ontogenetic time.

With regard to this problem, various positions have emerged in the history of biology, which I charted in chapter 4, each offering distinct perspectives and important considerations. Boundary skeptics, drawing from developmental and physiological evidence, reject the notion of a clear-cut distinction between organisms and their environments. They emphasize the recurrent interactions characterized by inseparable codetermination, ultimately advocating for a perspective that regards organisms and environments as commingled systems (sections 4.3 and 4.5.1). On the other hand, boundary pragmatists assert the pragmatic and contextual nature of the practices of "individuation," aligning with the specific demands of explanatory projects while remaining agnostic regarding the strict separation or unity of organisms and environments (section 4.2.5). Single boundary seekers take a different route by proposing specific and comprehensive criteria that serve to demarcate organisms from their surroundings. Their approach hinges on the identification of singular markers for defining the boundary between an organism and its environment (section 4.5.3). Meanwhile, boundary renegotiators endeavor to reconfigure the traditional segregation of what is considered "internal" and "external" to an organism (section 4.4).

Each of these positions, while offering valuable insights, falls short of providing a full resolution to the problem of organism-environment separation. As I see it, I think the issue of separation between a developing organism and its environment should be framed in terms of *self-individuation* (in the sense that Child and his organicist colleagues understood it; see section 4.2.1): How do organisms actively become different, to a certain extent, from their environing conditions? As a response to this, I argued for the necessity of advancing a novel perspective that not only incorporates the valuable elements of these existing views but also expands them in significant ways. In this context, I introduced the stance of *shifting boundaries defenders* (section 4.6), which draws on a diverse range of phylogenetic and ontogenetic case studies to offer a fresh approach to addressing this complex issue. According to my vantage point, the conventional notion of a single organism-environment boundary should be challenged. I posited the existence of multiple synchronically instantiated boundaries that correspond to ontologically diverse (biotic and abiotic) environmental factors. Considering the ontological diversity inherent in the (extrinsically defined) constituents of an organism's environment (an abstract class-as-many), it becomes apparent that construing a singular boundary proves inadequate. Such a singular boundary imposes an excessively weighty ontological burden on any entity or process seeking to serve as the sole determinant for demarcating an organism from its surroundings. Instead, empirical evidence suggests that different environmental components are distinctly parsed out by separate boundaries, each contributing to the delineation of an organism's bounds with its environment. These boundaries are not static but undergo dynamic shifts and negotiations throughout an organism's life cycle. They are subject to adjustments, breakdowns, and alterations in permeability and new emerge during ontogeny (e.g., through organogenesis or developmental processes such as metamorphosis).

Future research undertakings must build on the proposal of shifting boundaries by applying it to detailed, concrete empirical case studies. These studies should aim to pinpoint the processes and structures that facilitate the separation between developing organisms and their environments. By delving into specific empirical examples, we can gain a better understanding of how this separation is enacted and maintained throughout the ontogeny of organisms. Such empirical investigations will be instrumental in providing empirical grounding and substantiation for the shifting boundaries framework, further advancing our comprehension of the organism-environment relationship.

In chapter 5, I substantiated how different construals of reciprocity impact how the nature of organism-environment separation is theorized, and I argued that there are good epistemic and ontological reasons for drawing boundaries between these relata, their complex interwovenness notwithstanding. I maintained that a perspective of organism-environment reciprocity that is parsed in terms of reciprocal causation is apt for contemporary debates—for example, for investigating how specific environmental factors substantially alter developmental trajectories and how, in turn, organisms shape their surroundings with important developmental and evolutionary consequences.

If we follow Toepfer's (2011) general framing, reciprocal causation posits *mutual interaction* between organism and environment, and in contrast, ontological coconstitution is a view of reciprocity based on the *interdependence* of the relata in question. Which one is more adequate to understand the bidirectional bond forged between organism and environment? In sections 5.5.1 and 5.5.2, I argued that reciprocal causation, grounded in both metaphysical and epistemic considerations, and understood in a diachronic, temporally extended view, provides the most fitting perspective on organism-environment reciprocity. An organism's interactions with its environment entail a dynamic process wherein, at time *t*, the organism modifies its surroundings. Subsequently, at time t_{+1}, specific elements of the reconfigured environment reciprocally influence the organism, leading to changes in its states. Importantly, the modifications an organism imparts on its environment at time *t* are causally connected to how the environmental conditions that enveloped it had transformed at time t_{-1} (see also Baedke et al. 2021).

Conversely, when considering ontological coconstitution as a framework for understanding the reciprocity between organisms and their environments, there is a deeper level of intertwining. In this view, organisms and environments become so intricately linked that they form a unified and mutually reliant entity, often referred to as an "organism-environment system." Thus, attempting to meaningfully separate the organism from its environment becomes a challenging if not futile endeavor (section 4.3). Nonetheless, I argued that there is a significant oversight that many proponents of ontological coconstitution have overlooked: the metaphysical status of the organismal environment as a relatum. This oversight carries substantial implications for three critical issues that cast doubt on the adequacy of ontological coconstitution.

Category mistake. The status of the organismal environment is conflated or believed to be on par with the status of the organism itself. Such conflation can lead to conceptual confusion and hinder a precise understanding of the nature of the organism-environment relationship. An organismal environment is not a singular, unified entity (section 3.2), so the position of ontological coconstitution attempts to amalgamate incommensurate relata. To assert that an organism and its environment are interdependent parts or symmetrical components of an inseparable organism-environment unit seems to constitute a fundamental category mistake. The very idea of fusing an organism, which embodies a distinct, self-individuating concretum, with a relational class or aggregative abstractum characterized by radically different criteria of identification, identity, persistence, and countability raises many alarm bells. This incongruity puts stress on the paradoxical nature of ontological coconstitution as a metaphysical standpoint on organism-environment reciprocity, as it seems to neglect what an environment represents in the first place.

Mismatch in corresponding dependences. Another problem with ontological coconstitution is the mismatch in the corresponding dependences between organisms and their environments. This discrepancy can create inconsistencies in how we conceptualize the reciprocal influence of these entities. Take again the principle of reciprocal codetermination of Lewontin and Levins—this time in a longer extract from Lewontin's *The Triple Helix*:

> Just as there can be no organism without an environment, so there can be no environment without an organism. There is a confusion between the correct assertion that there is a physical world outside of an organism that would continue to exist in the absence of the species, and the incorrect claim that environments exist without species. The earth will precess on its axis and produce periodic glacial and interglacial ages, volcanoes will erupt, evaporation from oceans will result in rain and snow, independent of any living beings. But glacial streams, volcanic ash deposits, and pools of water are not environments. They are physical conditions from which environments may be built. An *environment* is something that surrounds or encircles, but for there to be a surrounding there must be something at the center to be surrounded. *The environment of an organism is the penumbra of external conditions that are relevant to it because it has effective interactions with those aspects of the outer world.* (Lewontin 2000, 48–49; first emphasis in original, second emphasis added)

If we read him carefully, Lewontin is actually arguing for a *causal reading* of the organismal environment, particularly highlighting the "penumbra of external conditions that are relevant to [an organism] because it has effective interactions with those aspects of the outer world" (Lewontin 2000, 48–49). Organismal environments, Lewontin would agree, are made up of the subset of causally effective abiotic and biotic factors that establish a relation with an organism, which correspond to the "effective environment," as I discussed in chapter 3 (section 3.1) and throughout this book. As emphasized by Lewontin, the biology of an organism undeniably plays a fundamental role in delineating its effective environment. I concur with this perspective, but it is crucial to clarify that this determination is not of a causal nature (as in "organisms cause their environments"); instead, it operates as a *structuring ontic condition.* The features and activities of organisms ontically establish what qualifies as their environments. In other words, organisms are *ontically indispensable* for the very institution of their environments (section 5.5.1), and in turn, they are *causally dependent* on environmental resources for their survival. However, this should not be confused to mean that organisms are causally indispensable for the extrinsic relationships that emerge between various abiotic and biotic components within that environment. These dealings are contingent upon and give rise to manifold causal interactions, distinct from the organism itself. Lewontin's view is perfectly compatible with the picture I presented in this book—and I think he has been misunderstood on several occasions by commentators—but we need to recognize that the principle of "reciprocal codetermination" that he and Richard Levins advanced is only symmetrical in appearance. Causal dependence (of organisms on their environments) and ontic indispensability (of organisms for environments) do not entail ontic inseparability. And because organisms hold ontological priority to establish the relata of this pairing, they can come apart from their environments: Organisms, through their traits and capacities, define which components will count as their environments; organismal environments require the context of particular token organisms to gain reference, and it is plausible to claim that, with a life cycle view in mind, not only one environment corresponds to an organism—environments are shifting, too.

Insensitivity to indexicality. Finally, ontological coconstitution often fails to account for the indexicality of organismal environments. In neglecting this aspect, ontological coconstitution may erroneously equate ontic indispensability with ontic inseparability, thereby overlooking the context-dependent nature of the organism-environment relationship. Environments, considered as classes-as-many, exhibit an ontic dependence on individual organisms at specific junctures of their life cycles to count as such. However, organisms are not wholly inseparable from these environments. Instead, organisms maintain multiple, dynamic boundaries with particular biotic and environmental components that serve as mediators, filters, or barriers (section 4.6). In stark contrast to organisms, environments never perform self-individuation. The assortment of heterogeneous factors encompassed as environmental components at a particular moment in an organism's life cycle is inherently relational, as discussed earlier. Nevertheless, we can discuss continuities and discontinuities within environments by virtue of indexicality. The constituents of the class-as-many are in a state of constant change, with some remaining relatively constant at different points in time. This dynamic nature represents a stark divergence from how organisms establish their bounds as relata.

In contrast to the three shortcomings faced by the view of ontological coconstitution, the perspective of reciprocal causation emerges as a more promising candidate for capturing organism-environment reciprocity on ontological grounds. Most notably, it does not necessitate both relata to be commensurate in the same stringent sense. In the framework of reciprocal causation, causes and effects are not obliged to belong to the same category of entities, whether they be abstract or concrete. With some conceptual refinement and nuanced articulation, reciprocal causation harmonizes seamlessly with the notion of an organismal environment as an indexical class-as-many. When we assert that an organism modifies its environment—for example, through niche-constructing activities—we mean that the organism alters at least one of the components within the relational class we refer to as "its environment" at a specific moment in its life cycle. A niche construction event would never imply changes to the entirety of factors encompassed within any given environment at a given time. Instead, it modifies specific environmental components, and these altered components may then trigger changes within the organism. Similarly, alterations to an organism brought about by environmental changes at a particular time would not involve the entirety of the environment but, rather, result from the involvement of specific causally effective environmental factors. The transformed developmental system, in turn, may be in a position to further modify the same or perhaps different components within the environment, creating a chain of reciprocal causation. Decisively, the factors driving causation on the environmental side do not have to be consistent throughout the entire time series as per the indexical nature of environments. The composition of members within the class "environment" may shift over time, depending on the context of an organism's ontogeny, without interrupting the diachronic history of interconnected causal loops between the organism and its ever-changing surroundings, thus allowing us to trace these causal relationships into the future.

Furthermore, through the concept of reciprocal causation, we can continue to acknowledge the existential requirements of organisms to be constitutively embedded in their surroundings to keep on living and the relational definition of environments (the ontic indispensability of organisms as counter-relata), in line with the reciprocal codetermination principle advocated by Lewontin and Levins. In light of these considerations, I think reciprocal causation provides a more robust framework for understanding organism-environment reciprocity than does ontological coconstitution.

My analysis of reciprocity construals revealed that the organism-environment relationship is far from being symmetrical, and the asymmetries encapsulated in the mutual dependencies of these relata are not the only ones. In chapter 6 (section 6.5), I argued that organismal agency stands as one of the foundational asymmetries in this pairing. Organisms are agents with intrinsic purposiveness that can perform goal-directed actions and shape their affordance landscapes, with potential causal consequences downstream to diverse environmental components. As agents, organisms embody bounded loci of causal influence across a range of actual and counterfactual scenarios, and as a consequence, explanations involving organismal activities have a high degree of invariance and proportionality (see section 6.5.5). In contrast, environments are causally dispersed and only have causal consequences on the development, behavior, ecology, or physiology of token organisms, but not bona fide agency. For this, they score poorly with respect to the above-described explanatory standards qua causal relata. Moreover, organismal environments do not exhibit intrinsic purposiveness or autonomous normative standards: They invariably exist in ontic dependency with organisms (section 6.6.2).

While organisms and environments engage in reciprocal causation over developmental time, with potential ecological and evolutionary consequences stemming from these interactions, the paring lacks a state of perfect causal symmetry. Through the convergence of these insights on one of the perennial problems in biology and philosophical reflections about life (i.e., organismal teleology and purposiveness; see sections 6.2 and 6.3), we find ourselves once again confronting a central philosophical conclusion that has permeated this entire book: The organism-environment relationship is characteristically asymmetrical.

As should be clear by now, when we speak of organisms and environments, we are invoking two profoundly different relata, and this constrains the kind of relationship they can sustain.

Now revisit, if you will, the image that opened this book: the snow fly, gracefully navigating the landscapes of the northern hemisphere during the throes of winter (section 1.1). What kind of relationship is instantiated between a snow fly and its environment, if we grant that these are relata that can be fruitfully juxtaposed? How would biologists go on to conceptualize this dyad?

My answer is the following: The organism-environment relationship instantiates a *causal relationship* between two interdependent relata that can be fruitfully spelled out in terms of "reciprocal causation." Organisms and environments are indeed involved in diachronically extended cycles of causation. This is also a bidirectional relationship: They mutually depend on one another, but in very different senses; organisms are causally dependent on their environing conditions, and organisms, in turn, are ontically indispensable for carving out those components of the world that will be allotted to an environment. Although this is a causal relationship, it is a profoundly asymmetrical one (e.g., causal reciprocity should not be conflated with causal symmetry): It juxtaposes two radically different relata—a self-individuating, agential concretum (i.e., an organism) that acts as a bounded locus of causation (even though it has shifting, dynamic boundaries throughout its life cycle) and an (ontologically heterogeneous, dynamic) indexical aggregative abstractum (i.e., an organismal environment) with dispersed causal consequences and no shared normativity. The snow fly is an agent embedded in its plural conditions of existence—which we abstract and call an "organismal environment" at a particular moment of its life cycle—shaping and being shaped by the affordances it encounters, not least when it sheds some of its legs to endure the unforgiving cold.

In this book, I have concentrated solely on the organism-environment relationship within the ontology and epistemology of biological science. It would be important to extend these inquiries to other disciplines and research areas in which the organism-environment relationship is equally central (e.g., the scientific study of cognition). It remains to be seen if my reconstructions hold outside the epistemic sphere of biology.

With regard to the epistemology of life sciences, the environment concept, and the organism-environment relationship, are some of the most powerful thinking tools of the biologist's toolkit, bestowing intelligibility and explanatory expediency upon their endeavors. As I argued in chapter 3, the conceptualization of the environment as a singular entity (e.g., as a sortal or class-as-one) is ontologically problematic. Nevertheless, it does offer certain epistemic advantages, as it enables biologists to comprehend the manifold array of surrounding conditions within which organismal ontogenies unfold. This simplification involves the use of two singular terms placed in juxtaposition: "organism" and "environment." This an abstract yet highly portable dichotomy. As Pearce (2010) rightly noted, starting from the latter half of the nineteenth century, "the concept of organism–environment interaction . . . became popular because it was an *abstract, portable dichotomy* produced by the conjunction of two singular terms" (242; emphasis added).

In chapter 2 (section 2.3), we witnessed how the organism-environment relationship was used from early on as a framing device for multiple inquiries across different biological disciplines. This epistemic malleability and exportability from disciplines and contexts (e.g., modeling, experimental manipulations, explanations, and theory building) is still part of what makes the talk of organism-environment interactions and organism-environment couplings so successful in biology. Although there have been many conceptual and technological changes in the life sciences, biologists still resort to the organism-environment pairing as a cornerstone of their thinking, but usually without recognizing it as such or taking it for granted. This book has taken a step forward to try to clarify this epistemic and heuristic building block of biological theory and praxis.

Furthermore, the concept of environment as a singular relatum aligns with the practical requirements of empirical studies conducted within biological science. When an organism's environment is examined in an experimental context, for example, there is no pressing need to deconstruct all its variables in chorus. Instead, it is assumed that those elements or processes left unaltered or uncontrolled are exerting a consistent, albeit unspecified, influence on the organisms under investigation. Similarly, in explanatory frameworks elucidating developmental or ecological phenomena, biologists can employ the term "environment" as a surrogate for a comprehensive collection of heterogeneous causal factors. When warranted, this allows them to make significant assertions without having to explicitly indicate the precise alterations occurring within the complex web of biotic and abiotic (environmental) processes.

Likewise, the overarching concept of organismal environment serves as a navigational beacon for biologists, directing their attention toward more granular components and factors worthy of further investigation. In essence, its use in singular linguistic form acts as a heuristic guide, facilitating future research by implying areas of interest and inquiry. What's more, the singularity of the term "environment" simplifies communication across diverse epistemic communities within biology—for instance, avoiding encumbrances of naming all the factors that are effectively affecting an organism at time *t* or throughout a complex time series (e.g., a portion of a life cycle).

Central concepts that we have explored in this book, such as phenotypic plasticity and niche construction, chiefly rely on the possibility of apprehending the entirety of abiotic and biotic circumstances that a developing organism faces as a single entity. In this sense, the environment concept plays fundamental epistemic roles to foster theorizing and model building. NCT or plasticity-led evolution hypotheses would be impossible to suggest if organismal environments could not be epistemically treated as singular entities. Its fine-grained ontological improperness notwithstanding, getting rid of this construal would be a loss for biological (especially evolutionary) thinking. Both NCT and phenotypic plasticity-led evolution open spaces for research and explanations that could move evolutionary biology forward.

As I said, biologists tacitly presume and accept the reach of the organism-environment relationship in their investigations, and this has profound consequences for their practices and theory-building endeavors. Many of the issues and problems addressed in this book have practical upshots. For instance, the problem of organism-environment separation is a salient scientific and philosophical issue on ontological, epistemological, and methodological grounds: For instance, it affects the construction of models and the apportioning of causal contributions of particular entities in scientific explanations, as well as the kind of scientific representations, from conceptual to mathematical models, that can be built for each of these relata; and it impacts the day-to-day practices of biologists in how they segregate and measure variables, and how certain experiments and interventions are set up.

As emphasized in chapter 4 (specifically in section 4.3.1), perspectives that subscribe to the inseparability of organisms and their environments encounter a limitation in addressing the individuation of biological entities and hinder scientific practice. Simply asserting that an organism and its environment constitute an inseparable unit does not provide a definitive solution for the individuation of such a system—that is, the process of distinguishing it from other systems. Individuation becomes imperative to discern between the proximal and distal environments of a specific system and to establish a meaningful framework for recognizing the boundaries and distinct characteristics of that system. This highlights another advantage of adopting an &HPS approach, as it allows the mobilization of past arguments to influence ongoing debates. In this context, in this book I draw upon Joseph Needham's objection to John Scott Haldane's position of ontological coconstitution and organism-environment inseparability, rekindling the relevance of this objection in discussions involving contemporary defenders of these positions. By bridging the past and present, an &HPS lens enriches our ability to critically engage with and contribute to scientific and philosophical dialogues. Biological practices and scientific explanations that correctly apportion causal contributions require organism-environment separation, so this is not only an ontological issue that concerns arm-chair philosophy; as I mentioned in chapter 4, there are also some scientific questions that rely on bona fide organism-environment separation (see section 4.5.2).

The notion of an organismal environment is an indispensable epistemic tool for grappling with the multifaceted conditions that influence the process of development and ecological interactions and constrain evolutionary pathways. In attempting to visualize the expansive complexity encapsulated within the notion of an environment, one is confronted with a rich tapestry of interrelated factors and processes that collectively influence the life of organisms. Think back, for instance, to figure 3.4. Rather than attempting to encompass the entirety of this multilayared web in one swoop, biologists judiciously abstract these elements and aggregate them into a singular relatum—an environment (see section 3.3.2). This abstraction serves epistemic purposes. Working with two singular relata in juxtaposition, the organism and its environment, offers a pragmatic and manageable framework for understanding the myriad factors that surround living systems. It simplifies the formidable task of addressing the diverse and intricate components that collectively constitute an organism's (shifting) surroundings.

In a broader context, I argued that the environment concept is one of the most important *devices of understanding* that biologists have in their theoretical toolkits. Mason and Langenheim, two important inspirations in the writing of this book, articulated their perspective along these lines:

They claimed that biologists use the idea of "environmental factor" as an abstract construct to linguistically consolidate the complex one-to-many relationship that exists between an organism and its surroundings. Scientists do this by simplifying it into a more intelligible one-to-one relationship: the connection between the organism and its environment (Mason and Langenheim 1957, 334).

By construing it as a composite, higher-order object (section 3.3.2), an organismal environment is a receptacle of causal variables that can be included in different models and approaches without spelling out everything that is comprised within it. It works as a sieve for experimental interventions, and as I have argued, it is a common ingredient of causal explanations. Indeed, it is challenging to envision biology without the aid of such a potent explanatory and operational notion. The concept of organismal environment, as a theoretical construct, is firmly entrenched in biological science and shows no signs of relinquishing its status as a cornerstone of understanding in this field. Environments, as I have said before, are here to stay.

However, in spite of its many useful epistemic advantages, there are many reasons why we should avoid the *pernicious reification* of an organismal environment. For one, the ontological heterogeneity of its components precludes simplistic causal reasoning; in fact, it would be misleading to treat an organismal environment as a single, unified cause. Second, an organismal environment is not free of indexicality; it depends on particular ontogenetic contexts to become intelligible. Therefore, treating "an environment" as a concrete thing is an ontic mistake given its dynamic, relational character. The metaphysical partitioning that the environment performs with respect to the causal elements that might potentially or actually affect an organism leaves out from its provinces many elements of the world that exist in close (spatiotemporal and causal) vicinity with the selected elements, and this is yet again another sign of *abstraction*.

A central problem with the treatments of some abstractions, heuristically and explanatory powerful as surely these are for particular scientific fields, hinges on the danger of ontologically overinterpreting them and thus turning abstracta—otherwise useful in how scientists grasp, communicate, and generate knowledge—into concrete entities, which are taken to represent reality faithfully. The fallacy of misplaced concreteness, which Whitehead warned us about, happens when we have mistaken our abstractions for concrete realities and have become so enamored with them—so enthralled by them—that we can no longer tell them apart. Critical reflexivity is still needed, especially for concepts such as "organismal environment" that both scientists and philosophers usually take for granted. Environments are not captured by transparent concepts and deserve careful metaphysical and epistemic analyses.

All in all, the organism-environment relationship furnishes a scientific understanding of one of the central features of life and its evolution: Developing organisms are embedded in their environing worlds, actively participating in their shaping while being shaped by them. It is perhaps a simple message, but one that we ought not to forget ever again.

Notes

Chapter 1

1. For discussion of the organism-environment relationship in evolutionary adaptation, see, among others, Brandon (1990); and Lloyd (2021). A widespread position regarding this relationship vis-à-vis adaptation is encapsulated in a famous dictum by evolutionary biologist George Williams: "Adaptation is always asymmetrical; organisms adapt to their environment, never vice versa" (Williams, 1992, 484). Aaby and Ramsey (2022) refer to this standpoint on the organism-environment relationship in evolutionary biology as "asymmetrical externalism." For background on the form-function dichotomy and how it has unfolded in biological thought, see Russell (1916); Amundson (2005); Laubichler and Maienschein (2009); and Ochoa and Barahona Echeverría (2014).

2. In that passage, Odling-Smee was referring to Herbert Simon's work on bounded rationality, which underscores the cognitive limitations of decision-making processes that are grounded in organism-environment interactions (see Simon 1956, 1982).

3. Charles Wolfe has argued that "the concept of organism is a *porte-manteau* concept, on which one hangs various, rather disparate research programs" and "which plays a series of roles—sometimes overt, sometimes masked—throughout the history of biology, and frequently in very 'valuative' or normative ways, also shifting between the biological and the social" (Wolfe 2014a, 158).

4. Godfrey-Smith (1996a, 3) has famously contended that the evolutionary function of cognition is "to enable the agent to deal with environmental complexity," and thus understanding the environment is crucial for naturalizing the mind. Moreover, numerous scholars have argued that dealing with cognition necessitates a sine qua non appreciation of the dynamic interplay between living beings and their surroundings. The contemporary trend of situating cognition within the framework of dynamic organism-environment interactions is encapsulated in the movement of "4E cognition"—an abbreviation for embodied, embedded, enacted, and extended

cognition. Alas, discussions of the organism-environment relationship revolving around 4E cognition are beyond the scope of this book. Only concepts stemming from ecological psychology, specifically the notion of "affordance," will make an appearance in chapter 6 when I address the theme of organismal agency.

5. Of course, doing &HPS has much older roots, and a somewhat novel label should not make us forget that a long-standing epistemic space is being (partially) carved out by it. To give an example, different traditions of combining history and the philosophy of science have been common in the German and French contexts (e.g., Marburg neo-Kantianism and the tradition of historical epistemology, respectively), and these have overlaps and bridges, but also differences, with current Anglophone &HPS approaches (see, e.g., Friedman and Nordmann 2006; Massimi 2009; Vagelli 2019).

6. The members of the Organicist Movement that took root in the interwar years can also be understood as bona fide *philosophers of biology* (on this, see Nicholson and Gawne 2015).

Chapter 2

1. For discussion of the place of organisms in Kant's critical philosophy, see, among others, Löw (1980); Steigerwald (2010); and Mensch (2013). I will briefly hash out Kant's views on organisms in chapter 5 (section 5.2.3) and chapter 6 (sections 6.2.4 and 6.5.3).

2. For example, in influential works such as Buffon's *Histoire naturelle générale et particulière* (1749–1789), Charles Bonnet's *Considérations sur les corps organisés* (1762), and Georges Cuvier's *Leçons d'anatomie comparée* (1800–1805), terms like *corps organique*, *corps organisé*, and *corps vivant* were primarily put in motion to denote individual living organisms. In contrast, authors such as Ducrotay de Blainville, Johannes Müller, Richard Owen, Karl Ernst von Baer, Auguste Comte, and Claude Bernard used the term "organism" in the technical sense to pick up "living beings" in their works belonging to different orientations, from physiology to embryology to milieu theories (see Cheung 2006, 331, 335–339).

3. By this, I do not intend to imply that the organism-as-individual construal is the only one biologists espouse. Different evolutionary and ecological perspectives have been discussed and assumed as well (see, e.g., Bouchard and Huneman 2013 and chapters therein).

4. For instance, without going very far back in time, the eighteenth century stands out as a crucial period that Zammito (2018) has called "the gestation of (German) biology," beginning with the debate on organisms involving Georg Ernst Stahl and Gottfried Leibniz and culminating in the contributions of developmental morphology by, among others, Carl Friedrich Kielmeyer and Friedrich Schelling (see also

Cheung 2014). Furthermore, reflections on living entities, even in the nineteenth century, were not limited to biology; they also extended to several philosophical traditions. Within the German context, for example, proponents of neo-Kantianism, phenomenology, and *Lebensphilosophie* shared substantial connections with the nascent life sciences, but they were also engaged in autonomous philosophical explorations concerning the nature of life (see Bianco 2019; Reiß 2022).

5. For historical work pursuing that project, see, among others, Selcer (2018) and Warde et al. (2018). For philosophical discussions, see Casetta (2023).

6. For instance, Michaëlsson (1939) relates *ambiance* to the present participle of the Latin verb *ambire*, frequently used by authors such as Plautus, Seneca, and Pliny, and claims that adjectival forms (*ambiant, ambient*) can be found in the sixteenth century. Moreover, *aria ambiente* appears in Italian as early as 1603 (Casetta 2023, 21), and the noun *ambiente*, which dates back to the seventeenth century, is prevalent in this language. Michaëlsson (1939) suggests that it is plausible that the Italian usage played a role in the emergence of the corresponding French term. Similarly, in Spanish, *ambiente* has been documented since 1692. For a recent survey, see Chien (2007).

7. For a discussion of the metaphysics of the organismal environment concept, please refer to chapter 3 (section 3.2). Compare this with Jakob von Uexküll's concept of *Umwelt*, covered in chapter 4 (section 4.3.3) and chapter 5 (section 5.2.5).

8. Lamarck was pushing against the view of French anatomist Xavier Bichat stating that all influences external to an organism tend only to destroy it. For Bichat, to live is "to defend oneself against death, as a beleaguered city defends itself against destruction, or a square of infantry against the squadron of cavalry that charges it" (Entralgo 1948, 56).

9. On Cuvier's notion of *conditions d'existence*, see, among others, Grene (2001) and Reiss (2009, chapter 5). For a classic discussion of how Darwin reframed Cuvierian conditions of existence to fit his theory of natural selection, see Ospovat (1981).

10. Darwin emphasizes in various sections of *On the Origin of Species* that the interactions between organisms constitute the most important biological relationship. Mentions are found not only in chapter III but also prominently at the outset of chapter IX on the imperfections of the geological record (e.g., Darwin 1859, 279). In numerous instances, Darwin contends that abiotic alterations are not a prerequisite for the occurrence of natural selection (82). Additionally, he downplays the direct influence of external factors like climate and food in generating new variations. Such ideas are discernible in the introduction, the renowned chapter IV, as well as chapters V, IX, and XI. To underpin his position, he appeals to biogeographical evidence that questions the impact of abiotic external conditions. On the one hand, he posits that identical surroundings do not invariably house the same species, and conversely, identical species can be observed in disparate settings across separate

spatial locations. On the other hand, he highlights that species distribution ranges are primarily delimited by competition with other living organisms, sometimes even more so than by adaptations to specific climates (Darwin 1859, 140).

11. For instance, in a very famous passage of Malthusian inspiration, Darwin stressed that "as more individuals are produced than can possibly survive, there must in every case be a struggle for existence, either one individual with another of the same species, or with the individuals of distinct species, or with the *physical conditions of life*" (Darwin 1859, 63–64; emphasis added). Furthermore, when Darwin discusses the "two great laws" that, for him, have shaped all organic beings (i.e., the unity of type and the conditions of existence), it is clear that he has a plurality in mind: "Natural selection acts by either now adapting the varying parts of each being to its *organic and inorganic conditions of life*; or by having adapted them during long-past periods of time: the adaptations being aided in some cases by use and disuse, being slightly affected by the direct action of the *external conditions of life*, and being in all cases subjected to the several laws of growth" (206; emphases added). Another passage where Darwin brings up "conditions of life" as a generalizing instance of the plural environing settings of organisms occurs in chapter VIII ("Hybridism"; see Darwin 1859, 267).

12. Martineau's translation of *milieu* was certainly not the first occurrence of the term "environment" in the English language. This notwithstanding, the history of a word is not necessarily the history of a particular concept, and this is a case in point. Benson (2022, 9–10) correctly points out that, while variations of "environment" did sporadically appear in English texts as early as the early eighteenth century, their widespread adoption in anything resembling their modern meanings did not occur until the latter half of the nineteenth century. Importantly, in 1827, the English neologism "environment" appeared in the writings of the British essayist and historian Thomas Carlyle as a result of his engagement with the works of Goethe. Carlyle's passage, found in *Miscellanies* (1827), where "environment" first appears—a translation of *Umgebung*—is, in fact, a translation of Goethe's *Dichtung und Wahrheit* (book XIII; for discussion, see Spitzer 1942, 204; Jessop 2012, 708–709; see also Benson 2020, 10). It is still unclear among scholars if Martineau later borrowed the term "environment" from Carlyle in his translation of Comte or if she coined it independently (Pearce 2010, 248).

13. For example, Spencer talked about physical, biological, and social environments in his works, and from the latter half of the nineteenth century onward, these notions gained widespread acceptance in several fields despite being ambiguously construed and involving ontologically heterogeneous factors and processes. As illustrations of the concept of social environment in action, including some problematic directions, see Baldwin (1897), 64–71; Wallace (1913); and Morgan (1919), 76–77; for additional discussions, see Taylan (2018, 2021). Spencer also influentially referred to society as an "organism" (see, e.g., Elwick 2003). In a Spencerian

worldview, if societies are indeed like organisms, it makes sense to postulate social environments analogous to regular organismal environments.

14. "Heredity" is another one of those condensing abstractions, merging all purported processes that ensure the similarity between parents and offspring, which were effectively introduced into biological practice and theorizing around the same decades of the nineteenth century (López-Beltrán 1994). For accounts of the different threads (e.g., medical, legal) that converged in the crystallization of the term "heredity," see López Beltrán (2004) and Müller-Wille and Brandt (2016).

15. For historical and philosophical examinations of the Baldwin effect, see Weber and Depew (2003); Ceccarelli (2019); Radick (2024); and Loison (2024). For additional discussions on the Baldwin effect and its reception in twentieth-century biology, see chapter 6 (section 6.4).

16. In early-twentieth-century biology, this motif, which will be revisited in successive chapters (especially section 6.3.2), revolved around a physiological comprehension of adaptation.

17. The anonymous author was echoing the work of sociologist and polymath Patrick Geddes and biologist Arthur Thomson. In their coauthored book, *Evolution*, Geddes and Thomson claimed, "Every creature is a bundle of adaptations. Indeed, as Weismann says of the whale, 'When we take away the adaptations, what have we left?'" (Geddes and Thomson 1911, 97). And with regard to behavior, "Living creatures are agents; they thrust as well as parry; they act on their surroundings, modifying them; they are ever seeking out a new environment, and conquering them" (Geddes and Thomson 1911, 198).

18. This section partially draws from Baedke and Fábregas-Tejeda (2023).

19. For historical studies of the British context of organicism, see, for instance, Nicholson and Gawne (2014) and Peterson (2016). For historical narratives of the development of organicism in the United States, see Esposito (2016). For historical studies of German-speaking organicism and holism, see, among others, Harrington (1996); Lawrence and Weisz (1998); Rieppel (2016); Müller (2017a); and Fábregas-Tejeda et al. (2021). For discussions of Italian organicism, see Tamborini (2022, 2023).

20. For the sake of simplicity, the polysemous term "mechanism" will not be explained in this book, but see Nicholson (2012) for a disambiguation. A brief discussion on "vitalism" and its recent, nuanced historiography will be articulated in chapter 6 (section 6.2.2).

21. Before Delage, the term "organicism" had already been used but with an opposite semantic field: In early-nineteenth-century French medicine, *organicisme* was associated with the Paris School, in contrast to the Montpellier School of vitalists (see Raynaud 1998; Wolfe 2009). Organicists, in this older usage, were reductionists

who focused on the role of specific organs (e.g., for accounting for disease) and not whole organisms. I thank Charles Wolfe for making me aware of this.

22. Prior to Haldane's writings, only a limited number of biologists invoked Delage's notion of "organicism" (see, e.g., Calderwood 1893, 492; for discussion, see also Cumston 1926, 361–364).

23. It should be mentioned that, outside of biology, organism-centered and holistic perspectives also emerged in other fields, especially in medicine and psychology (see, e.g., Harrington 1996; Ash 1998).

24. Due to this heterogeneity, some scholars, such as Baedke (2019), favor the term "organism-centered perspectives" over the plural "organicisms." This nomenclature not only offers a more inclusive characterization but also avoids a potentially overly British-centric categorization. While I concur with this critical stance, in this book, I employ the terms "organicism(s)" and "holism(s)" for the sake of conceptual simplicity in the exposition of ideas. It is important to note, however, that this choice is made without the assertion that "British organicism" serves as the quintessential or foundational archetype for other organism-centered perspectives that emerged during the early twentieth century.

25. Interestingly enough, even the contentious issue of the inheritance of acquired characteristics was broached in the Vivarium by taking the environment as a waging point (see, e.g., Przibram 1917).

Chapter 3

1. For discussion of the semantic, epistemic, and ontological diversity of what are presumed to be "environments," and what count as such in past and present life sciences, see Benson (2020); Pontarotti et al. (2022); Baedke and Buklijas (2023); and Fábregas-Tejeda (2024a).

2. "Effective environment" should not be confused here with the meaning that evolutionary biologist Leigh Van Valen gave to this term in his theoretical work on the interplay between species, extinction, and the so-called Red Queen hypothesis. There, the "effective environment" of an organism (qua type) translates to its adaptive zone, in addition to the effects of other organisms within that adaptive zone (see Van Valen 1971, 1973). There is similarity, however, between the notion of effective environment I am focusing on and how some ecologists construe the term "habitat." For example, the well-known plant ecologist Arthur George Tansley defined "habitat" as "the sum total of environmental conditions actually working on the individual plant" (Tansley 1920, 121; see also Weaver and Clements 1929, 163). It is also not uncommon, then, to find scientists that use "environment" in its widest, universal sense, while restricting "habitat" to allude to "place" in the existential sense outlined above.

3. In the past, ecologists have also construed the contrast between environment simpliciter and effective environment in less sharp, categorical terms (see, e.g., Allee and Park 1939, 167; see also Haskell 1940, 6).

4. This was a point that I already stressed in the previous chapter. A notorious aspect of the historical development of the environment concept from the late nineteenth century onwards was to cardinally construe it as inherently relational. So, to be sure, the relationality character of the environment of an organism is not something specific to its juxtaposition to this counter-relatum. In fact, I think, alongside other scholars, that the relational aspect is a projectable and sine qua non property of environments. To be an environment is to be an environment of *something*. On this, Benson (2020, 12) asserts, "Without that fundamental relationality the concept of environment loses much of its distinctiveness."

5. There are, of course, many more adjectives, adnouns, and appellations that biologists have come up with to refer to the environing conditions of organisms. It goes without saying that I will not attempt to cover them all here. I will restrict my exposition to just a couple of examples—by no means representative of entire disciplines or subfields—to illustrate some general philosophical theses about how the organism-environment pairing is customarily construed within the life sciences.

6. This is something that scientists themselves have argued several times (see, e.g., Niven 1983; Prothero 2013, 161; Andrews 2017, 242; Tomasello 2022, 130). See also the discussion of Jakob von Uexküll's ideas in chapter 4 (section 4.3.3). Importantly, this should not be taken to mean that, within biology, other epistemic constructs of the environment (e.g., a shared, collective environment that simultaneously affects different individual organisms) are not possible. In fact, preeminently, the idea of a "population environment" is common conceptual currency across ecology and evolutionary biology. I will discuss this idea in detail, as well as the rise and fall of "individual environments" in twentieth-century biology, in chapter 5 (section 5.3.4).

7. As I have mentioned, some scholars might want to argue that the corollary of the relationality thesis is much more general and cuts through all uses of the environment concept—in other words, that there are no environments if nothing is being environed, be it a molecule, a cell, an organism, a population, a community, a city, a bridge, or whatever.

8. I will come back to reciprocal codetermination and how it bears on the issue of organism-environment reciprocity and, in particular, on the general view of ontological coconstitution between organism and environment in future chapters. As we will see, I think there are principled reasons not to treat reciprocal codetermination as implying the cognate thesis of organism-environment ontological inseparability. I will offer some arguments in support of this view in chapter 5 (section 5.5.1).

9. Two well-known philosophical defenses of this view can be found in Jonas (1966) and Maturana and Varela (1972). I consider that framing the organism-environment

relationship in terms of the persistence and the self-maintenance of the former relatum is a useful starting point, but much more can be said about this pairing (see chapter 6 on organismal agency).

10. On how adaptivity is connected with organismal agency, see chapter 6 (section 6.5.4).

11. It must be said, however, that even though there is a general recognition of the de facto dynamicity of organismal environments, this is not always a tenet that is baked into all biological models or explanations that deal with them. The dynamicity of environments can be bracketed off or be treated as an expendable epistemic principle in scientific practice. This is partially accounted for by technical constraints and its "high dimensionality" (Murugan et al. 2021, 2).

12. For the problem at hand, I think that a practice-only approach—centered on studying closely how scientists talk and think about environments—won't serve as the best philosophical fulcrum for understanding the environment as the relatum of an organism. For instance, if we compared the uses of "environment" (sensu lato) in different biological models, as Huneman (2022) has recently done, the outcome of philosophical explication becoming really knotty—if not plainly impossible—should not surprise us at all. This is indeed to be expected. In contrast to this position, and instead of attempting to understand the myriad uses of the environment concept throughout the biological sciences at once, I think that we can be more precise in the questions we are trying to tackle as philosophers: "What is an environment (qua relatum) for an organism? What kind of entity is an organismal environment?" These are questions that, as I show here, are amenable to philosophical explication.

13. The revitalization of metaphysical analyses within the philosophy of science should be disentangled from another philosophical project—although sometimes they go hand-in-hand—namely, "scientific metaphysics." This project aims at concocting a general ontology or naturalized worldview of what exists based exclusively on what is considered by the current array of the sciences. In this view, metaphysics should be conducted as another branch of science inasmuch as the latter is (professedly) best positioned to tell us about how the world actually is. Simply put, what's good for the goose is good for the gander (see, e.g., Ladyman et al. 2009). For a critical assessment, see Polger (2022).

14. As Triviño (2022, 4) mentions, these two different modes of transacting in the metaphysics of biology are not dichotomously opposed; rather, they resolve into a "question of emphasis" for the issue under evaluation. Recently, Cristina Villegas and Vanessa Triviño (2023) argued that a third category could be added—namely, "Metaphysics *from* Biology"—which aims to enrich the traditional toolkit of the metaphysician by the development of new notions that are necessary to fathom complex biological realities.

15. I submit that my work, although navigating both modes, is somewhat more closely aligned to the pole of metaphysics *for* biology given that I am not analyzing particular biological theories or practices: The organismal environment concept features transversally in the repertoire of several of them.

16. Andrewartha (1970) introduced the neologism "malentities" to refer to what Browning (1962) had previously called "hazards" as a fifth class of component of the environment of organisms that could expand the classification of Andrewartha and Birch (1954).

17. That being said, it is important to clarify that not all accounts of sortals require a counting criterion (see, e.g., Wiggins 2001).

18. For discussion of ostensive definitions, see Kotarbinska (1960). Importantly, in a broader interpretation, "what is essential for ostensive definition is not the pointing gesture but the result which we want to obtain thereby. The aim is to distinguish in our surroundings a certain definite object and to draw to it the attention of the addressee of the definition in question" (Kotarbinska 1960, 2). For the issue at stake, the question remains: How do we distinguish organismal environments from broader surroundings?

19. Please note that, as a general methodological approach, scientists consider a presumed shared "population environment" and then make the inference that all token organisms, in the case of Goff et al. (2020), frog larvae living in the same pond, are affected by it. As a counterpoint to this, in the last few years a new trend in ecology to measure "individualized niches" has emerged (for discussion, see Trappes et al. 2022). Issues about individual environments will be covered in chapter 5 of this book.

20. This is surely an oversimplification, as seedlings and mature plants also exhibit diverse movements (Dumais and Forterre 2012, 472; see also Hart 1990). These movements, which are highly flexible and adaptable, might also signal changes in what constitutes the microenvironments in which plants are embedded at particular instances of their life cycle.

21. The complementary issue of how organisms causally affect their surroundings (e.g., through different types of niche construction) will be discussed in chapters 5 and 6.

22. In contrast to what I present here, some philosophers of language would prefer to restrict indexicality to those expressions that vary in reference or content in relation to *utterances* or other forms of speech acts. As I am more interested in the metaphysical side of indexicals, I won't go further with what happens when we introduce the side of utterers, nor I will broach related issues such as the proper rules of use of indexicals (for discussion, see Ezcurdia 2014).

23. As philosophers of biology have discussed, this might be easy for entities that conform, in broad strokes, to the vertebrate Bauplan, but things get complicated when critters such as siphonophores, plants, aspens, lichens, or colonial life forms are considered. This is sometimes framed as one of the important components of the "biological individuality" debate, and it is one that extends very deeply into the history of biological science (see, e.g., overviews in Wilson 2005, chapter 1; Clarke 2010; Lidgard and Nyhart 2017). For discussion of how to distinguish (and not conflate) "organismality" from the broader concept of "biological individuality," see Prieto (2023, 2024).

24. Although Mason and Langenheim's article suffers from many of the strong commitments and deficiencies of logical empiricism, I still think it has important things to say to us in the twenty-first century. Trained as biologists, Mason and Langeheim devoted substantial time to philosophical topics throughout their careers (see, e.g., Mason and Langenheim 1961; Mason 1957).

25. After his early writings, Russell changed his mind and considered that classes were "logical fictions." The theory of classes that Russell developed alongside Alfred North Whitehead in the *Principia Mathematica* provided a notation to represent them but denied that, strictly speaking, there are such things as classes (see Whitehead and Russell 1910, 196; for an analysis, see Boër 1973). This is a case where useful constructions do not need to be ontologically sanctioned.

26. Discussions of abstraction and understanding cannot be circumvented in any epistemology of organismal environments, but these by no means exhaust everything that is epistemically interesting about them. More work needs to be done to arrive at a proper and comprehensive epistemology of organismal environments.

27. However, also because of this, and especially when environmental components are treated merely as "background conditions" or "noise disturbances" to be kept at bay in order to highlight, for instance, genetic contributions to organismal phenotypes, their developmental and evolutionary relevance gets screened off by how the experiments were set up in the first place (for discussion, see Sultan 2017).

28. Scholars usually contrast abstraction with idealization, claiming that the latter, although also related to the degree of detail of a description, involves introducing simplifying misrepresentations or distortions of the target system. This, nonetheless, delivers epistemic goods to their users and serves manifold roles in scientific practice. For classic accounts, see Jones (2005) and Godfrey-Smith (2009); for recent discussions, see Potochnik (2020) and Levy (2021).

29. A proviso should be added here: I do not intend to exhaust all epistemic dimensions of abstraction or claim that my analysis has accomplished a comprehensive characterization. For example, abstraction, qua complex, situated epistemic process, involves scaffolded interactions between developing different scientific skills and deploying them (e.g., through visualizations, techniques, or artifacts; see Martínez

and Huang 2011). I will only capsulize certain aspects of abstraction, focusing on its outcomes and products, for the purposes of framing an organismal environment qua relatum in biological practice.

30. This example is taken from Baedke and Fábregas-Tejeda (2023).

31. Contrast this with the ontological interpretation of an organismal environment as a single entity: "If one treats the environment as a class-as-one, complexity upon complexity begins to accrue to the class" (Mason and Langenheim 1957, 333).

Chapter 4

1. This section and the introduction to section 4.2 partially draw from Fábregas-Tejeda (under review).

2. This is an instance of what Pradeu (2010) has called "phenomenal individuation," according to which our everyday experience as humans allows us to incontrovertibly distinguish biological individuals like organisms. Suffice it to say that this approach faulters when common sense is confronted with entities that do not conform to the vertebrate Bauplan.

3. This is reminiscent of two understandings of organismal environments that we reviewed in the previous chapter (section 2.1)—that is, environment as external world and environment as effective environment.

4. A similar debate is ongoing in the philosophy of mind and the philosophy of the cognitive sciences—for example, as it pertains to different discussions in 4E approaches to cognition and the free energy principle (see, e.g., Corris 2020; Werner 2021; Kirchhoff and Kiverstein 2021). How cognition features in, or intersects with, the discussions of this chapter, and particularly with respect to my proposal of shifting organism-environment boundaries during life cycles (section 4.5), is something worth exploring in the future but which nevertheless falls outside the purview of the present work.

5. On the recent debate on biological individuality, see for instance, Clarke (2010); Pradeu (2016); and Prieto (2024). For an overview of how "organisms" have been demarcated from other biological individuals, see Prieto (2023). For discussions of environments in the direction mentioned, see Pontarotti et al. (2022) and Baedke and Buklijas (2023).

6. Here, the scope of the problem will be mainly restricted to multicellular organisms that undergo development. In that sense, throughout this chapter "organism" will be shorthand for "multicellular developing organism." Even though I will also cover some unicellular eukaryotes (e.g., in section 4.6), bacteria and archaea will be absent from these discussions. For organisms belonging to those groups, the cell membrane is usually postulated as a bona fide mediating boundary with their

surroundings, and a dynamic one indeed throughout their life cycles. My constriction is pragmatic in origin and deals with length constraints: Although I think many aspects of what I argue for in this chapter also apply to unicellular eukaryotes, bacteria, and archaea, a more careful treatment should be devoted to those organisms.

7. For a discussion of flowing streams, whirlpools, and vortexes as metaphors of organismal development in organicist biology, see Nicholson (2018).

8. As an example of the reception of Prenant's book, see Dobzhansky (1949).

9. On what explains this transition in Uexküll's thinking (i.e., a shift from a mechanicist organismal ontology to a subject-based one), see Bastard Rico (2021).

10. This section partially draws from Baedke et al. (2020).

11. Sections 4.5 and 4.6 partially draw from Fábregas-Tejeda (under review).

12. This is the sense in which I speak of concreteness, in contrast to abstractness, in this chapter. With this, I do not intend to imply that the abstract accounts of boundary seekers are devoid of material content; I only want to stress that, in their accounts, the causal relata of the processes alluded to are hard to specify and follow throughout particular ontogenies because they require high degrees of abstraction.

13. My focus in this chapter is on how the interfaces between organisms and environments are set during ontogeny, not on providing a full-fledged metaphysical framework to account for the diachronic identity of these relata as they undergo different changes in time (for discussions of the organismal environment and diachronic identity, see section 3.2.3). Certainly, a conceptual framework that aids by winnowing organism-environment limits is a necessary component of an account of organismal diachronic identity, but the latter is a much more expansive metaphysical problem that cannot be undertaken here. For recent discussions of the topic, see Meincke and Dupré (2020).

Chapter 5

1. A preliminary version of this classification was originally introduced in Baedke et al. (2021). For a different historical taxonomy and paired visual representations, see Di Paolo (2020).

2. "Concomitant reaction" (sensu Baedke et al. 2021) refers to a simultaneous event that brings about changes in both an organism and its surrounding environment. This concept bears resemblance to John Stuart Mill's view of the "law of coexistence" (see Mill 1846, 43; for one example, see Raerinne and Baedke 2015). In turn, the idea of "transaction," advanced by pragmatists John Dewey and Arthur Bentley, is an epistemological stance that also incorporates scientific observers studying the couplings between organism and environment, assuming "no pre-knowledge of

either organism or environment . . . with full freedom reserved for their developing examination" (Dewey and Bentley 1946a, 536). The boundaries of organisms and their environments are negotiated as outcomes of inquiry (see also Dewey and Bentley 1946b, 1949). For background on Dewey's ideas on the organism-environment relationship, see Pearce (2014b, 2020).

3. This move of the expansion—or distortion, as some scholars will surely interpret it—of Kantian reciprocity was not unique to early-twentieth-century discussions about life. Other authors extended it beyond the living world. Whitehead (1929) contended, "The relation of part to whole has the special reciprocity associated with the notion of organism . . . ; but *this relation reigns throughout nature*" (185; emphasis added).

4. This mutual Bernardian influence does not mean that Haldane and Henderson were in agreement all the time. In fact, they were often at odds on issues such as the nature of organismal organization and the limits of mechanistic explanations (see, e.g., Henderson 1915, 381; 1918, 576).

5. The second half of this section partially draws from materials previously published in Baedke et al. (2020, 2021) and Fábregas-Tejeda and Vergara-Silva (2022).

6. Generally speaking, the positions of organism-environment reciprocal causation and ontological coconstitution were not exclusive of early-twentieth-century biologists. In fact, several philosophers and psychologists around those years embraced them in their respective domains and deliberations. For instance, Pearce (2020, chapter 4) has shown that both reciprocal causation and ontological coconstitution—he calls them "the reciprocal causes view" and "the dual aspects view," respectively—were heavily debated in British idealism and American pragmatism in connection to discussions of Darwin, Spencer, and Hegel.

7. It must be said that not all organicists agreed with the validity of this type of experimental research that probed into the causal influences of particular environmental factors on developing organisms (see, e.g., the strong skepticism voiced by Ritter [1916, 464]).

8. For Livingston, however, this causal reciprocity should compel biologists to loosen up a staunch boundary between organism and environment, recognizing "the purely artificial character of our conventional distinction between the organism on the one hand and its environment on the other. *The innumerable series of chains of causes and effects*, wherein the effect of one cause is itself the cause of another effect, and so forth, *reach into and through the plant* without reference to its spatial limitations" (Livingston 1912, 214; emphases added).

9. This designation is usual in historiographic work, but it should be noted that the members of the group, while actively meeting, used the expression "Biotheoretical Gathering(s)" (Abir-Am 1987, 12).

10. Likewise, organicist myrmecologist William Morton Wheeler had once claimed, echoing Whitehead, that "the organism may be said to seek, and in many cases even to make, its own environment" (Wheeler 1905, 538).

11. The introduction of this section and section 5.3.1 partially draw from materials previously published in Fábregas-Tejeda and Vergara-Silva (2022) and Baedke and Fábregas-Tejeda (2023).

12. Although the so-called gene's-eye view of evolution can be traced back to the Modern Synthesis, this standpoint should be more precisely characterized as a distinct, extreme variant within it, which was more diverse and heterogeneous, as recent historiographic research shows (see Delisle 2011; Ågren 2023, 555–568).

13. For a broad-strokes historical overview of how other epistemic roles ascribed to organisms got displaced, see Baedke and Fábregas-Tejeda (2023, 134–137). However, it is important to note that not all proponents of the Modern Synthesis, such as Dobzhansky and Bernhard Rensch, aligned themselves exclusively with either the gene-centric or organism-centric perspective regarding the chief units of evolution. Some of them held broader, and at times even holistic, views on the matter (see Delisle 2008).

14. For the epistemic justification of adopting the gene's-eye view of evolution, see Ågren (2021). Depew (2017) covers the relationship between some architects of the Modern Synthesis and ontogenetic thinking. On the demotion of the epistemic status of the organism, see Walsh (2015, chapters 1–3).

15. Dobzhansky elaborated in another publication, "The role of environment in evolution is more subtle than was realized in the past. The organism does not suffer passively changes produced by external agents. In the production of mutations, environment acts as a trigger mechanism, but it is, of course, decisive in natural selection. However, natural selection does not 'change' the organism; it merely provides the opportunity for the organism to react to changes in the environment by adaptive transformations. *The reactions may or may not occur, depending upon the availability of genetic materials* supplied by the mutation and recombination processes" (Dobzhanksy 1950b, 2010; emphasis added).

16. In the words of Stebbins, "If a particular change in the environment takes place, a population may respond to it either by evolving in a new direction, that is, by carrying out an adaptive shift, or by becoming extinct. Which response will take place depends largely upon the composition of the gene pool at the time when the environmental change appears. The role of mutations in determining transspecific evolutionary change depends upon the basis of the entire spectrum of effects that they produce, upon the adaptive advantage or disadvantage of mutant individuals, and upon *the role of the mutant gene in population-environment interactions*" (Stebbins 1974, 4–5; emphasis added).

17. Interestingly, Nicholson chose to phrase some of his insights on reciprocity by switching the explanatory relatum back to organisms: "Many organisms alter the chemical and physical nature of their environment, and, generally, the increase in density of such organisms causes the character of the environment to become progressively less suitable for themselves. This clearly is a form of competition and provides the mechanism necessary for balance" (Nicholson 1933, 140). For a historical reconstruction of Nicholson's views, see Kimler (1986).

18. In contrast to the statements of Brandon and colleagues, my analysis (section 2.2.2) shows that, in Darwin's original work, there was no concept of "population environment" (or other environment concept, for that matter), but rather, that of "conditions of existence" of individual organisms.

19. J. B. S. Haldane, perhaps influenced by his father, asserted that this was only an epistemic stance: "It is however important to remember that [this distinction] is an abstraction. The physiologist finds his main difficulty in making it spatially. Thus he sometimes finds it convenient to regard the blood plasma as part of the organism, sometimes as part of the environment. The geneticist is a little unsure where to make it in time" (Haldane 1936, 349).

20. This subsection partially draws from Baedke et al. (2021).

21. For a broader analysis of how various concepts that originally referred to individual organisms were later reinterpreted by Modern Synthesis–inspired thinking in terms of populations, see Rama (2024).

22. Sections 5.4.1, 5.4.2, and 5.5.2 partially draw from Baedke et al. (2021). Section 5.4.1 partially builds from the historiography of Fábregas-Tejeda and Vergara-Silva (2022) and adapts material from Fábregas-Tejeda and Vergara-Silva (2018b).

23. In contrast, the view of mutual structural fitting, advanced by Henderson and his followers in the first decades of the twentieth century has not received new theoretical impulses from contemporary biologists.

24. This might not be true in all countries and research communities. For instance, in the United Sates of America, many evolutionary biologists do not recognize the aptness of reciprocal causation as a modeling tool (Hazelwood 2023). Likewise, NCT has received many criticisms since its onset. For example, some scientists have claimed that organism-environment reciprocity, as understood by NCT, is nothing new in evolutionism (see Brodie 2005).

25. For scholars such as Walsh (2012), physical niche construction and experiential niche construction are not kinds of niche construction but, rather, different interpretations of what niche construction entails for development and evolution.

26. The way Trappes and colleagues define the "individualized niche" might be reminiscent of the effective organismal environment (see section 3.1), but there are

some differences that should be made explicit. The idea of an effective organismal environment is broader, as it refers to the sum of causal factors affecting the development of a token organism in a given point in time, and the individualized niche only recovers a subset of those factors that are required for the individual's survival and reproductive success. In other words, the idea of the organismal environment and that of the individualized niche partition differently the heterogeneous plurality of elements that comprise an organism's environing conditions. In that regard, they are two distinct abstractions (see section 3.3).

27. Remember that, in stressing this, I am only considering causal interactions between token developing organisms and their indexical environments at particular moments of their life cycles (following the relationality desideratum considered in chapter 3 and the metaphysical assessment presented therein).

28. I think that even cases of experiential niche construction can be framed through a reciprocal causation lens (see Baedke et al. 2021). There is nothing proprietary to external niche construction that renders an exclusive causal construal and discards it for the case of experiential niche construction.

29. For a defense of organisms as bona fide sortals, see Okasha (forthcoming).

30. In this book, I will remain agnostic regarding which extant theory of causation (e.g., regularity, counterfactual, interventionist, and process theories) would be the best ontological (and epistemic) fit for thinking about organism-environment reciprocal causation. Further detailed work would be needed to spell out this matter and select between distinct possibilities.

31. I could be much more precise here: Walsh (2022) specifically claims that the role the concept of environment plays in models of population dynamics is that of an abstraction. But that is not equivalent to committing to the organismal environment as an abstraction, as I have indeed argued for in this book. Going further, I do believe that all invocations of environments, and not only the organismal environment, are products of abstraction processes—very useful and epistemically powerful for the sciences, but abstractions in the end (see Fábregas-Tejeda 2024a).

32. As we will see in the next chapter, Walsh (2015) endorses a view according to which organisms and their environments together constitute a system of affordances—the *affordance landscape*. The features of the affordance landscape cannot be apportioned into those elements provided exclusively by the environment and those contributed by the organism. But that is not the same as the organism and the environment reciprocally constituting one another. So, strictly speaking, Walsh does not count ontological coconstitution as the best construal of organism-environment reciprocity.

33. This does not mean that I give the nod to reciprocal causation as a flawless construal. Here, I am only adjudicating between these two widespread views. Perhaps

we would need a different or refined view of organism-environment reciprocity in the future.

34. In that passage, McDougall is citing Haldane (1936, 149).

Chapter 6

1. This section and section 6.3 partially draw from Fábregas-Tejeda (2024b). This section also adapts material from Fábregas-Tejeda et al. (2024).

2. Vitalism, a philosophy of life that is often misunderstood and maligned, has undergone critical reevaluation by scholars in recent years. Scholars have given a more nuanced picture (see Chen 2024) and debunked the presumption that it was a homogeneous and unified position (see Donohue and Wolfe 2023 and chapters therein). Wolfe (2011) has contributed to this revalorization by distinguishing between two kinds of vitalism: "functional vitalism" and "substantival vitalism." The former reflects a desire to comprehend life's complexity while acknowledging its unique characteristics and phenomena, seeking to model and understand life without reducing it to mechanistic processes. On the other hand, substantival vitalism posits the existence of specific entities or forces that transcend causal relations studied by traditional physical science. Importantly, then, vitalism encompasses varying degrees of metaphysical commitments (Bognon-Küss et al. 2018).

3. It should be mentioned explicitly that not all instances of the heuristic approach are Kantian in a committed philosophical sense. In other words, the Kantian view is an instance of the heuristic approach, but not all heuristic views on organismal teleology are Kantian. For discussion, see Gambarotto and Nahas (2022).

4. Kant's views on organisms are complex and have been the subject of exegetical disagreements between specialized scholars. As Steigerwald (2010) says,

> Kant's conception of living organisms as 'natural purposes' is profoundly amphibious and does complex work in the *Critique of the Power of Judgment* by enlisting both concepts of nature and concepts of reason in its formation. Kant recognized that living organisms unsettle us. Our study of these unique natural products demonstrates their remarkable organization and regularity, but they remain forms of natural order that are contigent in terms of the mechanisms of nature. We can account for the possibility only by appealing to a concept of reason, namely, the concept of purpose. But organisms do not fit within the domain of concepts of reason; they are formed by the concept of natural processes rather than on the basis of an extrinc idea. Recognized as natural products, yet judged to be purposive. (291)

5. *The Study of Living Things* was a normative book about the proper method that biology should take, and Russell devoted the rest of his career applying it to some of the greatest problems of biology in his next books—namely, the problems of development and heredity and the study of animal behaviour (Russell 1930, 1934). A synthesis of his life's work, with numerous empirical examples, can be found in Russell (1945).

6. From the 1930s onward, however, a shift in Lillie's thinking occurred: He increasingly started to defend the causal influence of psychic factors as the explanation of the directiveness and creativeness of life. Like many of his colleagues at the University of Chicago (see Steffes 2007), Lillie began embracing Whitehead's process philosophy and became a committed panexperientialist. In his 1945 book, the leitmotivs of Lillie's work are still there, but the dynamic equilibration between organism and environment, and his old organicist standpoint, completely gave way to the power of the mental as the ultimate source of explanation for living phenomena (Lillie 1945).

7. This subsection draws from Baedke and Fábregas-Tejeda (2023), as well as from Fábregas-Tejeda and Baedke (2023b).

8. In the biological and philosophy of biology literature, we mostly see authors rejecting teleomentalism explicitly (e.g., Sultan et al. 2022). Interestingly, however, some authors have argued against anthropocentric views of cognition and intentionality and have postulated a continuum of intentional dynamics to capture and flesh out the goal-directed activities of organisms (see Sims 2021). Exploring the possible interfaces between organismal agency and cognition in this nonanthropocentric view falls outside the purview of this book (but see, e.g., Harrison 2023).

9. Within the philosophy of biology, a debate has emerged in the last decades concerning contrasting "causalist" and "statisticalist" interpretations of natural selection. Covering them—or trying to adjudicate between them—would distract from the main focus of this book. For overviews of this debate, see Walsh (2007); Ramsey (2013); Otsuka (2016); and Walsh et al. (2017).

10. There is large corpus on theories of function in the philosophy of biology, with "selected effects" being one of the most prominent ones. The proponents of this standpoint argue that a function is not only the explanatory factor behind the past selection and favoring of a specific structure but also serves as a causal explanans for the existence of the trait itself. In other words, an organismal trait exists because it performs the function that contributed to its selection (see, among other classic references, Millikan 1989; Neander 1991; for a recent overview, see Christie et al. 2022). Regardless of the validity of these theories, the discussions of biological functions are pursuing a different target, and confusions should be avoided on this regard: They are not concerned with *organismal internal teleology* (see below). In functional explanations, purposes or functions are ascribed to individual *traits* of organisms, whereas in agential explanations, purposes are attributed to the *organisms* themselves as integrated *wholes*. The debate on organismal agency is about internal teleology, not external teleology. For a discussion of why functional accounts cannot fully deal with the problem of organismal agency, see Walsh (2008); McLaughlin (2000); and Nahas (2024b). For these reasons, I will not say more about the functions debate in this book.

11. This subsection partially draws from Fábregas-Tejeda and Baedke (2023b).

12. Alas, examining "cell agency" lies beyond the scope of this book. Interested readers may consult, for instance, Newman (2022).

13. Likewise, I will not address two significant challenges highlighted by Barham (2012) regarding the project aimed at naturalizing teleology and normativity. These challenges are commonly known as the "scope problem" and the "ground problem." The former involves determining the appropriate extent to which our concept of normative agency can be applied across the phylogenetic board. The latter consists in providing an explanation for the phenomenon of normative agency within the framework of our existing scientific knowledge of nature. On different answers to the scope problem, see Burge (2009); Fulda (2017); and Tomasello (2022).

14. Desmond and Huneman (2020) frame intrinsic "purposiveness eliminativism" in the name "Neo-Fisherian option," in which organismal behavior is "analyzed along the lines of the 'maximizing agent analogy', where organisms behave in such a way that maximizes their inclusive fitness. The underlying assumption is called the 'phenotypic gambit' . . . , which holds that the choice of a phenotype by the organism mirrors the allele dynamics that underlie evolution. In this way, natural selection is taken to design organisms so that they make decisions similar to what, as it were, natural selection would do if it were making the decision" (45). This view is conceptually distinct from the teleonomy approach, which was covered in section 6.4.

15. There are, of course, other interesting standpoints to explain or broach goal-directed behaviors but that do not rely on the centrality of the organism-environment dyad and are applied broadly to different relata (see, e.g., Babcock and McShea 2021).

16. It must be said that, in their early writings, Varela and Maturana were skeptical of attributing goals and teleological dynamics to living systems (e.g., Varela and Maturana 1972, 381–382). Not until the end of his career would Varela's attitude toward organismal teleology change (see Weber and Varela 2002).

17. It is not my intention here to solve the debate over the ontology of affordances. For good overviews of contrasting positions, see Chemero (2003) and Heras-Escribano (2019, 61–89).

18. For the sake of conciseness and simplicity, other criteria will not be discussed here. For discussion, see Ylikoski and Kuorikoski (2010) and Baedke et al. (2020).

19. I do not deny that sometimes a less coarse-grained description of a behavior (below the level of the whole organism) might be useful or even necessary in certain scientific projects. My claim is only that behavior happens at the level of the organism-in-its-environment, and this is why thinking about causal relations at

this level yields the right proportionality in the sort of causal explanations I am describing.

20. I am not disregarding collective agency or the possibility of group normativity altogether. However, purportedly transferring the intrinsic normativity from within a single organism (or from a group or population with aligned goals) to the environment of another organism is a very different issue. This is what I am disputing in these pages.

References

"(1) Experimental Zoologie (2) Mechanism, Life and Personality." 1914. *Nature* 94 (2347): 193–195.

Aaby, Bendik Hellem. 2021a. "Organism-Environment Interactions in Evolutionary Theory." PhD thesis, KU Leuven. https://lirias.kuleuven.be/retrieve/620296.

Aaby, Bendik Hellem. 2021b. "The Ecological Dimension of Natural Selection." *Philosophy of Science* 88 (5): 1199–1209.

Aaby, Bendik Hellem, and Hugh Desmond. 2021. "Niche Construction and Teleology: Organisms as Agents and Contributors in Ecology, Development, and Evolution." *Biology & Philosophy* 36 (5): 47. https://doi.org/10.1007/s10539-021-09821-2.

Aaby, Bendik Hellem, and Grant Ramsey. 2022. "Three Kinds of Niche Construction." *British Journal for the Philosophy of Science* 73 (2): 351–372.

Abbott, James Francis. 1911. "Recent Progress in Some Fields of Experimental Zoology. I." *Transactions of the American Microscopical Society* 30 (3): 217–233.

Abir-Am, Pnina. 1982. "The Discourse of Physical Power and Biological Knowledge in the 1930s: A Reappraisal of the Rockefeller Foundation's 'Policy' in Molecular Biology." *Social Studies of Science* 12 (3): 341–382.

Abir-Am, Pnina G. 1987. "The Biotheoretical Gathering, Trans-Disciplinary Authority and the Incipient Legitimation of Molecular Biology in the 1930s: New Perspective on the Historical Sociology of Science." *History of Science* 25 (1): 1–70.

Abir-Am, Pnina G. 1991. "The Philosophical Background of Joseph Needham's Work in Chemical Embryology." In *A Conceptual History of Modern Embryology*, edited by Scott F. Gilbert, 159–180. Boston: Springer.

Abouheif, Ehab, Marie-Julie Favé, Ana Sofia Ibarrarán-Viniegra, Maryna P. Lesoway, Ab Matteen Rafiqi, and Rajendhran Rajakumar. 2014. "Eco-Evo-Devo: The Time Has Come." In *Ecological Genomics: Ecology and the Evolution of Genes and Genomes*, edited by Christian R. Landry and Nadia Aubin-Horth, 107–125. Dordrecht, The Netherlands: Springer.

Abrams, Marshall. 2009. "What Determines Biological Fitness? The Problem of the Reference Environment." *Synthese* 166 (1): 21–40.

Affifi, Ramsey. 2020. "Engaging the Adaptive Subject: Learning Evolution Beyond the Cell Walls." *Biological Theory* 15 (3): 121–135.

Agar, Wilfred Eade. 1913. "Transmission of Environmental Effects from Parent to Offspring in *Simocephalus vetulus.*" *Philosophical Transactions of the Royal Society of London. Series B, Containing Papers of a Biological Character* 203 (294–302): 319–350.

Agarwal, K. S. 2008. *Fundamentals of Ecology*. New Delhi: APH Publishing.

Ågren, J. Arvid. 2021. *The Gene's-Eye View of Evolution*. Oxford: Oxford University Press.

Ågren, J. Arvid. 2023. "Genes and Organisms in the Legacy of the Modern Synthesis." In *Evolutionary Biology: Contemporary and Historical Reflections Upon Core Theory*, edited by Thomas E. Dickins and Benjamin J. A. Dickins, 555–568. Cham, Switzerland: Springer.

Aguilera, Oscar, Agustín F. Fernández, Alberto Muñoz, and Mario F. Fraga. 2010. "Epigenetics and Environment: A Complex Relationship." *Journal of Applied Physiology* 109 (1): 243–251.

Akbar, Siddiq, Lei Gu, Yunfei Sun, Lu Zhang, Kai Lyu, Yuan Huang, and Zhou Yang. 2022. "Understanding Host-Microbiome-Environment Interactions: Insights from *Daphnia* as a Model Organism." *Science of The Total Environment* 808: 152093. https://doi.org/10.1016/j.scitotenv.2021.152093.

Alassia, Fiorela. 2022. "¿Es posible una ontología procesual de las entidades bioquímicas? Consideraciones a partir del caso de los receptores celulares y la señalización celular." *Estudios de filosofía* 65: 153–175.

Alexandrov, Daniel, and Elena Aronova. 2004. "Russian Theoretical Biology between Heresy and Orthodoxy: Georgii Shaposhnikov and His Experiments on Plant Lice." In *Darwinian Heresies*, edited by Abigail Lustig, Michael Ruse, and Robert J. Richards, 14–47. Cambridge, UK: Cambridge University Press.

Allee, Warder Clyde. 1941. "Integration of Problems Concerning Protozoan Populations with Those of General Biology." *American Naturalist* 75 (760): 473–487.

Allee, Warder Clyde, Orlando Park, Alfred E. Emerson, Thomas Park, and Karl P. Schmidt. 1949. *Principles of Animal Ecology*. Philadelphia: Saunders.

Allee, Warder Clyde, and Thomas Park. 1939. "Concerning Ecological Principles." *Science* 89 (2304): 166–169.

Allen, Colin, and Marc Bekoff. 1995. "Function, Natural Design, and Animal Behavior: Philosophical and Ethological Considerations." In *Perspectives in Ethology 11: Behavioral Design*, edited by N. S. Thompson, 1–46. New York: Plenum Press.

Allen, Colin, and Jacob Neal. 2020. "Teleological Notions in Biology." In *Stanford Encyclopedia of Philosophy*. Article published March 20, 1996; last modified February 26, 2020. https://plato.stanford.edu/entries/teleology-biology/.

Allen, Garland E. 1967. "J. S. Haldane: The Development of the Idea of Control Mechanisms in Respiration." *Journal of the History of Medicine and Allied Sciences* 22 (4): 392–412.

Allen, Garland E. 2005. "Mechanism, Vitalism and Organicism in Late Nineteenth and Twentieth-Century Biology: The Importance of Historical Context." *Studies in History and Philosophy of Science Part C: Studies in History and Philosophy of Biological and Biomedical Sciences* 36 (2): 261–283.

Allen, Garland E. 2008. "Rebel with Two Causes: Hans Driesch." In *Rebels, Mavericks, and Heretics in Biology*, edited by Oren Harman and Michael Dietrich, 37–64. New Haven, CT: Yale University Press.

Allison, A. C. 1964. "Polymorphism and Natural Selection in Human Populations." *Cold Spring Harbor Symposia on Quantitative Biology* 29: 137–149.

Altamirano, Marco. 2016. *Time, Technology and Environment: An Essay on the Philosophy of Nature*. Edinburgh: Edinburgh University Press.

Amidon, Kevin S. 2008. "Adolf Meyer-Abich, Holism, and the Negotiation of Theoretical Biology." *Biological Theory* 3 (4): 357–370.

Amundson, Ron. 1994. "John T. Gulick and the Active Organism: Adaptation, Isolation, and the Politics of Evolution." In *Darwin's Laboratory: Evolutionary Theory and Natural History in the Pacific*, edited by Roy MacLeod and Philip F. Rehbock, 110–139. Honolulu: University of Hawai'i Press.

Amundson, Ron. 2005. *The Changing Role of the Embryo in Evolutionary Thought: Roots of Evo-Devo*. Cambridge, UK: Cambridge University Press.

Andrewartha, Herbert George. 1957. "The Use of Conceptual Models in Population Ecology." *Cold Spring Harbor Symposia on Quantitative Biology* 22: 219–236.

Andrewartha, Herbert George. 1970. *Introduction to the Study of Animal Populations*, 2nd ed. London: Chapman and Hall.

Andrewartha, Herbert George, and Louis Charles Birch. 1954. *The Distribution and Abundance of Animals*. Chicago: University of Chicago Press.

Andrews, John H. 2017. "The Environment." In *Comparative Ecology of Microorganisms and Macroorganisms*, edited by John H. Andrews, 241–282. New York: Springer.

Andrews, Mel. 2021. "The Math Is Not the Territory: Navigating the Free Energy Principle." *Biology & Philosophy* 36 (3): 30. https://doi.org/10.1007/s10539-021-09807-0.

Angyal, Andras. 1941. *Foundations for a Science of Personality*. New York: The Commonwealth Fund.

"The Animal and Its Environment: A Text-Book of the Natural History of Animals." 1924. *Nature* 114: 152–153.

Antonovics, Janis, Norman C. Ellstrand, and Robert N. Brandon. 1988. "Genetic Variation and Environmental Variation: Expectations and Experiments." In *Plant Evolutionary Biology*, edited by Leslie D. Gottlieb and Subodh K. Jain, 275–303. Dordrecht, The Netherlands: Springer.

Appel, Toby A. 1987. *The Cuvier-Geoffroy Debate: French Biology in the Decades before Darwin*. Oxford: Oxford University Press.

Arabatzis, Theodore. 2017. "What's in It for the Historian of Science? Reflections on the Value of Philosophy of Science for History of Science." *International Studies in the Philosophy of Science* 31 (1): 69–82.

Arthur, Wallace. 2004. *Biased Embryos and Evolution*. Cambridge, UK: Cambridge University Press.

Ash, Mitchell G. 1998. *Gestalt Psychology in German Culture, 1890–1967: Holism and the Quest for Objectivity*. Cambridge, UK: Cambridge University Press.

Atkinson, J. W. 1985. "E. G. Conklin on Evolution: The Popular Writings of an Embryologist." *Journal of the History of Biology* 18 (1): 31–50.

Austin, Christopher J. 2016. "The Ontology of Organisms: Mechanistic Modules or Patterned Processes?" *Biology & Philosophy* 31 (5): 639–662.

Ayala, Francisco J. 1970. "Teleological Explanations in Evolutionary Biology." *Philosophy of Science* 37 (1): 1–15.

Babcock, Gunnar, and Daniel W. McShea. 2021. "An Externalist Teleology." *Synthese* 199 (3): 8755–8780.

Badyaev, Alexander V., and Tobias Uller. 2009. "Parental Effects in Ecology and Evolution: Mechanisms, Processes and Implications." *Philosophical Transactions of the Royal Society B: Biological Sciences* 364 (1520): 1169–1177.

Baedke, Jan. 2013. "The Epigenetic Landscape in the Course of Time: Conrad Hal Waddington's Methodological Impact on the Life Sciences." *Studies in History and Philosophy of Science Part C: Studies in History and Philosophy of Biological and Biomedical Sciences* 44 (4, Part B): 756–773.

Baedke, Jan. 2018. *Above the Gene, Beyond Biology: Toward a Philosophy of Epigenetics*. Pittsburgh: University of Pittsburgh Press.

Baedke, Jan. 2019. "O Organism, Where Art Thou? Old and New Challenges for Organism-Centered Biology." *Journal of the History of Biology* 52 (2): 293–324.

Baedke, Jan. 2021. "What's Wrong with Evolutionary Causation?" *Acta Biotheoretica* 69 (1): 79–89.

Baedke, Jan. 2025. *The Organism*. Cambridge, UK: Cambridge University Press.

Baedke, Jan, Alexander Böhm, and Stefan Reiners-Selbach. 2024. "From Kant to Holism: The Decline of Neo-Kantianism and the Rise of Theoretical Biology." In *New Perspectives on Neo-Kantianism and the Sciences*, edited by Helmut Pulte, Jan Baedke, Daniel Koenig, and Gunther Nickel, 253–277. New York: Routledge.

Baedke, Jan, and Christina Brandt. 2022. "Between the Wars, Facing a Scientific Crisis: The Theoretical and Methodological Bottleneck of Interwar Biology." *Journal of the History of Biology* 55 (2): 209–217.

Baedke, Jan, and Tatjana Buklijas. 2023. "Where Organisms Meet the Environment: Introduction to the Special Issue 'What Counts as Environment in Biology and Medicine: Historical, Philosophical and Sociological Perspectives.'" *Studies in History and Philosophy of Science* 99: A4–A9.

Baedke, Jan, and Alejandro Fábregas-Tejeda. 2023. "The Organism in Evolutionary Explanation: From Early Twentieth Century to the Extended Evolutionary Synthesis." In *Evolutionary Biology: Contemporary and Historical Reflections upon Core Theory*, edited by Thomas E. Dickins and Benjamin J. A. Dickins, 121–150. Cham, Switzerland: Springer International Publishing. https://doi.org/10.1007/978-3-031-22028-9_8.

Baedke, Jan, Alejandro Fábregas-Tejeda, and Abigail Nieves Delgado. 2020. "The Holobiont Concept before Margulis." *Journal of Experimental Zoology Part B: Molecular and Developmental Evolution* 334 (3): 149–155. https://doi.org/10.1002/jez.b.22931.

Baedke, Jan, Alejandro Fábregas-Tejeda, and Guido I. Prieto. 2021. "Unknotting Reciprocal Causation between Organism and Environment." *Biology & Philosophy* 36 (5): 48. https://doi.org/10.1007/s10539-021-09815-0.

Baedke, Jan, Alejandro Fábregas-Tejeda, and Francisco Vergara-Silva. 2020. "Does the Extended Evolutionary Synthesis Entail Extended Explanatory Power?" *Biology & Philosophy* 35 (1): 20. https://doi.org/10.1007/s10539-020-9736-5.

Baedke, Jan, and Scott F. Gilbert. 2021. "Evolution and Development." In *Stanford Encyclopedia of Philosophy*. Article published July 8, 2020. https://plato.stanford.edu/archives/fall2021/entries/evolution-development/.

Baldwin, James Mark. 1897. *Social and Ethical Interpretations in Mental Development: A Study in Social Psychology*. London: Macmillan.

Balgooyen, Thomas G. 1973. "Toward a More Operational Definition of Ecology." *Ecology* 54 (6): 1199–1200.

Ballard, Edward G. 1955. "On the Nature and Use of Dialectic." *Philosophy of Science* 22 (3): 205–213.

Barberousse, Anouk, Michel Morange, and Thomas Pradeu. 2009. "Introduction." In *Mapping the Future of Biology: Evolving Concepts and Theories*, edited by Anouk Barberousse, Michel Morange, and Thomas Pradeu, 1–14. Dordrecht: Springer Netherlands.

Barham, James. 2012. "Normativity, Agency, and Life." *Studies in History and Philosophy of Science Part C: Studies in History and Philosophy of Biological and Biomedical Sciences* 43 (1): 92–103.

Barker, George Frederick. 1880. "Some Modern Aspects of Life Question." *Science* (10): 112–118.

Barker, Gillian, Eric Desjardins, and Trevor Pearce, eds. 2014a. *Entangled Life: Organism and Environment in the Biological and Social Sciences*. Dordrecht, The Netherlands: Springer.

Barker, Gillian, Eric Desjardins, and Trevor Pearce, eds. 2014b. "Introduction: Perspectives on Entangled Life." In *Entangled Life: Organism and Environment in the Biological and Social Sciences*, edited by Gillian Barker, Eric Desjardins, and Trevor Pearce, 1–9. Dordrecht, The Netherlands: Springer.

Bartholomew, George A. 1958. "The Role of Physiology in the Distribution of Terrestrial Vertebrates." In *Zoogeography*, edited by Carl L. Hubbs, 81–95. Washington, DC: American Association for the Advancement of Science.

Baskett, Marissa L. 2012. "Integrating Mechanistic Organism-Environment Interactions into the Basic Theory of Community and Evolutionary Ecology." *Journal of Experimental Biology* 215 (6): 948–961.

Bastard Rico, Juan Alberto. 2021. "El cambio del concepto de *milieu* al de *Umwelt* en el marco de la crítica uexküllliana al mecanicismo en biología." *Revista colombiana de filosofía de la ciencia* 21 (43): 43–68.

Bastian, Henry Charlton. 1870. "Facts and Reasonings Concerning the Heterogenous Evolution of Living Things." *Nature* 2 (35): 170–177.

Bateson, Patrick. 1979. "How Do Sensitive Periods Arise and What Are They For?" *Animal Behaviour* 27: 470–486.

Bateson, Patrick. 2005. "The Return of the Whole Organism." *Journal of Biosciences* 30 (1): 31–39.

Bateson, William. 1894. *Materials for the Study of Variation: Treated with Especial Regard to Discontinuity in the Origin of Species*. New York: Macmillan.

Bauer, Silke, Judy Shamoun-Baranes, Cecilia Nilsson, Andrew Farnsworth, Jeffrey F. Kelly, Don R. Reynolds, Adriaan M. Dokter, et al. 2019. "The Grand Challenges of Migration Ecology That Radar Aeroecology Can Help Answer." *Ecography* 42 (5): 861–875.

Beckner, Morton. 1969. *The Biological Way of Thought*. Berkeley: University of California Press.

Beisbart, Claus, Helmut Pulte, and Thomas Reydon. 2019. "Editorial: Fifty Years Journal for General Philosophy of Science." *Journal for General Philosophy of Science* 50 (1): 1–8.

Beller, Mara. 1992. "The Genesis of Bohr's Complementarity Principle and the Bohr-Heisenberg Dialogue." In *The Scientific Enterprise: The Bar-Hillel Colloquium: Studies in History, Philosophy, and Sociology of Science*, vol. 4, edited by Edna Ullmann-Margalit, 273–293. Dordrecht, The Netherlands: Springer.

Benson, Etienne S. 2020. *Surroundings: A History of Environments and Environmentalisms*. Chicago: University of Chicago Press.

Benson, Keith R. 1989. "Biology's 'Phoenix': Historical Perspectives on the Importance of the Organism." *American Zoologist* 29 (3): 1067–1074.

Berg, Hein van den. 2013. "The Wolffian Roots of Kant's Teleology." *Studies in History and Philosophy of Science Part C: Studies in History and Philosophy of Biological and Biomedical Sciences* 44 (4, Part B): 724–734.

Berg, Hein van den, and Boris Demarest. 2020. "Axiomatic Natural Philosophy and the Emergence of Biology as a Science." *Journal of the History of Biology* 53 (3): 379–422.

Bernal, John Desmond. 1967. *The Origin of Life*. Cleveland: World Publishing.

Bernard, Claude. 1878. *Leçons sur les phénomènes de la vie, communs aux animaux et aux végétaux*. Paris: J.-B. Baillière et Fils.

Bernard, Claude. 1927. *An Introduction to the Study of Experimental Medicine*. New York: Abelard- Schuman.

Bertalanffy, Ludwig von. 1932. *Theoretische Biologie*. Berlin: Gebrüder Borntraeger.

Bertalanffy, Ludwig von. 1933. *Modern Theories of Development: An Introduction to Theoretical Biology*. Oxford: Oxford University Press.

Bertalanffy, Ludwig von. 1951. "Towards a Physical Theory of Organic Teleology: Feedback and Dynamics." *Human Biology* 23 (4): 346–361.

Bertalanffy, Ludwig von. 1953. *Biophysik des Fließgleichgewichts: Einführung in die Physik offener Systeme und ihre Anwendung in der Biologie*. Wiesbaden, Germany: Vieweg+Teubner Verlag.

Beyler, Richard H. 1996. "Targeting the Organism: The Scientific and Cultural Context of Pascual Jordan's Quantum Biology, 1932–1947." *Isis* 87 (2): 248–273.

Bianco, Giuseppe. 2019. "Philosophies of Life." In *The Cambridge History of Modern European Thought*, vol. 2. *The Twentieth Century*, edited by Peter E. Gordon and Warren Breckman, 153–175. Cambridge, UK: Cambridge University Press.

Bich, Leonardo. 2019. "The Problem of Functional Boundaries in Prebiotic and Inter-Biological Systems." In *Systemics of Incompleteness and Quasi-Systems*, edited by Gianfranco Minati, Mario R. Abram, and Eliano Pessa, 295–302. Cham, Switzerland: Springer.

Bich, Leonardo, and Argyris Arnellos. 2012. "Autopoiesis, Autonomy and Organizational Biology: Critical Remarks on Life After Ashby?" *Cybernetics and Human Knowing* 19 (4): 75–103.

Bich, Leonardo, and William Bechtel. 2022. "Control Mechanisms: Explaining the Integration and Versatility of Biological Organisms." *Adaptive Behavior* 30 (5): 389–407.

Bich, Leonardo, Matteo Mossio, Kepa Ruiz-Mirazo, and Alvaro Moreno. 2016. "Biological Regulation: Controlling the System from Within." *Biology & Philosophy* 31 (2): 237–265.

Bierens de Haan, J. A. 1947. "Animal Psychology and the Science of Animal Behaviour." *Behaviour* 1 (1): 71–80.

Billings, William Dwight. 1952. "The Environmental Complex in Relation to Plant Growth and Distribution." *Quarterly Review of Biology* 27 (3): 251–265.

Blanchard, Thomas. 2020. "Explanatory Abstraction and the Goldilocks Problem: Interventionism Gets Things Just Right." *British Journal for the Philosophy of Science* 71 (2): 633–663.

Blandino, Giovanni. 1969. *Theories on the Nature of Life*. New York: Philosophical Library.

Blitz, David. 1992. *Emergent Evolution: Qualitative Novelty and the Levels of Reality*. Dordrecht, The Netherlands: Springer.

Bock, Walter J. 1980. "The Definition and Recognition of Biological Adaptation." *American Zoologist* 20 (1): 217–227.

Bodenheimer, Friedrich Simon. 1928. "Welche Faktorem regulieren die Individuenzahl einer Insektenart in der Natur?" *Biologisches Zentralblatt* 48: 714–739.

Bodenheimer, Friedrich Simon, and M. Schiffer. 1952. "Mathematical Studies in Animal Populations." *Acta Biotheoretica* 10 (1): 23–56.

Boër, Steven E. 1973. "Russell on Classes as Logical Fictions." *Analysis* 33 (6): 206–208.

Bognon-Küss, Cécilia, Bohang Chen, and Charles T. Wolfe. 2018. "Metaphysics, Function and the Engineering of Life: The Problem of Vitalism." *Kairos Journal of Philosophy & Science* 20 (1): 113–140.

Bognon-Küss, Cécilia, and Charles T. Wolfe. 2019. "The Idea of 'Philosophy of Biology before Biology': A Methodological Provocation." In *Philosophy of Biology before Biology*, edited by Cécilia Bognon-Küss and Charles T. Wolfe, 4–23. London: Routledge.

Böhm, Alexander, Stefan Reiners-Selbach, Jan Baedke, Alejandro Fábregas Tejeda, and Daniel J. Nicholson. 2022. "What Was Theoretical Biology? A Topic-Modelling Analysis of a Multilingual Corpus of Monographs and Journals, 1901–1971." Digital Humanities im deutschsprachigen Raum, Postdam. https://doi.org/10.5281/zenodo.6322557.

Böker, Hans. 1935. "Artumwandlung durch Umkonstruktion, Umkonstruktion durch aktives Reagieren der Organismen." *Acta Biotheoretica* 1 (1–2): 17–34.

Böker, Hans. 2021. "Species Transformation through Reconstruction: Reconstruction through Active Reaction of Organisms. Translated by Alexander Böhm and Jan Baedke." *Biological Theory* 16 (2): 114–122.

Bollati, V., and A. Baccarelli. 2010. "Environmental Epigenetics." *Heredity* 105 (1): 105–112.

Botz-Bornstein, Thorsten. 2020. *Micro and Macro Philosophy: Organicism in Biology, Philosophy, and Politics*. New York: Brill.

Bouchard, Frederic, and Philippe Huneman, eds. 2013. *From Groups to Individuals: Evolution and Emerging Individuality*. Cambridge, MA: MIT Press.

Bourne, Gilbert C. 1910. "Problems of Animal Morphology." *Science* 32 (830): 729–742.

Bowler, Peter J. 2014. "Herbert Spencer and Lamarckism." In *Herbert Spencer: Legacies*, edited by Mark Francis and Michael Taylor, 203–221. New York: Routledge.

Bradley, Ben. 2022. "Natural Selection According to Darwin: Cause or Effect?" *History and Philosophy of the Life Sciences* 44 (2): 13. https://doi.org/10.1007/s40656-022-00485-z.

Bradshaw, Anthony David. 1965. "Evolutionary Significance of Phenotypic Plasticity in Plants." In *Advances in Genetics*, edited by E. W. Caspari and J. M. Thoday, 13:115–55. Academic Press.

Brandon, Robert N. 1981. "Biological Teleology: Questions and Explanations." *Studies in History and Philosophy of Science Part A* 12 (2): 91–105.

Brandon, Robert N. 1990. *Adaptation and Environment*. Princeton, NJ: Princeton University Press.

Brandon, Robert N. 1992. "Environment." In *Keywords in Evolutionary Biology*, edited by Evelyn Fox Keller and Elisabeth Anne Lloyd, 81–86. Cambridge, MA: Harvard University Press.

Brandon, Robert N. 2012. "The Concept of the Environment in Evolutionary Theory." In *The Environment: Philosophy, Science, and Ethics*, edited by William P. Kabasenche, Michael O'Rourke, and Matthew H. Slater, 19–35. Cambridge, MA: MIT Press.

Brentari, Carlo. 2015. *Jakob von Uexküll: The Discovery of the Umwelt between Biosemiotics and Theoretical Biology*. Dordrecht, The Netherlands: Springer.

Brierley, William B. 1921. "On a Form of *Botrytis cinerea*, with Colourless Sclerotia." *Philosophical Transactions of the Royal Society of London. Series B* 210 (372–381): 83–114.

Brodie, Edmund D. 2005. "Caution: Niche Construction Ahead." *Evolution* 59 (1): 249–251.

Brooks, Daniel S. 2019. "Conceptual Heterogeneity and the Legacy of Organicism: Thoughts on the Life Organic." *History and Philosophy of the Life Sciences* 41 (2): 24. https://doi.org/10.1007/s40656-019-0263-0.

Brooks, Daniel S. 2021. "Adaptive Design, Contingency, and Ontological Principles for Limited Beings." *Philosophy of Science* 88 (5): 871–881.

Brown, J. 1883. "Leaves and Their Environment." *Nature* 28 (707): 55.

Browning, T. O. 1962. "The Environments of Animals and Plants." *Journal of Theoretical Biology* 2 (1): 63–68.

Bruni, Nadia C., Nancy G. Dengler, and Jane P. Young. 1996. "Leaf Developmental Plasticity of *Ranunculus flabellaris* in Response to Terrestrial and Submerged Environments." *Canadian Journal of Botany* 74 (6): 823–837.

Bunge, Mario. 1992. "System Boundary." *International Journal of General Systems* 20 (3): 215–219.

Burge, Tyler. 2009. "Primitive Agency and Natural Norms." *Philosophy and Phenomenological Research* 79 (2): 251–278.

Burton, Alan C. 1939. "The Properties of the Steady State Compared to Those of Equilibrium as Shown in Characteristic Biological Behavior." *Journal of Cellular and Comparative Physiology* 14 (3): 327–349.

Buskell, Andrew. 2019. "Reciprocal Causation and the Extended Evolutionary Synthesis." *Biological Theory* 14 (4): 267–279.

Butler, Stella V. F. 1988. "Centers and Peripheries: The Development of British Physiology, 1870–1914." *Journal of the History of Biology* 21 (3): 473–500.

Buxton, Patrick Alfred. 1923. *Animal Life in Deserts: A Study of the Fauna in Relation to the Environment*. London: E. Arnold.

Cain, Stanley A. 1944. *Foundation of Plant Geography*. New York: Harper & Brothers.

Calderwood, W. L. 1893. "Applied Natural History." *Nature* 47 (1221): 492–494.

Canguilhem, Georges. (1952) 2008. *Knowledge of Life*. New York: Fordham University Press.

Cannon, Walter B. 1929. "Organization for Physiological Homeostasis." *Physical Review* 9 (3): 399–431.

Caponi, Gustavo. 2022. "The Darwinian Turn in the Understanding of Biological Environment." *Biological Theory* 17 (1): 20–35.

Carlson, Elof Axel. 1971. "An Unacknowledged Founding of Molecular Biology: H. J. Muller's Contributions to Gene Theory, 1910–1936." *Journal of the History of Biology* 4 (1): 149–170.

Caron, Joseph A. 1988. "'Biology' in the Life Sciences: A Historiographical Contribution." *History of Science* 26 (3): 223–268.

Carrillo, Natalia, and Sergio Martínez. 2023. "Scientific Inquiry: From Metaphors to Abstraction." *Perspectives on Science*, 31 (2): 233–261.

Casetta, Elena. 2023. *Filosofia dell'ambiente*. Bologna, Italy: Il mulino.

Caskey, Rena. 1930. "Animal Adaptations to Environmental Influences." *Bios* 1 (1): 52–63.

Ceccarelli, David. 2019. "Between Social and Biological Heredity: Cope and Baldwin on Evolution, Inheritance, and Mind." *Journal of the History of Biology* 52 (1): 161–194.

Chang, Hasok. 2008. "Contingent Transcendental Arguments for Metaphysical Principles." *Royal Institute of Philosophy Supplements* 63: 113–133.

Chang, Hasok. 2012. "Beyond Case-Studies: History as Philosophy." In *Integrating History and Philosophy of Science: Problems and Prospects*, edited by Seymour Mauskopf and Tad Schmaltz, 109–124. Dordrecht, The Netherlands: Springer.

Chang, Hasok. 2021. "Presentist History for Pluralist Science." *Journal for General Philosophy of Science* 52 (1): 97–114.

Chappell, Timothy. 1997. "Introduction. Respecting Nature—Environmental Thinking in the Light of Philosophical Theory." In *The Philosophy of the Environment*, 1–18. Edinburgh: Edinburgh University Press.

Charlesworth, Brian. 1971. "Selection in Density-Regulated Populations." *Ecology* 52 (3): 469–474.

Chemero, Anthony. 2003. "An Outline of a Theory of Affordances." *Ecological Psychology* 15 (2): 181–195.

Chen, Bohang. 2018. "A Non-Metaphysical Evaluation of Vitalism in the Early Twentieth Century." *History and Philosophy of the Life Sciences* 40 (3): 50. https://doi.org/10.1007/s40656-018-0221-2.

Chen, Bohang. 2023. "A Historico-Logical Re-Assessment of Hans Driesch's Vitalism." In *Vitalism and Its Legacy in Twentieth Century Life Sciences and Philosophy*, edited by Christopher Donohue and Charles T. Wolfe, 49–65. Cham, Switzerland: Springer International Publishing.

Chen, Bohang. 2024. *On the Riddle of Life: A Historico-Logical Study of Vitalism*. Cham, Switzerland: Springer International Publishing.

Cheung, Tobias. 2000. *Die Organisation des Lebendigen: die Entstehung des biologischen Organismusbegriffs bei Cuvier, Leibniz und Kant*. Frankfurt, Germany: Campus

Cheung, Tobias. 2006. "From the Organism of a Body to the Body of an Organism: Occurrence and Meaning of the Word 'Organism' from the Seventeenth to the Nineteenth Centuries." *British Journal for the History of Science* 39 (3): 319–339.

Cheung, Tobias. 2010. "What Is an 'Organism'? On the Occurrence of a New Term and Its Conceptual Transformations 1680–1850." *History and Philosophy of the Life Sciences* 32 (2–3): 155–194.

Cheung, Tobias. 2014. *Organismen. Agenten zwischen Innen- und Außenwelten 1780–1860*. Bielefeld, Germany: Transcript Verlag.

Cheung, Tobias. 2021. *Das physiologische Dispositiv der Moderne: Agentenmodelle systemischer Interaktionsformen 1800–1930*. Würzburg, Germany: Königshausen & Neumann.

Chien, Jui-Pi. 2007. "Umwelt, Milieu(x), and Environment: A Survey of Cross-Cultural Concept Mutations." *Semiotica* 2007 (167): 65–89.

Child, Charles Manning. 1902. "Studies on Regulation." *Archiv für Entwicklungsmechanik der Organismen* 15 (2): 187–237.

Child, Charles Manning. 1906a. "Contributions toward a Theory of Regulation." *Archiv für Entwicklungsmechanik der Organismen* 20 (3): 380–426.

Child, Charles Manning. 1906b. "The Relation between Functional Regulation and Form-Regulation." *Journal of Experimental Zoology* 3 (4): 559–582.

Child, Charles Manning. 1908. "The Physiological Basis of Restitution of Lost Parts." *Journal of Experimental Zoology* 5 (4): 485–502.

Child, Charles Manning. 1911a. "The Regulatory Processes in Organisms." *Journal of Morphology* 22 (2): 171–222.

Child, Charles Manning. 1911b. "Studies on the Dynamics of Morphogenesis and Inheritance in Experimental Reproduction. II. Physiological Dominance of Anterior over Posterior Regions in the Regulation of Planaria Dorotocephala." *Journal of Experimental Zoology* 11 (3): 187–220.

Child, Charles Manning. 1913. "Studies on the Dynamics of Morphogenesis and Inheritance in Experimental Reproduction. VI. The Nature of the Axial Gradients in Planaria and Their Relation to Antero-Posterior Dominance, Polarity and Symmetry." *Archiv für Entwicklungsmechanik der Organismen* 37 (1): 108–158.

Child, Charles Manning. 1914. "Susceptibility Gradients in Animals." *Science* 39 (993): 73–76.

Child, Charles Manning. 1915. *Individuality in Organisms.* Chicago: University of Chicago Press.

Child, Charles Manning. 1916. "The Basis of Physiological Individuality in Organisms." *Science* 43 (1111): 511–523.

Child, Charles Manning. 1927. "The Individual and Environment from a Physiological Viewpoint." In *The Child, the Clinic and the Court*, edited by Jane Addams et al., 126–155. New York: New Republic.

Chiu, Lynn. 2019. "Decoupling, Commingling, and the Evolutionary Significance of Experiential Niche Construction." In *Evolutionary Causation: Biological and Philosophical Reflections*, edited by Tobias Uller and Kevin Lala, 299–322. Cambridge, MA: MIT Press.

Chiu, Lynn, and Scott F. Gilbert. 2020. "Niche Construction and the Transition to Herbivory: Phenotype Switching and the Organization of New Nutritional Modes." In *Phenotypic Switching*, edited by Herbert Levine, Mohit Kumar Jolly, Prakash Kulkarni, and Vidyanand Nanjundiah, 459–482. Cambridge, MA: Academic Press.

Christie, Joshua R., Carl Brusse, Pierrick Bourrat, Peter Takacs, and Paul E. Griffiths. 2022. "Are Biological Traits Explained by Their 'Selected Effect' Functions?" *Australasian Philosophical Review* 6 (4): 335–359.

Churchill, Frederick B. 1969. "From Machine-Theory to Entelechy: Two Studies in Developmental Teleology." *Journal of the History of Biology* 2 (1): 165–185.

Clark, Andrew D., Dominik Deffner, Kevin Laland, John Odling-Smee, and John Endler. 2020. "Niche Construction Affects the Variability and Strength of Natural Selection." *American Naturalist* 195 (1): 16–30.

Clarke, Ellen. 2010. "The Problem of Biological Individuality." *Biological Theory* 5 (4): 312–325.

Clements, Frederic E. 1928. *Plant Succession and Indicators.* New York: H. W. Wilson.

Coen, Deborah R. 2006. "Living Precisely in Fin-de-Siècle Vienna." *Journal of the History of Biology* 39 (3): 493–523.

Coleman, William. 1977. *Biology in the Nineteenth Century: Problems of Form, Function and Transformation.* Cambridge, UK: Cambridge University Press.

Coleman, William. 1985. "The Cognitive Basis of the Discipline: Claude Bernard on Physiology." *Isis* 76 (1): 49–70.

Collins, James P. 1986. "Evolutionary Ecology and the Use of Natural Selection in Ecological Theory." *Journal of the History of Biology* 19 (2): 257–288.

Colton, Harold Sellers. 1908. "Some Effects of Environment on the Growth of Lymnæa Columella Say." *Proceedings of the Academy of Natural Sciences of Philadelphia* 60 (3): 410–448.

Conklin, Edwin Grant. 1934. "A Generation's Progress in the Study of Evolution." *Science* 80 (2068): 147–156.

Conklin, Edwin Grant. 1944. "Ends as Well as Means in Life and Evolution." *Transactions of the New York Academy of Sciences* 6 (4 Series II): 125–136.

Conklin, Edwin Grant. 1929. "Problems of Development." *American Naturalist* 63 (684): 5–36.

Conklin, Edwin Grant. 1937. "What Shapes Our Ends?" *American Scholar* 6 (2): 225–235.

Cooper, William S. 1926. "The Fundamentals of Vegetational Change." *Ecology* 7 (4): 391–413.

Corning, Peter A., Stuart A. Kauffman, Denis Noble, James Alan Shapiro, Richard Irwin Vane-Wright, and Addy Pross. 2023. *Evolution "on Purpose": Teleonomy in Living Systems.* Cambridge, MA: MIT Press.

Corris, Amanda. 2020. "Defining the Environment in Organism-Environment Systems." *Frontiers in Psychology* 11: 1285. https://www.frontiersin.org/articles/10.3389/fpsyg.2020.01285.

Cortés-García, David, and Arantza Etxeberria Agiriano. 2023. "Ontologies in Evolutionary Biology: The Role of the Organism in the Two Syntheses." In *Life and Mind: New Directions in the Philosophy of Biology and Cognitive Sciences,* edited by José Manuel Viejo and Mariano Sanjuán, 185–205. Cham, Switzerland: Springer.

Cross, Stephen J., and William R. Albury. 1987. "Walter B. Cannon, L. J. Henderson, and the Organic Analogy." *Osiris* 3: 165–192.

Cumston, Charles Greene. 1926. *An Introduction to the History of Medicine from the Time of the Pharaohs to the of the XVIIth Century.* London: Kegan Paul, Trench, Trubner & Co.

Dahn, Ryan. 2019. "Big Science, Nazified? Pascual Jordan, Adolf Meyer-Abich, and the Abortive Scientific Journal Physis." *Isis* 110 (1): 68–90.

Danchakoff, Vera. 1917. "Differentiation by Segregation and Environment in the Developing Organism." *American Naturalist* 51 (607): 419–428.

Darwin, Charles. 1859. *On the Origin of Species by Means of Natural Selection, or Preservation of Favoured Races in the Struggle for Life*. London: John Murray.

Darwin, Charles. 1875. *The Variation of Animals and Plants under Domestication*, 2nd ed., vol. 2. London: John Murray.

Davidson, Robert A., and Rosalie A. Dunn. 1967. "A Correlation Approach to Certain Problems of Population-Environment Relations." *American Journal of Botany* 54 (5): 529–538.

Dawkins, Richard. 1986. *The Blind Watchmaker*. New York: W. W. Norton.

De Bont, Raf. 2010. "Organisms in Their Milieu: Alfred Giard, His Pupils, and Early Ethology, 1870–1930." *Isis* 101 (1): 1–29.

Delisle, Richard G. 2008. "Expanding the Framework of the Holism/Reductionism Debate in Neo-Darwinism: The Case of Theodosius Dobzhansky and Bernhard Rensch." *History and Philosophy of the Life Sciences* 30 (2): 207–226.

Delisle, Richard G. 2011. "What Was Really Synthesized during the Evolutionary Synthesis? A Historiographic Proposal." *Studies in History and Philosophy of Science Part C: Studies in History and Philosophy of Biological and Biomedical Sciences* 42 (1): 50–59.

Dennett, Daniel Clement. 1987. *The Intentional Stance*. Cambridge, MA: MIT Press.

Depew, David J. 2017. "Natural Selection, Adaptation, and the Recovery of Development." In *Challenging the Modern Synthesis: Adaptation, Development, and Inheritance*, edited by Philippe Huneman and Denis Walsh, 37–67. Oxford: Oxford University Press.

Desmond, Hugh, and Philippe Huneman. 2020. "The Ontology of Organismic Agency: A Kantian Approach." In *Natural Born Monads: On the Metaphysics of Organisms and Human Individuals*, edited by Andrea Altobrando and Pierfrancesco Biasetti, 33–64. Berlin: De Gruyter.

Desmond, Hugh, and Philippe Huneman. 2022. "The Integrated Information Theory of Agency." *Behavioral and Brain Sciences* 45: e45. https://doi.org/10.1017/S0140525X21002004.

Dewey, John, and Arthur F. Bentley. 1946a. "Interaction and Transaction." *Journal of Philosophy* 43 (19): 505–517.

Dewey, John, and Arthur F. Bentley. 1946b. "Transactions as Known and Named." *Journal of Philosophy* 43 (20): 533–551.

Dewey, John, and Arthur Fisher Bentley. 1949. *Knowing and the Known*. Boston: Beacon Press.

Dickins, Tom E., and R. A. Barton. 2013. "Reciprocal Causation and the Proximate-Ultimate Distinction." *Biology & Philosophy* 28 (5): 747–756.

Di Paolo, Ezequiel A. 2005. "Autopoiesis, Adaptivity, Teleology, Agency." *Phenomenology and the Cognitive Sciences* 4 (4): 429–452.

Di Paolo, Ezequiel A. 2020. "Picturing Organisms and Their Environments: Interaction, Transaction, and Constitution Loops." *Frontiers in Psychology* 11: 1912. https://www.frontiersin.org/articles/10.3389/fpsyg.2020.01912.

Dobzhansky, Theodosius. 1937. *Genetics and the Origin of Species*. New York: Columbia University Press.

Dobzhansky, Theodosius. 1949. "Marxist Biology, French Style." *Journal of Heredity* 40 (3): 78–79.

Dobzhansky, Theodosius. 1950a. "Evolution in the Tropics." *American Scientist* 38 (2): 208–221.

Dobzhansky, Theodosius. 1950b. "Heredity, Environment, and Evolution." *Science* 111 (2877): 161–166.

Dobzhansky, Theodosius. 1950c. "Mendelian Populations and Their Evolution." *American Naturalist* 84 (819): 401–418.

Dobzhansky, Theodosius. 1951. *Genetics and the Origin of Species*, 3rd ed., revised. New York: Columbia University Press.

Dobzhansky, Theodosius. 1955. "A Review of Some Fundamental Concepts and Problems of Population Genetics." *Cold Spring Harbor Symposia on Quantitative Biology* 20: 1–15.

Dobzhansky, Theodosius. 1957a. "Mendelian Populations as Genetic Systems." In *Population Studies: Animal Ecology and Demography. Cold Spring Harbor Symposia on Quantitative Biology*, vol. XXII, 285–393. Cold Spring Harbor, NY: The Biological Laboratory.

Dobzhansky, Theodosius. 1957b. "What Is 'Environment'?: Reply to Mr. Morrill." *American Naturalist* 91 (859): 269–271.

Dobzhansky, Theodosius, Francisco J. Ayala, George Ledyard Stebbins, and James W. Valentine. 1977. *Evolution*. San Francisco: W. H. Freeman.

Dobzhansky, Theodosius, and B. Spassky. 1963. "Genetics of Natural Populations. XXXIV. Adaptive Norm, Genetic Load and Genetic Elite in *Drosophila pseudoobscura*." *Genetics* 48 (11): 1467–1485.

Dobzhansky, Theodosius, and Bruce Wallace. 1953. "The Genetics of Homeostasis in Drosophila." *Proceedings of the National Academy of Sciences* 39 (3): 162–171.

Donohue, Christopher, and Charles T. Wolfe, eds. 2023. *Vitalism and Its Legacy in Twentieth Century Life Sciences and Philosophy*. Cham, Switzerland: Springer International Publishing.

Douglas, Claude Gordon. 1936. "John Scott Haldane, 1860–1936." *Obituary Notices of Fellows of the Royal Society* 2 (5): 114–139.

Drack, Manfred, Wilfried Apfalter, and David Pouvreau. 2007. "On the Making of a System Theory of Life: Paul A Weiss and Ludwig von Bertalanffy's Conceptual Connection." *Quarterly Review of Biology* 82 (4): 349–373.

Driesch, Hans. 1892. "Entwicklungsmechanische Studien. I. Der Werth der Beiden Ersten Furchungszellen in der Echinodermenentwicklung. Experimentelle Erzeugung von Theft- und Doppelbildungen." *Zeitschrift für wissenschaftliche Zoologie* 53: 160–184.

Driesch, Hans. 1900. "Studien über das Regulationsvermögen der Organismen." *Archiv für Entwicklungsmechanik der Organismen* 10 (2): 411–434.

Driesch, Hans. 1908. *The Science and Philosophy of the Organism*. Aberdeen: Aberdeen University.

Driesch, Hans. 1914. *The History and Theory of Vitalism*. London: Macmillan.

Drude, Oscar. 1906. "The Position of Ecology in Modern Science." In *International Congress on the Arts and Sciences*, edited by Howard J. Rogers, 177–190. New York: The University Alliance.

Dumais, Jacques, and Yoël Forterre. 2012. "'Vegetable Dynamicks': The Role of Water in Plant Movements." *Annual Review of Fluid Mechanics* 44 (1): 453–478.

Dupré, John. 2021. *The Metaphysics of Biology*. Cambridge, UK: Cambridge University Press.

Dürken, Bernhard. (1932) 2016. *Experimental Analysis of Development*. Abingdon: Routledge.

Edsall, John T. 1972. "Blood and Hemoglobin: The Evolution of Knowledge of Functional Adaptation in a Biochemical System." *Journal of the History of Biology* 5 (2): 205–257.

Ehrenreich, Ian M., and David W. Pfennig. 2016. "Genetic Assimilation: A Review of Its Potential Proximate Causes and Evolutionary Consequences." *Annals of Botany* 117 (5): 769–779.

Ehrle, Elwood B. 1965. "The Historico-Philosophical Background of Evolutionary Thought." *Evolution* 19 (4): 575–577.

Elwick, James. 2003. "Herbert Spencer and the Disunity of the Social Organism." *History of Science* 41 (1): 35–72.

Endler, John A. 1989. "Organisms and Environment." *BioScience* 39 (5): 330–331.

England, Sam J., Katie Lihou, and Daniel Robert. 2023. "Static Electricity Passively Attracts Ticks onto Hosts." *Current Biology* 33 (14): 3041–3047.E4. https://doi.org/10.1016/j.cub.2023.06.021.

Entralgo, Pedro Laín. 1948. "Sensualism and Vitalism in Bichat's 'Anatomie Générale.'" *Journal of the History of Medicine and Allied Sciences* 3 (1): 47–64.

"Entwicklungsbiologie und Ganzheit." 1938. *Nature* 141 (3569): 535–535.

Esposito, Maurizio. 2013. "Heredity, Development and Evolution: The Unmodern Synthesis of E. S. Russell." *Theory in Biosciences* 132 (3): 165–180.

Esposito, Maurizio. 2014. "Problematic 'Idiosyncrasies': Rediscovering the Historical Context of D'Arcy Wentworth Thompson's Science of Form." *Science in Context* 27 (1): 79–107.

Esposito, Maurizio. 2016. *Romantic Biology, 1890–1945*. New York: Routledge.

Esposito, Maurizio. 2017. "The Organismal Synthesis: Holistic Science and Developmental Evolution in the English-Speaking World, 1915–1954." In *The Darwinian Tradition in Context: Research Programs in Evolutionary Biology*, edited by Richard G. Delisle, 219–241. Cham, Switzerland: Springer International Publishing.

Esposito, Maurizio. 2018. "Sobre el uso y significado de la historia en filosofía de la ciencia." *Revista Colombiana de Filosofía de la Ciencia* 18 (37): 91–117.

Esposito, Maurizio. 2024. "'Agency' in the History of the Life Sciences: A Historico-Philosophical Account." In *The Riddle of Organismal Agency: New Historical and Philosophical Reflections*, edited by Alejandro Fábregas-Tejeda, Gregory Radick, Guido I. Prieto, and Jan Baedke, 23–40. Abingdon: Routledge.

Etxeberria, Arantza. Forthcoming. "Environment(s), Autonomy and (A)Symmetries." In *Outonomy: Fleshing Out Autonomy Beyond the Individual*, edited by Xabier E. Barandiaran and Arantza Etxeberria. Springer.

Etxeberria, Arantza, and Jon Umerez. 2006. "Organismo y organización en la biología teórica: ¿vuelta al organicismo?" *Ludus Vitalis* 14 (26): 3–38.

Ezcurdia, Maite. 2014. *Los indéxicos y la semántica de Kaplan*. Mexico City: Instituto de Investigaciones Filosóficas—Universidad Nacional Autónoma de México.

Fábregas-Tejeda, Alejandro. 2024a. "Research Environments *Vis-à-Vis* Biological Environments: Ontological Parallels, Epistemic Parallax, and Metaphilosophical

Parallelization." *European Journal for Philosophy of Science* 14 (3): 46. https://doi.org/10.1007/s13194-024-00603-7.

Fábregas-Tejeda, Alejandro. 2024b. "Charting Contrasting Stances on Organismal Purposiveness and Agency in Early Twentieth-Century Biology." In *The Riddle of Organismal Agency: New Historical and Philosophical Perspectives*, edited by Alejandro Fábregas-Tejeda, Jan Baedke, Guido I. Prieto, and Gregory Radick, 41–60. Oxon: Routledge.

Fábregas-Tejeda, Alejandro. Under review. "Re-Negotiating Organism-Environment Separation."

Fábregas-Tejeda, Alejandro, and Jan Baedke. 2023a. "Organisms and the Causes and Consequences of Selection: A Reply to Vidya et al." In *Evolutionary Biology: Contemporary and Historical Reflections upon Core Theory*, edited by Thomas E. Dickins and Benjamin J. A. Dickins, 159–169. Cham, Switzerland: Springer International Publishing. https://doi.org/10.1007/978-3-031-22028-9_10.

Fábregas-Tejeda, Alejandro, and Jan Baedke. 2023b. "Teleology, Organisms, and Genes: A Commentary on Haig." In *Evolutionary Biology: Contemporary and Historical Reflections upon Core Theory*, edited by Thomas E. Dickins and Benjamin J. A. Dickins, 249–264. Cham, Switzerland: Springer International Publishing. https://doi.org/10.1007/978-3-031-22028-9_15.

Fábregas-Tejeda, Alejandro, Jan Baedke, Guido I. Prieto, and Gregory Radick. 2024. "Organismal Agency: A Persistent Riddle in the History and Philosophy of Biology." In *The Riddle of Organismal Agency: New Historical and Philosophical Reflections*, edited by Alejandro Fábregas-Tejeda, Jan Baedke, Guido I. Prieto, and Gregory Radick, 1–20. Abingdon: Routledge.

Fábregas-Tejeda, Alejandro, Abigail Nieves Delgado, and Jan Baedke. 2021. "Revisiting Hans Böker's 'Species Transformation Through Reconstruction: Reconstruction Through Active Reaction of Organisms' (1935)." *Biological Theory* 16 (2): 63–75. https://doi.org/10.1007/s13752-020-00370-7.

Fábregas-Tejeda, Alejandro, and Mariano Martín-Villuendas. 2023. "What Is the Philosophy of Organismal Biology?" *ArtefaCToS. Revista de estudios sobre la ciencia y la tecnología* 12 (1): 5–25. https://doi.org/10.14201/art2023121525.

Fábregas-Tejeda, Alejandro, and Grant Ramsey. 2024. "Driftability and Niche Construction." *Synthese* 204: 162. https://doi.org/10.1007/s11229-024-04815-5

Fábregas-Tejeda, Alejandro, and Francisco Vergara-Silva. 2018a. "Hierarchy Theory of Evolution and the Extended Evolutionary Synthesis: Some Epistemic Bridges, Some Conceptual Rifts." *Evolutionary Biology* 45 (2): 127–139. https://doi.org/10.1007/s11692-017-9438-3.

Fábregas-Tejeda, Alejandro, and Francisco Vergara-Silva. 2018b. "The Emerging Structure of the Extended Evolutionary Synthesis: Where Does Evo-Devo Fit In?" *Theory in Biosciences* 137 (2): 169–184. https://doi.org/10.1007/s12064-018-0269-2.

Fábregas-Tejeda, Alejandro, and Francisco Vergara-Silva. 2022. "'Man-Made Futures': Conrad Hal Waddington, Biological Theory, and the Anthropocene." *Azimuth: Philosophical Coordinates in Modern and Contemporary Age* 19 (1): 35–56. https://doi.org/10.1400/289171.

Fabris, Flavia. 2019. "Conrad Hal Waddington (1905–1975)." In *Evolutionary Developmental Biology: A Reference Guide*, edited by Laura Nuño de la Rosa and Gerd B. Müller, 299–313. Cham, Switzerland: Springer

Falconer, Douglas Scott. 1952. "The Problem of Environment and Selection." *American Naturalist* 86 (830): 293–298.

Falconer, Douglas Scott. 1960. *Introduction to Quantitative Genetics*. Edinburgh: Oliver and Boyd.

Farmer, J. Bretland. 1903. "On Stimulus and Mechanism as Factors in Organisation." *New Phytologist* 2 (10): 217–225.

Feldman, Fred. 1973. "Sortal Predicates." *Noûs* 7 (3): 268–282.

Ferry, Ronald M. 1942. "Lawrence Joseph Henderson." *Science* 95 (2465): 316–318.

Fisher, Ronald A. 1930. *The Genetical Theory of Natural Selection*. Oxford: Clarendon Press.

"Fisheries Scientist." 1954. *The Glasgow Herald*. August 27, 1954.

Foucault, Michel. 1970. *The Order of Things: An Archaeology of the Human Sciences*. New York: Pantheon Books.

Fox, Howard. 1883. "Influence of 'Environment' upon Plants." *Nature* 27 (692): 315.

Fox Keller, Evelyn. 1990. "Physics and the Emergence of Molecular Biology: A History of Cognitive and Political Synergy." *Journal of the History of Biology* 23 (3): 389–409.

Fox Keller, Evelyn. 2001. "Beyond the Gene but Beneath the Skin." In *Cycles of Contingency: Developmental Systems and Evolution*, edited by Susan Oyama, Paul Griffiths, and Russell D. Gray, 299–312. Cambridge: MIT Press.

Francescotti, Robert. 2012. "Understanding the Intrinsic/Extrinsic Distinction." *Metascience* 21 (1): 91–94.

Francis, Mark. 2014. *Herbert Spencer and the Invention of Modern Life*. New York: Routledge.

Frankfurt, Harry G. 1978. "The Problem of Action." *American Philosophical Quarterly* 15 (2): 157–162.

Frazer, Persifor. 1890. "The Persistence of Plant and Animal Life Under Changing Conditions of Environment." *American Naturalist* 24 (282): 517–529.

Friebe, Cord. 2008. "Kant's Ontology of Organisms." In *Purposiveness: Teleology Between Nature and Mind*, edited by Luca Illetterati and Francesca Michelini, 59–74. Frankfurt: Ontos.

Friedman, Michael, and Alfred Nordmann, eds. 2006. *The Kantian Legacy in Nineteenth-Century Science*. Cambridge, MA: MIT Press.

Fromhage, Lutz, and Alasdair I. Houston. 2022. "Biological Adaptation in Light of the Lewontin-Williams (a)Symmetry." *Evolution* 76 (7): 1619–1624.

Fry, Iris. 1996. "On the Biological Significance of the Properties of Matter: L. J. Henderson's Theory of the Fitness of the Environment." *Journal of the History of Biology* 29 (2): 155–196.

Fulda, Fermín C. 2017. "Natural Agency: The Case of Bacterial Cognition." *Journal of the American Philosophical Association* 3 (1): 69–90.

Fulda, Fermín C. 2023. "Agential Autonomy and Biological Individuality." *Evolution & Development* 25 (6): 353–370.

Fusco, Giuseppe, and Alessandro Minelli. 2010. "Phenotypic Plasticity in Development and Evolution: Facts and Concepts." *Philosophical Transactions of the Royal Society B: Biological Sciences* 365 (1540): 547–556.

Fusco, Giuseppe, and Alessandro Minelli. 2019. *The Biology of Reproduction*. Cambridge, UK: Cambridge University Press.

Gallegos Ordorica, Sergio Armando. 2016. "The Explanatory Role of Abstraction Processes in Models: The Case of Aggregations." *Studies in History and Philosophy of Science Part A* 56: 161–167.

Gambarotto, Andrea. 2017. *Vital Forces, Teleology and Organization: Philosophy of Nature and the Rise of Biology in Germany*. Cham, Switzerland: Springer.

Gambarotto, Andrea. 2019. "Teleology: A Case Study in iHPS." In *The Past, Present, and Future of Integrated History and Philosophy of Science*, edited by Emily Herring, Kevin Matthew Jones, Konstantin S. Kiprijanov, and Laura M. Sellers, 150–166. London: Routledge.

Gambarotto, Andrea, and Auguste Nahas. 2022. "Teleology and the Organism: Kant's Controversial Legacy for Contemporary Biology." *Studies in History and Philosophy of Science* 93: 47–56.

Gambarotto, Andrea, and Auguste Nahas. 2023. "Nature and Agency: Towards a Post-Kantian Naturalism." *Topoi* 42: 767–780. https://doi.org/10.1007/s11245-023-09882-w.

García, Vivette. 2005. "Explicar sin techo, sin piso: El desarrollo sin muros." *Revista colombiana de filosofía de la ciencia* VI (12–13): 9–31.

García-Roger, Eduardo M., Esther Lubzens, Diego Fontaneto, and Manuel Serra. 2019. "Facing Adversity: Dormant Embryos in Rotifers." *Biological Bulletin* 237 (2): 119–144.

Gause, Georgii Frantsevich. 1934. *The Struggle for Existence*. Baltimore: Williams and Wilkins.

Gause, Georgii Frantsevich. 1937. "Experimental Populations of Microscopic Organisms." *Ecology* 18 (2): 173–179.

Gawne, Richard, Kenneth Z. McKenna, and H. Frederik Nijhout. 2018. "Unmodern Synthesis: Developmental Hierarchies and the Origin of Phenotypes." *BioEssays* 40 (1): 1600265. https://doi.org/10.1002/bies.201600265.

Ge, Yue, Da-Zhi Wang, Jen-Fu Chiu, Susana Cristobal, David Sheehan, Frédéric Silvestre, Xianxuan Peng, et al. 2013. "Environmental OMICS: Current Status and Future Directions." *Journal of Integrated OMICS* 3 (2): 75–87.

Geddes, Patrick, and Arthur Thomson. 1911. *Evolution*. New York: Henry Holt.

Gelderloos, Carl. 2019. *Biological Modernism: The New Human in Weimar Culture*. Evanston, IL: Northwestern University Press.

Gerard, Jean-François, and Marie-Line Maublanc. 2023. "Nothing in the Environment Makes Sense Except in the Light of a Living System: Organisms, Their Relationships to the Environment, and Evolution." *Evolutionary Biology* 50 (1): 18–29.

Gerard, Ralph W. 1952. "Ralph Stayner Lillie: 1875–1952." *Science* 116 (3019): 496–497.

Gerber, Lucie. 2020. "The Art of Growing Old: Environmental Manipulation, Physiological Rhythms, and the Advent of Microcebus Murinus as a Primate Model of Aging." *History and Philosophy of the Life Sciences* 42 (2): 26. https://doi.org/10.1007/s40656-020-00321-2.

Geroulanos, Stefanos, and Todd Meyers. 2018. "The Organism and Its Environment: Integration, Interiority, and Individuality around 1930." In *The Human Body in the Age of Catastrophe: Brittleness, Integration, Science, and the Great War*, 161–206. Chicago: University of Chicago Press.

Ghiselin, Michael T. 1994. "Darwin's Language May Seem Teleological, but His Thinking Is Another Matter." *Biology and Philosophy* 9 (4): 489–492.

Ghiselin, Michael T. 1997. *Metaphysics and the Origin of Species*. New York: SUNY Press.

Gibson, Abraham, and Cindy Ermus. 2019. "The History of Science and the Science of History: Computational Methods, Algorithms, and the Future of the Field." *Isis* 110 (3): 555–566.

Gibson, James J. 1979. *The Ecological Approach to Visual Perception*. Boston: Houghton Mifflin.

Giglioni, Guido. 2013. "Jean-Baptiste Lamarck and the Place of Irritability in the History of Life and Death." In *Vitalism and the Scientific Image in Post-Enlightenment Life Science, 1800–2010*, edited by Sebastian Normandin and Charles T. Wolfe, 19–49. Dordrecht: Springer Netherlands.

Gilbert, Scott F. 2016. "Developmental Plasticity and Developmental Symbiosis: The Return of Eco-Devo." *Current Topics in Developmental Biology* 116: 415–433.

Gilbert, Scott F., Jan Sapp, and Alfred I. Tauber. 2012. "A Symbiotic View of Life: We Have Never Been Individuals." *Quarterly Review of Biology* 87 (4): 325–341.

Gillette, Roy. 1946. "The Directiveness of Organic Activities. E. S. Russell." *Quarterly Review of Biology* 21 (2): 181.

Gissis, Snait B. 2024. *Lamarckism and the Emergence of "Scientific" Social Sciences in Nineteenth-Century Britain and France*. Cham, Switzerland: Springer Nature.

Glymour, Bruce. 2011. "Modeling Environments: Interactive Causation and Adaptations to Environmental Conditions." *Philosophy of Science* 78 (3): 448–471.

Godfrey-Smith, Peter. 1996a. *Complexity and the Function of Mind in Nature*. Cambridge, UK: Cambridge University Press.

Godfrey-Smith, Peter. 1996b. "Précis of Complexity and the Function of Mind in Nature." *Adaptive Behavior* 4 (3–4): 453–465.

Godfrey-Smith, Peter. 2009a. "Abstractions, Idealizations, and Evolutionary Biology." In *Mapping the Future of Biology: Evolving Concepts and Theories*, edited by Anouk Barberousse, Michel Morange, and Thomas Pradeu, 47–56. Dordrecht: Springer Netherlands

Godfrey-Smith, Peter. 2009b. *Darwinian Populations and Natural Selection*. Oxford: Oxford University Press.

Goff, Cory B., Susan C. Walls, David Rodriguez, and Caitlin R. Gabor. 2020. "Changes in Physiology and Microbial Diversity in Larval Ornate Chorus Frogs Are Associated with Habitat Quality." *Conservation Physiology* 8 (1): coaa047.

Golding, Dominic, Katie L. Rupp, Anne Sustar, Brandon Pratt, and John C. Tuthill. 2023. "Snow Flies Self-Amputate Freezing Limbs to Sustain Behavior at Sub-Zero Temperatures." *Current Biology* 33 (21): 4549–4556.e3. https://doi.org/10.1016/j.cub.2023.09.002.

Goldstein, Kurt. (1934) 1995. *The Organism: A Holistic Approach to Biology Derived from Pathological Data in Man*. New York: Zone Books.

Goodman, Martin. 2008. *Suffer and Survive: The Extreme Life of J. S. Haldane*. Simon & Schuster UK.

Goodrich, Edwin Stephen. 1924. *Living Organisms: An Account of Their Origin & Evolution*. Oxford: Clarendon Press.

Goudge, Thomas Anderson. 1955. "What Is a Population?" *Philosophy of Science* 22 (4): 272–279.

Goudge, Thomas Anderson. 1961. *The Ascent of Life: A Philosophical Study of the Theory of Evolution*. London: George Allen and Unwin.

Graham, Michael. 1954. "E. S. Russell 1887–1954." *ICES Journal of Marine Science* 20 (2): 135–139.

Greenwood, Thomas. 1931. "The International Congress of the History of Science and Technology." *Nature* 128 (3219): 77–79.

Grene, Marjorie. 1961. "Statistics and Selection." *British Journal for the Philosophy of Science* 12 (45): 25–42.

Grene, Marjorie. 1968. *Approaches to a Philosophical Biology*. New York: Basic Books.

Grene, Marjorie. 1972. "Aristotle and Modern Biology." *Journal of the History of Ideas* 33 (3): 395–424.

Grene, Marjorie. 2001. "Darwin, Cuvier and Geoffroy: Comments and Questions." *History and Philosophy of the Life Sciences* 23 (2): 187–211.

Griffiths, Paul E., and Russell D. Gray. 1994. "Developmental Systems and Evolutionary Explanation." *Journal of Philosophy* 91 (6): 277–304.

Griffiths, Paul, and Russell D. Gray. 2001. "Darwinism and Developmental Systems." In *Cycles of Contingency: Developmental Systems and Evolution*, edited by Susan Oyama, Paul Griffiths, and Russell D Gray, 195–218. Cambridge, MA: MIT Press.

Grodwohl, Jean-Baptiste, Franco Porto, and Charbel N. El-Hani. 2018. "The Instability of Field Experiments: Building an Experimental Research Tradition on the Rocky Seashores (1950–1985)." *History and Philosophy of the Life Sciences* 40 (3): 45. https://doi.org/10.1007/s40656-018-0209-y.

Gruber, Jacob W. 1954. "Hexicology: A Note for the History of Ecology." *Ecology* 35 (3): 415–417.

Guay, Alexandre, and Thomas Pradeu. 2020. "Right Out of the Box: How to Situate Metaphysics of Science in Relation to Other Metaphysical Approaches." *Synthese* 197 (5): 1847–1866.

Gulick, John T. 1873. "On Diversity of Evolution Under One Set of External Conditions." *Zoological Journal of the Linnean Society* 11 (56): 496–505.

Gulick, John T. 1882. "On the Variation of Species as Related to Their Geographical Distribution, Illustrated by the Achatinellinæ." *Nature* 6 (142): 222–224.

Gulick, John T. 1890. "Divergent Evolution and the Darwinian Theory." *Annals and Magazine of Natural History* 5 (26): 156–166.

Gulick, John T. 1905. *Evolution, Racial and Habitudinal*. Washington, DC: Carnegie Institution of Washington.

Gulick, John Thomas. 1888. "Divergent Evolution Through Cumulative Segregation." *Journal of the Linnean Society of London, Zoology* 20 (120): 189–274.

Gunderman, Ricard B. 2005. "LILLIE, Ralph Stayner (1875–1952)." In *Dictionary of Modern American Philosophers*, edited by John R. Shook, 1473–1474. London: Thoemmes-Continuum.

Haber, Matt H., and Jay Odenbaugh. 2009. "The Edges and Boundaries of Biological Objects." *Biological Theory* 4 (3): 219–224.

Haeckel, Ernst. 1866a. *Generelle Morphologie der Organismen: allgemeine Grundzüge der organischen Formen-Wissenschaft, mechanisch begründet durch die von Charles Darwin reformirte Descendenz-Theorie. I. Allgemeine Anatomie der Organismen*. Berlin: Verlag von Georg Reimer.

Haeckel, Ernst. 1866b. *Generelle Morphologie der Organismen: allgemeine Grundzüge der organischen Formen-Wissenschaft, mechanisch begründet durch die von Charles Darwin reformirte Descendenz-Theorie. II. Allgemeine Entwickelungsgeschichte der Organismen*. Berlin: Verlag von Georg Reimer.

Haeckel, Ernst. 1868. *Natürliche Schöpfungsgeschichte*. Berlin: Verlag von Georg Reimer.

Haeckel, Ernst. 1870. "Über Entwicklungsgang und Aufgabe der Zoologie." *Jenaische Zeitschrift für Medizin und Naturwissenschaft* V (3): 353–370.

Hagen, Joel B. 2021. *Life Out of Balance: Homeostasis and Adaptation in a Darwinian World*. Tuscaloosa: University of Alabama Press.

Haldane, J. S. 1935. "The Physiology of Descartes and Its Modern Developments." *Acta Biotheoretica* 1 (1): 5–16. https://doi.org/10.1007/BF02324292.

Haldane, John Burdon Sanderson. 1936. "Some Principles of Causal Analysis in Genetics." *Erkenntnis* 6: 346–357.

Haldane, John Burdon Sanderson. 1946. "The Interaction of Nature and Nurture." *Annals of Eugenics* 13 (1): 197–205.

Haldane, John Burdon Sanderson. 1956. "The Relation Between Density Regulation and Natural Selection." *Proceedings of the Royal Society of London. Series B Biological Sciences* 145 (920): 306–308.

Haldane, John Scott. 1884. "Life and Mechanism." *Mind* os-IX (33): 27–47.

Haldane, John Scott. 1916. "The New Physiology." *Science* 44 (1140): 619–631.

Haldane, John Scott. 1917. *Organism and Environment as Illustrated by the Physiology of Breathing*. New Haven, CT: Yale University Press.

Haldane, John Scott. 1919. *The New Physiology and Other Addresses*. London: Charles Griffin.

Haldane, John Scott. 1921. *Mechanism, Life and Personality: An Examination of the Mechanistic Theory of Life and Mind*. London: John Murray.

Haldane, John Scott. 1927. "An Address on Biology and Medicine." *British Medical Journal* 1 (3463): 909–912.

Haldane, John Scott. 1929. *The Sciences and Philosophy*. London: Hodder and Stroughton.

Haldane, John Scott. 1931. *The Philosophical Basis of Biology*. London: Doubleday, Doran.

Haldane, John Scott. 1936. *The Philosophy of a Biologist*. Oxford: Clarendon Press.

Haldane, John Scott, Edward Stuart Russell, and Leslie Mackenzie. 1923. "IV—Symposium: The Relations Between Biology and Psychology." *Aristotelian Society Supplementary Volume* 3 (1): 56–94.

Haldane, John Scott, D'Arcy W. Thompson, P. Chalmers Mitchell, and L. T. Hobhouse. 1917. "Symposium: Are Physical, Biological and Psychological Categories Irreducible?" *Proceedings of the Aristotelian Society* 18: 419–478.

Haldane, Richard Burdon, and John Scott Haldane. 1883. "The Relation of Philosophy to Science." In *Essays in Philosophical Criticism*, edited by Andrew Seth and Richard Burdon Haldane, 41–66. London: Longmans, Green.

Hamburger, Viktor. 1997. "Wilhelm Roux: Visionary with a Blind Spot." *Journal of the History of Biology* 30 (2): 229–238.

Haraway, Donna Jeanne. 1976. *Crystals, Fabrics, and Fields: Metaphors of Organicism in Twentieth-Century Developmental Biology*. New Haven, CT: Yale University Press.

Harrington, Anne. 1996. *Reenchanted Science: Holism in German Culture from Wilhelm II to Hitler*. Princeton, NJ: Princeton University Press.

Harrison, David. 2023. "Mind the Matter: Active Matter, Basal Cognition, and the Making of Bio-Inspired Artificial Intelligence." PhD thesis, University of Cambridge.

Hart, J. W. 1990. *Plant Tropisms: And Other Growth Movements*. New York: Springer Science & Business Media.

Hartmann, Nicolai. 1950. *Philosophie der Natur: Abriss der speziellen Kategorienlehre.* Berlin: W. de Gruyter.

Harwood, Jonathan. 1996. "Weimar Culture and Biological Theory: A Study of Richard Woltereck (1877–1944)." *History of Science* 34 (3): 347–377.

Hawley, Amos H. 1973. "Ecology and Population." *Science* 179 (4079): 1196–1201.

Hazelwood, Caleb. 2023. "Reciprocal Causation and Biological Practice." *Biology & Philosophy* 38 (1): 5. https://doi.org/10.1007/s10539-023-09895-0.

Heikertinger, Franz. 1917. "Das Scheinproblem von der Zweckmäßigkeit im Organischen." *Biologisches Zentralblatt* 37:333–352.

Hein, Hilde. 1969. "Molecular Biology vs. Organicism: The Enduring Dispute Between Mechanism and Vitalism." *Synthese* 20 (2): 238–253.

Henderson, Lawrence J. 1908. "The Theory of Neutrality Regulation in the Animal Organism." *American Journal of Physiology* 21 (4): 427–448.

Henderson, Lawrence J. 1911. "A Critical Study of the Process of Acid Excretion." *Journal of Biological Chemistry* 9 (5): 403–424.

Henderson, Lawrence J. 1914. "The Functions of an Environment." *Science* 39 (1006): 524–527.

Henderson, Lawrence. J. 1915. "Review: [Untitled]." *Science* 42 (1081): 378–382.

Henderson, Lawrence J. 1922. "Orthogenesis from the Standpoint of the Biochemist." *American Naturalist* 56 (643): 97–104.

Henderson, Lawrence J. 1913a. *The Fitness of the Environment: An Inquiry into the Biological Significance of the Properties of Matter.* New York: Macmillan.

Henderson, Lawrence J. 1913b. "The Fitness of the Environment: An Inquiry into the Biological Significance of the Properties of Matter." *American Naturalist* 47 (554): 105–115.

Henderson, Lawrence J. 1913c. "The Regulation of Neutrality in the Animal Body." *Science* 37 (950): 389–395.

Henderson, Lawrence J. 1917a. *The Order of Nature.* Cambridge, MA: Harvard University Press.

Henderson, Lawrence J. 1917b. "Acidosis." *Science* 46 (1178): 73–83

Henderson, Lawrence J. 1918. "Mechanism, from the Standpoint of Physical Science." *Philosophical Review* 27 (6): 571–576.

Henderson, Lawrence J. 1927. "Introduction." In *An Introduction to the Study of Experimental Medicine,* edited by Claude Bernard, v–xii. New York: Abelard- Schuman.

Henderson, Lawrence J. 1928. *Blood: A Study in General Physiology*. New Haven, CT: Yale University Press.

Henslow, George. 1895. *The Origin of Plant Structures by Self-Adaptation to the Environment*. London: K. Paul, Trench, Trübner.

Heras-Escribano, Manuel. 2019. *The Philosophy of Affordances*. Cham, Switzerland: Palgrave Macmillan.

Hernández, Isaac, and Davide Vecchi. 2019. "The Interactive Construction of Biological Individuality Through Biotic Entrenchment." *Frontiers in Psychology* 10:2578. https://www.frontiersin.org/articles/10.3389/fpsyg.2019.02578.

Herring, Emily, and Gregory Radick. 2019. "Emergence in Biology: From Organicism to Systems Biology." In *The Routledge Handbook of Emergence*, edited by Sophie Gibb, Robin Findlay Hendry, and Tom Lancaster. Abingdon: Routledge.

Hertwig, Oscar. 1892. "Urmund und Spina Bifida: Eine vergleichend morphologische, teratologische Studie an missgebildeten Froscheiern." *Archiv für Mikroskopische Anatomie* 39: 353–489.

Hertwig, Oscar. 1894. *Zeit- und Straitfragen der Biologie. Heft I: Präformation oder Epigenese? Grundzüge einer Entwicklungs-Theorie der Organismen*. Jena, Germany: Gustav Fischer.

Hildebrand, Jayne. 2023. *Novel Environments: Science, Description, and Victorian Fiction*. Oxford: Oxford University Press.

Hill, Archibald Vivian. 1928. "The Diffusion of Oxygen and Lactic Acid Through Tissues." *Proceedings of the Royal Society of London Series B Containing Papers of a Biological Character* 104 (728): 39–96.

Hill, Archibald Vivian. 1930. "Colloid Science Applied to Biology. A General Discussion. Part I. Equilibrium in Protein Systems. Membrane-Phenomena in Living Matter: Equilibrium or Steady State." *Transactions of the Faraday Society* 26: 667–673.

Hodge, Michael Jonathan Sessions. 1971. "Lamarck's Science of Living Bodies." *British Journal for the History of Science* 5 (4): 323–352.

Hoernlé, Reinhold Friedrich Alfred. 1931. "Must Biological Processes Be Either Purposive or Mechanistic?" In *Proceedings of the Seventh International Congress of Philosophy: Held at Oxford, England, September 1–6, 1930*, edited by Gilbert Ryle, 44–46. Oxford: Oxford University Press.

Hoffmann-Kolss, Vera. 2010. *The Metaphysics of Extrinsic Properties*. Frankfurt: Ontos Verlag.

Holmes, Frederic L. 1963. "The 'Milieu Intérieur' and the Cell Theory." *Bulletin of the History of Medicine* 37 (4): 315–335.

Holmes, Frederic L. 1969. "Joseph Barcroft and the Fixity of the Internal Environment." *Journal of the History of Biology* 2 (1): 89–122.

Holmes, Frederic L. 1986. "Claude Bernard, The 'Milieu Intérieur', and Regulatory Physiology." *History and Philosophy of the Life Sciences* 8 (1): 3–25.

Hongbo, Shao, Chen Sixue, and Marian Brestic. 2015. "Environment-Living Organism's Interactions from Physiology to Genomics." *International Journal of Genomics* 2015: e270736. https://doi.org/10.1155/2015/270736.

Hooker, Henry D. 1919. "Behavior and Assimilation." *American Naturalist* 53 (629): 506–514.

Hora, Sunder Lal. 1930. "Animal Plasticity and Environment." *Nature* 126 (3177): 435–436.

Horowitz, Milton W. 1946. Review of *General Biology and Philosophy of Organism*, by Ralph Stayner Lillie. *American Journal of Psychology* 59 (2): 321–322.

Hubbs, Carl L. 1946. "General Biology and Philosophy of Organism. Ralph Stayner Lillie." *American Naturalist* 80 (792): 381–382.

Huneman, Philippe. 2010. "Assessing the Prospects for a Return of Organisms in Evolutionary Biology." *History and Philosophy of the Life Sciences* 32 (2–3): 341–371.

Huneman, Philippe. 2017. "Kant's Concept of Organism Revisited: A Framework for a Possible Synthesis Between Developmentalism and Adaptationism?" *Monist* 100 (3): 373–390.

Huneman, Philippe. 2022. "What Is It Like to Be an Environment? A Semantic and Epistemological Inquiry." *Biological Theory* 17 (1): 94–112.

Huneman, Philippe, and Charles T. Wolfe. 2010. "Introduction." *History and Philosophy of the Life Sciences* 32 (2–3): 147–154.

Huxley, Julian. 1912. *The Individual in the Animal Kingdom*. Cambridge, UK: Cambridge University Press.

Hyman, Libbie H. 1955. "Charles Manning Child." *Science* 121 (3151): 717–718.

Hyman, Libbie H. 1957. *Charles Manning Child, 1869—1954. A Biographical Memoir.* Washington, DC: National Academy of Sciences.

Illetterati, Luca, and Francesca Michelini. 2008. "Introduction." In *Purposiveness: Teleology Between Nature and Mind*, edited by Luca Illetterati and Francesca Michelini, 1–8. Frankfurt: Ontos.

"In Memoriam Prof. Dr. A. Meyer-Abich." 1971. *Acta Biotheoretica* 20 (3): 81–82.

Jacobs, Jonathan. 1986. "Teleology and Reduction in Biology." *Biology and Philosophy* 1 (4): 389–399.

Japyassú, Hilton F., and Kevin N. Laland. 2017. "Extended Spider Cognition." *Animal Cognition* 20 (3): 375–395.

Jaroš, Filip, and Carlo Brentari. 2022. "Organisms as Subjects: Jakob von Uexküll and Adolf Portmann on the Autonomy of Living Beings and Anthropological Difference." *History and Philosophy of the Life Sciences* 44 (3): 36. https://doi.org/10.1007/s40656-022-00518-7.

Jaroš, Filip, and Jiří Klouda, eds. 2021. *Adolf Portmann: A Thinker of Self-Expressive Life*. Cham, Switzerland: Springer International Publishing.

Jennings, Herbert Spencer. 1905. "The Method of Regulation in Behavior and in Other Fields." *Journal of Experimental Zoology* 2 (4): 473–494.

Jennings, Herbert Spencer. 1906. *Behavior of Lower Animals*. New York: Columbia University Press.

Jensen, Paul. 1907. *Organische Zweckmässigkeit, Entwicklung und Vererbung vom Standpunkte der Physiologie*. Jena, Germany: G. Fischer.

Jessop, Ralph. 2012. "Coinage of the Term Environment: A Word Without Authority and Carlyle's Displacement of the Mechanical Metaphor." *Literature Compass* 9: 708–720.

Johnston, Mark. 1992. "Constitution Is Not Identity." *Mind* 101 (401): 89–106.

Jonas, Hans. 1966. *The Phenomenon of Life: Toward a Philosophical Biology*. New York: Harper & Row.

Jonckers, L. H. M. 1973. "The Concept of Population in Biology." *Acta Biotheoretica* 22 (2): 78–108.

Jones, Andrew. 2023. *How Kant Matters for Biology: A Philosophical History*. Melksham, UK: University of Wales Press.

Jones, Martin R. 2005. "Idealization and Abstraction: A Framework." In *Idealization XII: Correcting the Model. Idealization and Abstraction in the Sciences*, edited by Martin R. Jones and Nancy Cartwright, 173–217. New York: Rodopi.

Jones, Nicholaos. 2018. "Strategies of Explanatory Abstraction in Molecular Systems Biology." *Philosophy of Science* 85 (5): 955–968.

Jones, R. Morley, and Kenneth Mather. 1958. "Interaction of Genotype and Environment in Continuous Variation. II. Analysis." *Biometrics* 14 (4): 489–498.

Jordan, David Starr, and Vernon Lyman Kellogg. 1907. *Evolution and Animal Life*. New York: D. Appelton.

Jordanova, Ludmilla Jane. 1984. *Lamarck*. Oxford: Oxford University Press.

Jost, Ludwig. 1907. *Lectures on Plant Physiology*. Oxford: Clarendon Press.

Kammerer, Paul. 1910. "Vererbung erzwungener Farbveränderungen." *Archiv für Entwicklungsmechanik der Organismen* 29 (3): 456–498.

Kant, Immanuel. (1790) 2000. *Critique of the Power of Judgment.* Cambridge, UK: Cambridge University Press.

Kauffman, Stuart, and Philip Clayton. 2006. "On Emergence, Agency, and Organization." *Biology and Philosophy* 21 (4): 501–521.

Kay, Lily E. 1993. *The Molecular Vision of Life: Caltech, the Rockefeller Foundation, and the Rise of the New Biology*. Oxford: Oxford University Press.

Kimler, William C. 1986. "Advantage, Adaptiveness, and Evolutionary Ecology." *Journal of the History of Biology* 19 (2): 215–233.

Kingsland, Sharon E. 1986. "Mathematical Figments, Biological Facts: Population Ecology in the Thirties." *Journal of the History of Biology* 19 (2): 235–256.

Kirchhoff, Michael D., and Julian Kiverstein. 2021. "How to Determine the Boundaries of the Mind: A Markov Blanket Proposal." *Synthese* 198 (5): 4791–4810.

Kirchhoff, Michael, and Julian Kiverstein. 2024. "Diachronic Constitution." *Manuscrito* 47 (1): 2022–0042.

Kirkpatrick, Mark. 1982. "Sexual Selection and the Evolution of Female Choice." *Evolution* 36 (1): 1–12.

Kissel, Marc, and Agustín Fuentes. 2021. "The Ripples of Modernity: How We Can Extend Paleoanthropology with the Extended Evolutionary Synthesis." *Evolutionary Anthropology: Issues, News, and Reviews* 30 (1): 84–98.

Kiverstein, Julian, and Michael Kirchhoff. 2023. "Dissolving the Causal-Constitution Fallacy: Diachronic Constitution and the Metaphysics of Extended Cognition." In *Situated Cognition Research: Methodological Foundations*, edited by Mark-Oliver Casper and Giuseppe Flavio Artese, 155–173. Cham, Switzerland: Springer.

Klebs, Georg. 1910. "Alterations of the Development and Forms of Plants as a Result of Environment." *Nature* 83 (2118): 414.

Klement, Kevin C. 2014. "Early Russell on Types and Plurals." *Journal for the History of Analytical Philosophy* 2 (6): 1–21.

Köchy, Kristian. 2019. "Uexküll's Legacy: Biological Reception and Biophilosophical Impact." In *Jakob von Uexküll and Philosophy: Life, Environments, Anthropology*, edited by Francesca Michelini and Kristian Köchy, 52–69. London: Routledge.

Koerner, Lisbet. 1999. *Linnaeus: Nature and Nation*. Cambridge, MA: Harvard University Press.

Körner, Stephan. 1955. *Kant*. Bristol, UK: Penguin Books.

Kotarbinska, Janina. 1960. "On Ostensive Definitions." *Philosophy of Science* 27 (1): 1–22.

Kovaka, Karen. 2015. "Biological Individuality and Scientific Practice." *Philosophy of Science* 82 (5): 1092–1103.

Krakauer, David C., Karen M. Page, and Douglas H. Erwin. 2009. "Diversity, Dilemmas, and Monopolies of Niche Construction." *American Naturalist* 173 (1): 26–40.

Krieger, Gerald J. 1998. "Transmogrifying Teleological Talk?" *History and Philosophy of the Life Sciences* 20 (1): 3–34.

Kroeker, Kristy J., Lauren E. Bell, Emily M. Donham, Umihiko Hoshijima, Sarah Lummis, Jason A. Toy, and Ellen Willis-Norton. 2020. "Ecological Change in Dynamic Environments: Accounting for Temporal Environmental Variability in Studies of Ocean Change Biology." *Global Change Biology* 26 (1): 54–67.

Kugler, Rolf. 1967. *Philosophische Aspekte der Biologie Adolf Portmanns*. Zürich: EVZ-Verlag.

Kulkarni, Saurabh S., Robert J. Denver, Ivan Gomez-Mestre, and Daniel R. Buchholz. 2017. "Genetic Accommodation via Modified Endocrine Signalling Explains Phenotypic Divergence Among Spadefoot Toad Species." *Nature Communications* 8 (1): 993. https://doi.org/10.1038/s41467-017-00996-5.

Kunz, Thomas H., Sidney A. Gauthreaux, Jr, Nickolay I. Hristov, Jason W. Horn, Gareth Jones, Elisabeth K. V. Kalko, Ronald P. Larkin, et al. 2008. "Aeroecology: Probing and Modeling the Aerosphere." *Integrative and Comparative Biology* 48 (1): 1–11.

Kuukkanen, Jouni-Matti. 2016. "Historicism and the Failure of HPS." *Studies in History and Philosophy of Science Part A* 55: 3–11.

Ladyman, James, Don Ross, John Collier, and David Spurrett. 2009. *Every Thing Must Go: Metaphysics Naturalized*. Oxford: Oxford University Press.

Laland, Kevin N., Blake Matthews, and Marcus W. Feldman. 2016. "An Introduction to Niche Construction Theory." *Evolutionary Ecology* 30 (2): 191–202.

Laland, Kevin N., John Odling-Smee, and John Endler. 2017. "Niche Construction, Sources of Selection and Trait Coevolution." *Interface Focus* 7 (5): 20160147. https://doi.org/10.1098/rsfs.2016.0147.

Laland, Kevin N. 2004. "Extending the Extended Phenotype." *Biology and Philosophy* 19 (3): 313–325.

Laland, Kevin N., and Michael J. O'Brien. 2011. "Cultural Niche Construction: An Introduction." *Biological Theory* 6 (3): 191–202.

Laland, Kevin N., F. John Odling-Smee, and Marcus W. Feldman. 1999. "Evolutionary Consequences of Niche Construction and Their Implications for Ecology." *Proceedings of the National Academy of Sciences* 96 (18): 10242–10247.

Laland, Kevin N., John Odling-Smee, and Marcus W. Feldman. 2019. "Understanding Niche Construction as an Evolutionary Process." In *Evolutionary Causation: Biological and Philosophical Reflection*, edited by Tobias Uller and Kevin N. Laland, 127–156. Cambridge, MA: MIT Press.

Laland, Kevin N., John Odling-Smee, William Hoppitt, and Tobias Uller. 2013. "More on How and Why: Cause and Effect in Biology Revisited." *Biology & Philosophy* 28 (5): 719–745.

Laland, Kevin N., Kim Sterelny, John Odling-Smee, William Hoppitt, and Tobias Uller. 2011. "Cause and Effect in Biology Revisited: Is Mayr's Proximate-Ultimate Dichotomy Still Useful?" *Science* 334 (6062): 1512–1516.

Laland, Kevin N., Tobias Uller, Marcus W. Feldman, Kim Sterelny, Gerd B. Müller, Armin Moczek, Eva Jablonka, and John Odling-Smee. 2015. "The Extended Evolutionary Synthesis: Its Structure, Assumptions and Predictions." *Proceedings of the Royal Society B: Biological Sciences* 282 (1813): 20151019. https://doi.org/10.1098/rspb.2015.1019.

Lamarck, Jean-Baptiste. 1802. *Recherches sur l'organisation des corps vivants et particulièrement sur son origine, sur la cause de son développement et des progrès de sa composition*. Paris: Maillard.

Lang, W. H. 1915. "Plant Morphology." *Science* 42 (1092): 780–791.

Lankester, Edwin Ray. 1870. "On the Use of the Term Homology in Modern Zoology, and the Distinction Between Homogenetic and Homoplastic Agreements." *Annals and Magazine of Natural History* 6 (31): 34–43.

Largent, Mark A. 1999. "Bionomics: Vernon Lyman Kellogg and the Defense of Darwinism." *Journal of the History of Biology* 32 (3): 465–488.

Laubichler, Manfred D. 2000. "The Organism Is Dead. Long Live the Organism!" *Perspectives on Science* 8 (3): 286–315.

Laubichler, Manfred D. 2006. "Allgemeine Biologie als selbständige Grundwissenschaft und die allgemeinen Grundlagen des Lebens." In *Der Hochsitz des Wissens: Das Allgemeine als wissenchaflicher Wert*, edited by Michael Hagner and Manfred D. Laubichler, 185–205. Zürich: Diaphanes.

Laubichler, Manfred D. 2017. "The Emergence of Theoretical and General Biology: The Broader Scientific Context for the Biologische Versuchsanstalt."

In *Vivarium: Experimental, Quantitative, and Theoretical Biology at Vienna's Biologische Versuchsanstalt*, edited by Gerhard H. Müller, 95–114. Cambridge, MA: MIT Press.

Laubichler, Manfred D., and Jane Maienschein, eds. 2009. *Form and Function in Developmental Evolution*. Cambridge, UK: Cambridge University Press.

Lawrence, Christopher, and George Weisz, eds. 1998. *Greater Than the Parts: Holism in Biomedicine, 1920–1950*. Oxford: Oxford University Press.

Le Conte, Joseph. 1891. "The Factors of Evolution." *Monist* 1 (3): 321–335.

Lennox, James G. 1992. "Teleology." In *Keywords in Evolutionary Biology*, edited by Elisabeth Lloyd and Evelyn Fox Keller, 324–333. Cambridge, MA: Harvard University Press.

Lennox, James G. 1993. "Darwin Was a Teleologist." *Biology and Philosophy* 8 (4): 409–421.

Lenoir, Timothy. 1989. *The Strategy of Life: Teleology and Mechanics in Nineteenth-Century German Biology*. Chicago: University of Chicago Press.

Lerner, Isadore Michael. 1950. *Population Genetics and Animal Improvement*. Cambridge, UK: Cambridge University Press.

Levene, Howard, Olga Pavlovsky, and Theodosius Dobzhansky. 1954. "Interaction of the Adaptive Values in Polymorphic Experimental Populations of *Drosophila pseudoobscura*." *Evolution* 8 (4): 335–349.

Levins, Richard. 1968. *Evolution in Changing Environments: Some Theoretical Explorations*. Princeton, NJ: Princeton University Press.

Levins, Richard, and Richard Lewontin. 1985. *The Dialectical Biologist*. Cambridge, MA: Harvard University Press.

Levit, Georgy S., and Uwe Hossfeld. 2019. "Ernst Haeckel in the History of Biology." *Current Biology* 29 (24): R1276– R1284.

Levy, Arnon. 2021. "Idealization and Abstraction: Refining the Distinction." *Synthese* 198 (24): 5855–5872.

Lewens, Tim. 2004. *Organisms and Artifacts: Design in Nature and Elsewhere*. Cambridge, MA: MIT Press.

Lewin, Kurt. 1922. *Der Begriff der Genese in Physik, Biologie und Entwicklungsgeschichte: Eine Untersuchung zur vergleichenden Wissenschaftslehre*. Berlin: Springer.

Lewis, D. 1954. "Gene-Environment Interaction: A Relationship Between Dominance, Heterosis, Phenotypic Stability and Variability." *Heredity* 8 (3): 333–356.

Lewontin, Richard C. 1955. "The Effects of Population Density and Composition on Viability in *Drosophila melanogaster.*" *Evolution* 9 (1): 27–41.

Lewontin, Richard, and Richard Levins. 1997. "Organism and Environment." *Capitalism Nature Socialism* 8 (2): 95–98.

Lewontin, Richard C. 1957. "The Adaptations of Populations to Varying Environments." *Cold Spring Harbor Symposia on Quantitative Biology* 22: 395–408.

Lewontin, Richard C. 1974. *The Genetic Basis of Evolutionary Change.* New York: Columbia University Press.

Lewontin, Richard C. 1983. "Gene, Organism, and Environment." In *Evolution: From Molecules to Men,* edited by D. Bendall, 273–285. Cambridge, UK: Cambridge University Press.

Lewontin, Richard C. 2000. *The Triple Helix: Gene, Organism, and Environment.* Cambridge, MA: Harvard University Press.

Lewontin, Richard C. 2001. "Gene, Organism, and Environment: A New Introduction." In *Cycles of Contingency: Developmental Systems and Evolution,* edited by Susan Oyama, Paul E. Griffiths, and Russell D. Gray, 55–57. Cambridge, MA: MIT Press.

Lidgard, Scott, and Lynn K. Nyhart, eds. 2017. *Biological Individuality: Integrating Scientific, Philosophical, and Historical Perspectives.* Chicago: University of Chicago Press.

Lightman, Bernard, ed. 2015. *Global Spencerism: The Communication and Appropriation of a British Evolutionist.* London: Brill.

Lillie, Ralph S. 1902. "On the Effects of Various Solutions on Ciliary and Muscular Movement in the Larvæ of Arenicola and Polygordius. II." *American Journal of Physiology* 7 (1): 25–55.

Lillie, Ralph S. 1908. "Momentary Elevation of Temperature as a Means of Producing Artificial Parthenogenesis in Starfish Eggs and the Conditions of Its Action." *Journal of Experimental Zoology* 5 (3): 375–428.

Lillie, Ralph S. 1913. Review of *The Fitness of the Environment,* by Lawrence J. Henderson. *Science* 38 (975): 337–342.

Lillie, Ralph S. 1915. "What Is Purposive and Intelligent Behavior from the Physiological Point of View?" *Journal of Philosophy, Psychology and Scientific Methods* 12 (22): 589–610.

Lillie, Ralph S. 1916. "The Theory of Anæsthesia." *Biological Bulletin* 30 (5): 311–366.

Lillie, Ralph S. 1918. "The Origin and Evolution of Life, on the Theory of the Action, Reaction and Interaction of Energy. By Henry Fairfield Osborn. New York, Charles Scribner's Sons. 1918. Pp. Xxxi+322. Price $3.00." *Science* 48 (1245): 472–474.

Lillie, Ralph S. 1920a. "The Nature of Protoplasmic and Nervous Transmission." *Journal of Physical Chemistry* 24 (3): 165–191.

Lillie, Ralph S. 1920b. "The Place of Life in Nature." *Journal of Philosophy, Psychology and Scientific Methods* 17 (18): 477–493.

Lillie, Ralph S. 1923. *Protoplasmic Action and Nervous Action.* Chicago: University of Chicago Press.

Lillie, Ralph S. 1924. "Reactivity of the Cell." In *General Cytology: A Textbook of Cellular Structure and Function for Students of Biology and Medicine*, edited by Edmund V. Cowdry, 167–233. Chicago: University of Chicago Press.

Lillie, Ralph S. 1945. *General Biology and Philosophy of Organism.* University of Chicago Press.

Lindsay, Alexander Dunlop. 1919. *The Philosophy of Immanuel Kant.* London: T. C. & E. C. Jack.

Liu, Daniel. 2019. "The Artificial Cell, the Semipermeable Membrane, and the Life That Never Was, 1864–1901." *Historical Studies in the Natural Sciences* 49 (5): 504–555.

Livingston, Burton E. 1913. "A New Problem in Adaptation." *Plant World* 16 (11): 315–318.

Livingston, Burton E. 1934. "Environments." *Science* 80 (2086): 569–576.

Livingston, Burton Edward. 1912. "A Schematic Representation of the Water Relations of Plants: A Pedagogical Suggestion." *Plant World* 15 (9): 214–218.

Lloyd, Elisabeth. 2021. *Adaptation.* Cambridge, UK: Cambridge University Press.

Locke, John. 1975. *An Essay Concerning Human Understanding*, edited by Peter H. Nidditch. *The Clarendon Edition of the Works of John Locke: An Essay Concerning Human Understanding.* Oxford: Oxford University Press

Lockman, Jeffrey J., Emily A. Lewis, and Nicholas E. Fears. 2018. "Ecology." In *The SAGE Encyclopedia of Lifespan Human Development*, edited by Marc H. Bornstein, 702–705. Thousand Oaks, CA: SAGE Publications.

Loeb, Jacques. 1909. "Experimental Study of the Influence of Environment on Animals." In *Darwin and Modern Science*, edited by A. C. Seward, 247–270. Cambridge, UK: Cambridge University Press.

Loeb, Jacques. 1912. *The Mechanistic Conception of Life: Biological Essays.* Chicago: University of Chicago Press.

Loeb, Jacques. 1919. *Forced Movements, Tropisms, and Animal Conduct.* Philadelphia and London: J. B. Lippincott.

Logan, Cheryl A., and Sabine Brauckmann. 2015. "Controlling and Culturing Diversity: Experimental Zoology Before World War II and Vienna's Biologische Versuchsanstalt." *Journal of Experimental Zoology Part A: Ecological Genetics and Physiology* 323 (4): 211–226.

Loison, Laurent. 2016. "Forms of Presentism in the History of Science. Rethinking the Project of Historical Epistemology." *Studies in History and Philosophy of Science Part A* 60: 29–37.

Loison, Laurent. 2024. *Beyond Lamarckism: Plasticity in Darwinian Evolution, 1890–1970*. London: Routledge.

López-Beltrán, Carlos. 1994. "Forging Heredity: From Metaphor to Cause, a Reification Story." *Studies in History and Philosophy of Science Part A* 25 (2): 211–235.

López-Beltrán, Carlos. 2004. *El sesgo hereditario: Ámbitos históricos del concepto de herencia biológica*. Mexico City: Univerisdad Nacional Autónoma de México.

Lotka, Alfred J. 1925. *Elements of Physical Biology*. Baltimore: Williams and Wilkins.

Love, Alan C. 2018. "Individuation, Individuality, and Experimental Practice in Developmental Biology." In *Individuation, Process and Scientific Practices*, edited by Otavio Bueno, Ruey-Lin Chen, and Melinda B. Fagan, 165–191. Oxford: Oxford University Press.

Löw, Reinhard. 1980. *Philosophie des Lebendigen: Der Begriff des Organischen bei Kant, sein Grund und seine Aktualität*. Frankfurt: Suhrkamp.

Lowe, Edward Jonathan. 1989. *Kinds of Being: A Study of Individuation, Identity, and the Logic of Sortal Terms*. Oxford: Blackwell.

Lowe, Winsor H., Thomas E. Martin, David K. Skelly, and H. Arthur Woods. 2021. "Metamorphosis in an Era of Increasing Climate Variability." *Trends in Ecology & Evolution* 36 (4): 360–375.

Lyons, Sherrie L. 2020. *From Cells to Organisms: Re-Envisioning Cell Theory*. Toronto: University of Toronto Press.

Macvicar, John Gibson. 1863. "The Theory of Terminal Fructification in the Simple Plant, of Ovules and Pollen, and of Spores." *Transactions of the Botanical Society of Edinburgh* 7 (1–4): 13–30.

Maelzer, D. A. 1965a. "A Discussion of Components of Environment in Ecology." *Journal of Theoretical Biology* 8 (1): 141–162.

Maelzer, D. A. 1965b. "Environment, Semantics, and System Theory in Ecology." *Journal of Theoretical Biology* 8 (3): 395–402.

Maienschein, Jane. 1988. "Whitman at Chicago: Establishing a Chicago Style of Biology?" In *The American Development of Biology*, edited by Ronald Rainger, Keith R.

Benson, and Jane Maienschein, 151–182. Philadelphia: University of Pennsylvania Press.

Maienschein, Jane. 1991. *Transforming Traditions in American Biology, 1880–1915*. Baltimore: Johns Hopkins University Press.

Maienschein, Jane. 2011. "'Organization' as Setting Boundaries of Individual Development." *Biological Theory* 6 (1): 73–79.

Maienschein, Jane, and Kate MacCord. 2022. *What Is Regeneration?* Chicago: University of Chicago Press.

Martin, Raymond. 2006. "Do Historians Need Philosophy?" *History and Theory* 45 (2): 252–260.

Martínez, Sergio F., and Xiang Huang. 2011. "Epistemic Groundings of Abstraction and Their Cognitive Dimension." *Philosophy of Science* 78 (3): 490–511.

Maslakova, Svetlana A. 2010. "Development to Metamorphosis of the Nemertean Pilidium Larva." *Frontiers in Zoology* 7 (1): 30. https://doi.org/10.1186/1742-9994-7-30.

Mason, Herbert L. 1957. "The Concept of the Flower and the Theory of Homology." *Madroño* 14 (3): 81–95.

Mason, Herbert L., and Jean H. Langenheim. 1957. "Language Analysis and the Concept 'Environment.'" *Ecology* 38 (2): 325–340.

Mason, Herbert L., and Jean H. Langenheim. 1961. "Natural Selection as an Ecological Concept." *Ecology* 42 (1): 158–165.

Massimi, Michela. 2009. "Philosophy and the Sciences After Kant." *Royal Institute of Philosophy Supplements* 65: 275–311.

Mather, Kenneth, and R. Morley Jones. 1958. "Interaction of Genotype and Environment in Continuous Variation. I. Description." *Biometrics* 14 (3): 343–359.

Mathews, Albert P. 1913. "Adaptation from the Point of View of the Physiologist." *American Naturalist* 47 (554): 90–104.

Matthews, Blake, Luc De Meester, Clive G. Jones, Bas W. Ibelings, Tjeerd J. Bouma, Visa Nuutinen, Johan van de Koppel, and John Odling-Smee. 2014. "Under Niche Construction: An Operational Bridge Between Ecology, Evolution, and Ecosystem Science." *Ecological Monographs* 84 (2): 245–263.

Matthews, John A. 2014. "Holocoenotic Environment." In *Encyclopedia of Environmental Change*, 526–526. Thousand Oaks, CA: SAGE Publications.

Maturana, Humberto R. 1975. "The Organization of the Living: A Theory of the Living Organization." *International Journal of Man-Machine Studies* 7 (3): 313–332.

Maturana, Humberto R., and Francisco J. Varela. 1972. *De máquinas y seres vivos. Autopoiesis: La organización de lo vivo.* Santiago de Chile: Editorial Universitaria.

Mayr, Ernst. 1959a. "Darwin and The Evolutionary Theory in Biology." In *Evolution and Anthropology: A Centennial Appraisal*, edited by Betty J. Meggers, 1–8. Washington, DC: Theo Gaus' Sons.

Mayr, Ernst. 1959b. "Reviewed Work: Darwin and the Darwinian Revolution by Himmelfarb Gertrude." *Scientific American* 201 (5): 209–216.

Mayr, Ernst. 1961. "Cause and Effect in Biology." *Science* 134 (3489): 1501–1506.

Mayr, Ernst. 1963. *Animal Species and Evolution.* Cambridge, MA: Harvard University Press.

Mayr, Ernst. 1964. "The Evolution of Living Systems." *Proceedings of the National Academy of Sciences* 51 (5): 934–941.

Mayr, Ernst. 1970. *Populations, Species, and Evolution: An Abridgment of Animal Species and Evolution.* Cambridge, MA: Harvard University Press.

Mayr, Ernst. 1972. "The Nature of the Darwinian Revolution." *Science* 176 (4038): 981–989.

Mayr, Ernst. 1982. *The Growth of Biological Thought: Diversity, Evolution, and Inheritance.* Cambridge, MA: Harvard University Press.

Mayr, Ernst. 1985. "Teleological and Teleonomic, a New Analysis." In *A Portrait of Twenty-Five Years: Boston Colloquium for the Philosophy of Science 1960–1985*, edited by Robert S. Cohen and Marx W. Wartofsky, 133–159. Dordrecht: Springer Netherlands.

Mayr, Ernst. 1993a. *One Long Argument: Charles Darwin and the Genesis of Modern Evolutionary Thought.* Cambridge, MA: Harvard University Press.

Mayr, Ernst. 1993b. "Proximate and Ultimate Causations." *Biology and Philosophy* 8 (1): 93–94.

Mayr, Ernst. 1998. "The Multiple Meanings of 'Teleological.'" *History and Philosophy of the Life Sciences* 20 (1): 35–40.

McDougall, William. 1936. "The Philosophy of J. S. Haldane." *Philosophy* 11 (44): 419–432.

McLaughlin, Peter. 2000. *What Functions Explain: Functional Explanation and Self-Reproducing Systems.* Cambridge, UK: Cambridge University Press.

McLaughlin, Peter. 2002. "Naming Biology." *Journal of the History of Biology* 35 (1): 1–4.

McVeigh, Ryan. 2021. "Organism and Environment in Auguste Comte." *History of the Human Sciences* 34 (3–4): 76–97.

Medawar, Peter B. 1954. "Problems of Life. An Evaluation of Modern Biological Thought. By Ludwig von Bertalanffy. London: Watts & Co. 1952." *Mind* LXIII (249): 105–108.

Meek, Alexander. 1920. "Environment and Reproduction." *Nature* 106 (2669): 532–533.

Meincke, Anne Sophie. 2018. "Autopoiesis, Biological Autonomy and the Process View of Life." *European Journal for Philosophy of Science* 9 (1): 5. https://doi.org/10.1007/s13194-018-0228-2.

Meincke, Anne Sophie. 2020. "Systems or Bodies? On How (Not) to Embody Autopoiesis." *Adaptive Behavior* 28 (2): 119–120.

Meincke, Anne Sophie, and John Dupré, eds. 2020. *Biological Identity: Perspectives from Metaphysics and the Philosophy of Biology*. London: Routledge.

Meloni, Maurizio. 2017. "Disentangling Life: Darwin, Selectionism, and the Postgenomic Return of the Environment." *Studies in History and Philosophy of Science Part C: Studies in History and Philosophy of Biological and Biomedical Sciences* 62: 10–19.

Meloni, Maurizio. 2019. *Impressionable Biologies: From the Archaeology of Plasticity to the Sociology of Epigenetics*. Abingdon, UK: Routledge.

Menatti, Laura, Leonardo Bich, and Cristian Saborido. 2022. "Health and Environment from Adaptation to Adaptivity: A Situated Relational Account." *History and Philosophy of the Life Sciences* 44 (3): 38. https://doi.org/10.1007/s40656-022-00515-w.

Mensch, Jennifer. 2013. *Kant's Organicism: Epigenesis and the Development of Critical Philosophy*. Chicago: University of Chicago Press.

Merrifield, F. 1905. "Experiments on Variations of Lepidoptera by Environment." *Nature* 72 (1878): 632–634.

Metcalf, Maynard M. 1906. "The Influence of the Plasticity of Organisms Upon Evolution." *Science* 23 (594): 786–787.

Meyer, Adolf. 1936. "Zwischen Scylla und Charybdis: Holistische Antikritik von Mechanismus und Vitalismus." *Acta Biotheoretica* 1 (3): 203–218.

Meyer-Abich, Adolf. 1934. *Ideen und Ideale der biologischen Erkenntnis: Beiträge zur Theorie und Geschichte der biologischen Ideologien*. Leipzig: J. A. Barth.

Meyer-Abich, Adolf. 1942. "Kant und das Biologische Denken." *Acta Biotheoretica* 6 (3): 185–211.

Meyer-Abich, Adolf. 1943. "Beiträge zur Theorie der Evolution der Organismen. I. Das typologische Grundgesetz und seine Folgerungen für Phylogenie und Entwicklungsphysiologie." *Acta Biotheoretica* 7 (1): 1–80.

Meyer-Abich, Adolf. 1948. *Naturphilosophie auf neuen Wegen*. Stuttgart, Germany: Hippokrates-Verlag Marquardt.

Meyer-Abich, Adolf. 1950. *Beiträge zur Theorie der Evolution der Organismen. II. Typensynthese durch Holobiose (Bibliotheca Biotheoretica 5)*. Leiden, The Netherlands: E. J. Brill.

Meyer-Abich, Adolf. 1955. "The Principle of Complementarity in Biology." *Acta Biotheoretica* 11 (2): 57–74.

Meyer-Abich, Adolf. 1964. *The Historico-Philosophical Background of the Modern Evolution-Biology*. Leiden, The Netherlands: E. J. Brill.

Michaëlsson, Karl. 1939. "Ambiance." *Studia Neophilologica* 12 (1): 91–119.

Michelini, Francesca, and Kristian Köchy, eds. 2019. *Jakob von Uexküll and Philosophy: Life, Environments, Anthropology*. New York: Routledge.

Michelini, Francesca, Matthias Wunsch, and Dirk Stederoth. 2018. "Philosophy of Nature and Organism's Autonomy: On Hegel, Plessner and Jonas' Theories of Living Beings." *History and Philosophy of the Life Sciences* 40 (3): 1–27.

Mill, John Stuart. 1846. *A System of Logic, Ratiocinative and Inductive*, vol. II, 2nd ed. London: John W. Parker.

Millikan, Ruth Garrett. 1989. "In Defense of Proper Functions." *Philosophy of Science* 56 (2): 288–302.

Millstein, Roberta L. 2008. "Distinguishing Drift and Selection Empirically: 'The Great Snail Debate' of the 1950s." *Journal of the History of Biology* 41 (2): 339–367.

Millstein, Roberta L. 2014. "How the Concept of Population Resolves Concepts of Environment." *Philosophy of Science* 81 (5): 741–755.

Milne, A. 1957. "Theories of Natural Control of Insect Populations." *Cold Spring Harbor Symposia on Quantitative Biology* 22: 253–271.

Milne, A. 1962. "On a Theory of Natural Control of Insect Population." *Journal of Theoretical Biology* 3 (1): 19–50.

Milocco, Lisandro, and Isaac Salazar-Ciudad. 2022. "A Method to Predict the Response to Directional Selection Using a Kalman Filter." *Proceedings of the National Academy of Sciences* 119 (28): e2117916119. https://doi.org/10.1073/pnas.2117916119.

Milocco, Lisandro, and Tobias Uller. 2023. "A Data-Driven Framework to Model the Organism–Environment System." *Evolution & Development* 25 (6): 439–450.

Minelli, Alessandro. 2003. *The Development of Animal Form: Ontogeny, Morphology, and Evolution.* Cambridge, UK: Cambridge University Press.

Mitman, Gregg, and Anne Fausto-Sterling. 1992. "Whatever Happened to Planaria? C. M. Child and the Physiology of Inheritance." In *The Right Tools for the Job: At Work in Twentieth-Century Life Sciences,* edited by Adele Clarke and Joan Fujimura, 172–197. Princeton, NJ: Princeton University Press.

Mivart, George. 1889. "Prof. Weismann's 'Essays.'" *Nature* 41 (1046): 38–41.

Moczek, Armin P. 2015. "Re-Evaluating the Environment in Developmental Evolution." *Frontiers in Ecology and Evolution* 3: 7. https://www.frontiersin.org/articles/10.3389/fevo.2015.00007.

Moczek, Armin P., Sonia Sultan, Susan Foster, Cris Ledón-Rettig, Ian Dworkin, H. Fred Nijhout, Ehab Abouheif, and David W. Pfennig. 2011. "The Role of Developmental Plasticity in Evolutionary Innovation." *Proceedings of the Royal Society B: Biological Sciences* 278 (1719): 2705–2713.

Monod, Jacques. (1970) 1971. *Chance and Necessity: An Essay on the Natural Philosophy of Modern Biology.* New York: Alfred A. Knopf.

Montévil, Maël, and Matteo Mossio. 2020. "The Identity of Organisms in Scientific Practice: Integrating Historical and Relational Conceptions." *Frontiers in Physiology* 11: 611. https://www.frontiersin.org/articles/10.3389/fphys.2020.00611.

Moreno, Alvaro. 2018. "On Minimal Autonomous Agency: Natural and Artificial." *Complex Systems* 27 (3): 289–313.

Moreno, Alvaro, and Matteo Mossio. 2015. *Biological Autonomy: A Philosophical and Theoretical Enquiry.* Dordrecht: Springer Netherlands.

Morgan, Conwy Lloyd. 1919. *Eugenics and Environment.* London: John Bale, Sons & Danielsson.

Morgan, Conwy Lloyd. 1929. "The Case for Emergent Evolution." *Philosophy* 4 (13): 23–38.

Morgan, Thomas Hunt. 1926. "Genetics and the Physiology of Development." *American Naturalist* 60 (671): 489–515.

Morgan, Thomas Hunt. 1927. *Experimental Embryology.* New York: Columbia University Press.

Morrill, Thomas A. 1957. "What Is 'Environment'?" *American Naturalist* 91 (859): 269.

Morrison, Margaret. 2002. "Modelling Populations: Pearson and Fisher on Mendelism and Biometry." *British Journal for the Philosophy of Science* 53 (1): 39–68.

Moss, Lenny, and Stuart A. Newman. 2015. "The grassblade beyond Newton: The Pragmatizing of Kant for Evolutionary-Developmental Biology." *Lebenswelt. Aesthetics and Philosophy of Experience* 7: 94–111. https://doi.org/10.13130/2240-9599/6686.

Mossio, Matteo, and Alvaro Moreno. 2010. "Organisational Closure in Biological Organisms." *History and Philosophy of the Life Sciences* 32 (2–3): 269–88.

Mossio, Matteo, and Leonardo Bich. 2017. "What Makes Biological Organisation Teleological?" *Synthese* 194 (4): 1089–1114.

Mossio, Matteo, Leonardo Bich, and Alvaro Moreno. 2013. "Emergence, Closure and Inter-Level Causation in Biological Systems." *Erkenntnis* 78 (2): 153–178.

Müller, Caroline, Barbara A. Caspers, Jürgen Gadau, and Sylvia Kaiser. 2020. "The Power of Infochemicals in Mediating Individualized Niches." *Trends in Ecology & Evolution* 35 (11): 981–989.

Müller, Gerd B. 2017a. "Biologische Versuchsanstalt: An Experiment in the Experimental Sciences." In *Vivarium. Experimental, Quantitative, and Theoretical Biology at Vienna's Biologische Versuchsanstalt*, edited by Gerd B. Müller, 3–18. Cambridge, MA: MIT Press.

Müller, Gerd B. 2017b. "The Substance of Form: Hans Przibram's Quest for Biological Experiment, Quantification, and Theory." In *Vivarium. Experimental, Quantitative, and Theoretical Biology at Vienna's Biologische Versuchsanstalt*, edited by Gerd B. Müller, 135–163. Cambridge, MA: MIT Press.

Müller, Gerd B., ed. 2017c. *Vivarium. Experimental, Quantitative, and Theoretical Biology at Vienna's Biologische Versuchsanstalt*. Cambridge, MA: MIT Press.

Müller, Gerhard H. 1994. "Wechselwirkung in the Life and Other Sciences: A Word, New Claims and a Concept Around 1800 . . . and Much Later." In *Romanticism in Science: Science in Europe, 1790–1840*, edited by Stefano Poggi and Maurizio Bossi, 1–14. Dordrecht: Springer Netherlands.

Muller, Herbert J. 1943. *Science and Criticism: The Humanistic Tradition in Contemporary Thought*. New Haven, CT: Yale University Press.

Müller-Wille, Staffan, and Christina Brandt, eds. 2016. *Heredity Explored: Between Public Domain and Experimental Science, 1850–1930*. Cambridge, MA: MIT Press.

Mumford, Stephen, and Matthew Tugby, eds. 2013. *Metaphysics and Science*. Oxford: Oxford University Press.

Murugan, Arvind, Kabir Husain, Michael J. Rust, Chelsea Hepler, Joseph Bass, Julian M. J. Pietsch, Peter S. Swain, et al. 2021. "Roadmap on Biology in Time Varying Environments." *Physical Biology* 18 (4): 1–33. https://doi.org/10.1088/1478-3975/abde8d.

Nahas, Auguste. 2024a. "Behavior, Purpose and Teleology Revisited: Locating Cybernetic Teleology in 20th Century Holism." In *The Riddle of Organismal Agency: New Historical and Philosophical Reflections*, edited by Alejandro Fábregas-Tejeda, Gregory Radick, Guido I. Prieto, and Jan Baedke, 77–93. Abingdon: Routledge.

Nahas, Auguste. 2024b. "Organisms as Agents: Beyond the Artifact Model of Teleology." PhD thesis, University of Toronto.

Nahas, Auguste, and Carl Sachs. 2023. "What's at Stake in the Debate over Naturalizing Teleology? An Overlooked Metatheoretical Debate." *Synthese* 201 (4): 142. https://doi.org/10.1007/s11229-023-04147-w.

Nassar, Dalia. 2016. "Analogical Reflection as a Source for the Science of Life: Kant and the Possibility of the Biological Sciences." *Studies in History and Philosophy of Science Part A* 58: 57–66.

Neander, Karen. 1991. "Functions as Selected Effects: The Conceptual Analyst's Defense." *Philosophy of Science* 58 (2): 168–184.

Needham, Joseph. 1928a. "Recent Developments in the Philosophy of Biology." *Quarterly Review of Biology* 3 (1): 77–91.

Needham, Joseph. 1928b. "Organicism in Biology." *Journal of Philosophical Studies* 3 (9): 29–40.

Needham, Joseph. 1930. *The Sceptical Biologist*. New York: W.W. Norton.

Needham, Joseph. 1936. *Order and Life*. Cambridge, UK: Cambridge University Press.

Newman, Stuart A. 2022. "Inherency and Agency in the Origin and Evolution of Biological Functions." *Biological Journal of the Linnean Society* 139 (4): 487–502. https://doi.org/10.1093/biolinnean/blac109.

Nichols, C. Reid, and Kaustubha Raghukumar. 2022. *Marine Environmental Characterization*. Cham, Switzerland: Springer Nature.

Nichols, George Elwood. 1924. "The Terrestrial Environment in Its Relation to Plant Life." In *Organic Adaptation to Environment*, edited by Malcolm Rutherford Thorpe, 1–43. New Haven, CT: Yale University Press.

Nicholson, Alexander John. 1933. "The Balance of Animal Populations." *Journal of Animal Ecology* 2 (1): 131–178.

Nicholson, Alexander John. 1954a. "Compensatory Reactions of Populations to Stresses, and Their Evolutionary Significance." *Australian Journal of Zoology* 2 (1): 1–8.

Nicholson, Alexander John. 1954b. "Experimental Demonstrations of Balance in Populations." *Nature* 173 (4410): 862–863.

Nicholson, Alexander John. 1960. "The Role of Population Dynamics in Natural Selection." In *Evolution After Darwin*, vol. I, edited by Sol Tax, 477–521. Chicago: University of Chicago Press.

Nicholson, Daniel J. 2012. "The Concept of Mechanism in Biology." *Studies in History and Philosophy of Science Part C: Studies in History and Philosophy of Biological and Biomedical Sciences* 43 (1): 152–163.

Nicholson, Daniel J. 2013. "Organisms ≠ Machines." *Studies in History and Philosophy of Science Part C: Studies in History and Philosophy of Biological and Biomedical Sciences* 44 (4): 669–678.

Nicholson, Daniel J. 2014. "The Return of the Organism as a Fundamental Explanatory Concept in Biology." *Philosophy Compass* 9 (5): 347–359.

Nicholson, Daniel J. 2018. "Reconceptualizing the Organism: From Complex Machine to Flowing Stream." In *Everything Flows: Towards a Processual Philosophy of Biology*, edited by Daniel J. Nicholson and John Dupré, 139–166. Oxford: Oxford University Press

Nicholson, Daniel J., and John Dupré, eds. 2018. *Everything Flows: Towards a Processual Philosophy of Biology*. Oxford: Oxford University Press.

Nicholson, Daniel J., and Richard Gawne. 2014. "Rethinking Woodger's Legacy in the Philosophy of Biology." *Journal of the History of Biology* 47 (2): 243–292.

Nicholson, Daniel J., and Richard Gawne. 2015. "Neither Logical Empiricism nor Vitalism, but Organicism: What the Philosophy of Biology Was." *History and Philosophy of the Life Sciences* 37 (4): 345–381.

Nickelsen, Kärin. 2017b. "Growth, Development, and Regeneration: Plant Biology in Vienna around 1900." In *Vivarium. Experimental, Quantitative, and Theoretical Biology at Vienna's Biologische Versuchsanstalt*, edited by Gerd B. Müller, 165–187. Cambridge, MA: MIT Press.

Nickelsen, Kärin. 2017a. "The Organism Strikes Back: Chlorella Algae and Their Impact on Photosynthesis Research, 1920s–1960s." *History and Philosophy of the Life Sciences* 39 (2): 9. https://doi.org/10.1007/s40656-017-0137-2.

Nicoglou, Antonine. 2011. "Defining the Boundaries of Development with Plasticity." *Biological Theory* 6 (1): 36–47.

Nicoglou, Antonine. 2018. "The Concept of Plasticity in the History of the Nature-Nurture Debate in the Early Twentieth Century." In *The Palgrave Handbook of Biology and Society*, edited by Maurizio Meloni, John Cromby, Des Fitzgerald, and Stephanie Lloyd, 97–122. London: Palgrave Macmillan UK.

Niven, B. S. 1983. "Two Different Animals May Not Have the Same Environment." *Journal of Theoretical Biology* 105 (2): 369–370.

Niven, B. S., and M. J. Liddle. 1994. "Towards a Classification of the Environment and the Community of *Quercus robur.*" *Journal of Vegetation Science* 5 (3): 317–326.

Nuño de la Rosa, Laura. 2010. "Becoming Organisms: The Organisation of Development and the Development of Organisation." *History and Philosophy of the Life Sciences* 32 (2–3): 289–315.

Nuño de la Rosa, Laura. 2023. "Agency in Reproduction." *Evolution & Development* 25 (6): 418–429.

Nuño de la Rosa, Laura, and Arantza Etxeberria. 2010. "¿Fue Darwin el «Newton de la brizna de hierba»? La herencia de Kant en la teoría darwinista de la evolución." *Endoxa* 24: 185–216.

Nyhart, Lynn K. 1995. *Biology Takes Form: Animal Morphology and the German Universities, 1800–1900.* Chicago: University of Chicago Press.

Nyhart, Lynn K. 2009. *Modern Nature: The Rise of the Biological Perspective in Germany.* Chicago: University of Chicago Press.

Nyhart, Lynn K., and Scott Lidgard. 2011. "Individuals at the Center of Biology: Rudolf Leuckart's Polymorphismus der Individuen and the Ongoing Narrative of Parts and Wholes. With an Annotated Translation." *Journal of the History of Biology* 44 (3): 373–443.

Ochoa, Carlos, and Ana Barahona. 2014. *El Jano de la morfología: De la homología a la homoplasia, historia, debates y evolución.* Mexico City: Centro de Estudios Filosóficos, Políticos y Sociales Vicente Lombardo Toledano.

Odling-Smee, F. John. 1988. "Niche-Constructing Phenotypes." In *The Role of Behavior in Evolution,* edited by Henry C. Plotkin, 73–132. Cambridge, MA: MIT Press.

Odling-Smee, F. John. 1994. "Niche Construction, Evolution and Culture." In *Companion Encyclopedia of Anthropology,* edited by Tim Ingold, 162–196. London: Routledge.

Odling-Smee, F. John, Kevin N. Laland, and Marcus W. Feldman. 1996. "Niche Construction." *American Naturalist* 147 (4): 641–648.

Odling-Smee, F. John, Kevin N. Laland, and Marcus W. Feldman. 2003. *Niche Construction: The Neglected Process in Evolution.* Princeton, NJ: Princeton University Press.

Odling-Smee, John. 2009. "Niche Construction in Evolution, Ecosystems and Developmental Biology." In *Mapping the Future of Biology: Evolving Concepts and Theories,* edited by Anouk Barberousse, Michel Morange, and Thomas Pradeu, 69–92. Dordrecht: Springer Netherlands.

Odling-Smee, John, Douglas H. Erwin, Eric P. Palkovacs, Marcus W. Feldman, and Kevin N. Laland. 2013. "Niche Construction Theory: A Practical Guide for Ecologists." *Quarterly Review of Biology* 88 (1): 3–28.

Odling-Smee, John, and Kevin N. Laland. 2011. "Ecological Inheritance and Cultural Inheritance: What Are They and How Do They Differ?" *Biological Theory* 6 (3): 220–230.

Okasha, Samir. 2018. *Agents and Goals in Evolution*. Oxford: Oxford University Press.

Okasha, Samir. Forthcoming. "On the Very Idea of Biological Individuality." *British Journal for the Philosophy of Science*. https://doi.org/10.1086/728048.

Okasha, Samir. 2024. "The Concept of Agent in Biology: Motivations and Meanings." *Biological Theory* 19: 6–10. https://doi.org/10.1007/s13752-023-00439-z.

Ortega Lozano, Ramón. 2022. "El cuerpo como un todo y no como una suma de partes: La propuesta holista de Walter B. Cannon." *Crítica* 54 (162): 29–55.

Ortega y Gasset, José. 1918. *Meditaciones del Quijote*. Madrid: Publicaciones de la Residencia de Estudiantes.

Orton, James Herbert. 1933. "Some Limiting Factors in the Environment of the Common Limpet, *P. vulgata*." *Nature* 131 (3315): 693–694.

Osborn, Henry Fairfield. 1895. "The Hereditary Mechanism and the Search for the Unknown Factors of Evolution." *American Naturalist* 29 (341): 418–439.

Ospovat, Dov. 1981. *The Development of Darwin's Theory: Natural History, Natural Theology, and Natural Selection, 1838–1859*. Cambridge, UK: Cambridge University Press.

Otsuka, Jun. 2015. "Using Causal Models to Integrate Proximate and Ultimate Causation." *Biology & Philosophy* 30 (1): 19–37.

Otsuka, Jun. 2016. "A Critical Review of the Statisticalist Debate." *Biology & Philosophy* 31 (4): 459–482.

Oyama, Susan. 2000a. "Causal Democracy and Causal Contributions in Developmental Systems Theory." *Philosophy of Science* 67: S332– S347.

Oyama, Susan. 2000b. *Evolution's Eye: A Systems View of the Biology-Culture Divide*. Durham, NC: Duke University Press.

Oyama, Susan. 2000c. *The Ontogeny of Information: Developmental Systems and Evolution*. Durham, NC: Duke University Press.

Oyama, Susan. 2006. "Boundaries and (Constructive) Interaction." In *Genes in Development. Re-Reading the Molecular Paradigm*, edited by Eva M. Neumann-Held and Christoph Rehmann-Sutter, 272–289. Durham, NC: Duke University Press.

Parascandola, John. 1971. "Organismic and Holistic Concepts in the Thought of L. J. Henderson." *Journal of the History of Biology* 4 (1): 63–113.

Parker, George Howard. 1924. "Organic Determinism." *Science* 59 (1537): 517–521.

Patten, Bernard C. 2001. "Jakob von Uexküll and the Theory of Environs." *Semiotica* 2001 (134): 423–443.

Paul, Sarah. 2020. *Philosophy of Action: A Contemporary Introduction*. New York: Routledge.

Pausas, Juli G., and William J. Bond. 2022. "Feedbacks in Ecology and Evolution." *Trends in Ecology & Evolution* 37 (8): 637–644.

Pearce, Trevor. 2010. "From 'Circumstances' to 'Environment': Herbert Spencer and the Origins of the Idea of Organism–Environment Interaction." *Studies in History and Philosophy of Science Part C: Studies in History and Philosophy of Biological and Biomedical Sciences* 41 (3): 241–252.

Pearce, Trevor. 2014a. "The Dialectical Biologist, circa 1890: John Dewey and the Oxford Hegelians." *Journal of the History of Philosophy* 52 (4): 747–777.

Pearce, Trevor. 2014b. "The Origins and Development of the Idea of Organism-Environment Interaction." In *Entangled Life: Organism and Environment in the Biological and Social Sciences*, edited by Gillian Barker, Eric Desjardins, and Trevor Pearce, 13–32. Dordrecht: Springer Netherlands.

Pearce, Trevor. 2020. *Pragmatism's Evolution: Organism and Environment in American Philosophy*. Chicago: University of Chicago Press.

Pearl, Raymond. 1901. "Studies on the Effects of Electricity on Organisms. II. The Reactions of Hydra to the Constant Current." *American Journal of Physiology* 5 (5): 301–320.

Pearl, Raymond. 1922. *The Biology of Death*. J. B. Lippincott.

Pearl, Raymond. 1925. *The Biology of Population Growth*. New York: A. A. Knopf.

Pearse, Arthur Sperry. 1922. "The Effects of Environment on Animals." *American Naturalist* 56 (643): 144–158.

Peirce, Charles Sanders. 1931. *The Collected Papers of C. S. Peirce*, vol. 1–6, edited by Charles Hartshorne and Paul Weiss. Cambridge, MA: Harvard University Press.

Peirce, George J. 1904. "Certain Undetermined Factors in Heredity and Environment." *American Naturalist* 38 (448): 285–293.

Perlman, Robert L. 2000. "The Concept of the Organism in Physiology." *Theory in Biosciences* 119 (3): 174–186. https://doi.org/10.1078/1431-7613-00015.

Perry, Ralph Barton. 1917. "Purpose as Tendency and Adaptation." *Philosophical Review* 26 (5): 477–495. https://doi.org/10.2307/2178045.

Peterson, Erik L. 2011. "The Excluded Philosophy of Evo-Devo? Revisiting C.H. Waddington's Failed Attempt to Embed Alfred North Whitehead's 'Organicism' in Evolutionary Biology." *History and Philosophy of the Life Sciences* 33 (3): 301–320.

Peterson, Erik L. 2016. *The Life Organic: The Theoretical Biology Club and the Roots of Epigenetics*. Pittsburgh: University of Pittsburgh Press.

Pfennig, David W., ed. 2021. *Phenotypic Plasticity & Evolution: Causes, Consequences, Controversies*. Boca Raton, FL: CRC Press.

Phillips, Denis C. 1970. "Organicism in the Late Nineteenth and Early Twentieth Centuries." *Journal of the History of Ideas* 31 (3): 413–432.

Piaget, Jean. (1967) 1971. *Biology and Knowledge: An Essay on the Relations Between Organic Regulations and Cognitive Processes*. Chicago: University of Chicago Press.

Pigliucci, Massimo. 1996. "How Organisms Respond to Environmental Changes: From Phenotypes to Molecules (and Vice Versa)." *Trends in Ecology & Evolution* 11 (4): 168–173.

Pigliucci, Massimo, Courtney J. Murren, and Carl D. Schlichting. 2006. "Phenotypic Plasticity and Evolution by Genetic Assimilation." *Journal of Experimental Biology* 209 (12): 2362–2367.

Pike, F. H., and E. L. Scott. 1915. "The Significance of Certain Internal Conditions of the Organism in Organic Evolution. First Paper. The Regulation of the Physico-Chemical Conditions of the Organism." *American Naturalist* 49 (582): 321–359.

Pimentel, David. 1961. "Animal Population Regulation by the Genetic Feed-Back Mechanism." *American Naturalist* 95 (881): 65–79.

Pimentel, David. 1968. "Population Regulation and Genetic Feedback." *Science* 159 (3822): 1432–1437.

Pittendrigh, Colin S. 1958. "Adaptation, Natural Selection, and Behavior." In *Behavior and Evolution*, edited by Anne Roe and George Gaylord Simpson, 390–416. New Haven, CT: Yale University Press.

"Plants Considered in Relation to Their Environment." 1886. *Nature* 33 (861): 607–609.

Platt, J. 1969. "Theorems on Boundaries in Hierarchical Systems." In *Hierarchical Structures*, edited by Lancelot Law Whyte, Albert G. Wilson, and Donna M. Wilson, 201–213. New York: Elsevier.

Platt, Robert B., W. D. Billings, David M. Gates, Charles E. Olmsted, Royal E. Shanks, and John R. Tester. 1964. "The Importance of Environment to Life." *BioScience* 14 (7): 25–29.

Plochmann, George Kimball. 1953. "D'Arcy Thompson: His Conception of the Living Body." *Philosophy of Science* 20 (2): 139–148.

Plotkin, Henry C. 1988. "Behavior and Evolution." In *The Role of Behavior in Evolution*, edited by Henry C. Plotkin, 1–17. Cambridge, MA: MIT Press.

Polger, Thomas W. 2022. "Naturalizing the Metaphysics of Science." *Philosophia* 50 (2): 659–670.

Pontarotti, Gaëlle, Antoine C. Dussault, and Francesca Merlin. 2022. "Conceptualizing the Environment in Natural Sciences: Guest Editorial." *Biological Theory* 17 (1): 1–3.

Pontarotti, Gaëlle, and Francesca Merlin. 2023. "From Exposome to Pathogenic Niche. Looking for an Operational Account of the Environment in Health Studies." In *Integrative Approaches in Environmental Health and Exposome Research: Epistemological and Practical Issues*, edited by Élodie Giroux, Francesca Merlin, and Yohan Fayet, 173–206. Cham, Switzerland: Springer International Publishing

Poorter, Hendrik, Fabio Fiorani, Roland Pieruschka, Tobias Wojciechowski, Wim H. van der Putten, Michael Kleyer, Uli Schurr, and Johannes Postma. 2016. "Pampered Inside, Pestered Outside? Differences and Similarities Between Plants Growing in Controlled Conditions and in the Field." *New Phytologist* 212 (4): 838–855.

Population Studies: Animal Ecology and Demography. Cold Spring Harbor Symposia on Quantitative Biology, vol. XXII. 1957. Cold Spring Harbor, NY: The Biological Laboratory.

Portides, Demetris. 2021. "Idealization and Abstraction in Scientific Modeling." *Synthese* 198 (24): 5873–5895.

Potochnik, Angela. 2020. *Idealization and the Aims of Science*. Chicago: University of Chicago Press.

Poulin, Robert, and Thomas H. Cribb. 2002. "Trematode Life Cycles: Short Is Sweet?" *Trends in Parasitology* 18 (4): 176–183.

Pradeu, Thomas. 2010. "What Is an Organism? An Immunological Answer." *History and Philosophy of the Life Sciences* 32 (2–3): 247–267.

Pradeu, Thomas. 2011. "A Mixed Self: The Role of Symbiosis in Development." *Biological Theory* 6 (1): 80–88.

Pradeu, Thomas. 2016. "The Many Faces of Biological Individuality." *Biology & Philosophy* 31 (6): 761–773.

Pradeu, Thomas. 2018. "Genidentity and Biological Processes." In *Everything Flows: Towards a Processual Philosophy of Biology*, edited by Daniel J. Nicholson and John Dupré, 96–112. Oxford: Oxford University Press.

Prenant, Marcel. 1938. *Biology and Marxism*. New York: International Publishers.

Prieto, Guido Ignacio. 2023. "'Organism' Versus 'Biological Individual': The Missing Demarcation." *ArtefaCToS. Revista de estudios sobre la ciencia y la tecnología* 12 (1): 27–54. https://doi.org/10.14201/art20231212754.

Prieto, Guido Ignacio. 2024. "Organisms and Biological Individuals: A Bio-Philosophical Inquiry." PhD thesis, Ruhr University Bochum.

Prieto, Guido Ignacio, and Alejandro Fábregas-Tejeda. 2022. "Richard Lewontin y la reciprocidad organismo-ambiente en la historia de la biología." *Ludus Vitalis* 29 (56): 31–38. https://revistas.uv.cl/index.php/ludusvitalis/article/view/3849.

Prothero, Donald R. 2013. *Bringing Fossils to Life: An Introduction to Paleobiology*. New York: Columbia University Press.

Przibram, Hans. 1908. "Die biologische Versuchsanstalt in Wien." *Zeitschrift für biologische Technik und Methodik* 1 (3): 234–264.

Przibram, Hans. 1917. "Die Umwelt des Keimplasmas." *Archiv für Entwicklungsmechanik der Organismen* 43 (1): 37–46.

Radick, Gregory. 2017. "Animal Agency in the Age of the Modern Synthesis: W.H. Thorpe's Example." *BJHS Themes* 2: 35–56.

Radick, Gregory. 2024. "The Baldwin Effect and the Potentialities for Thoughtful Darwinism Around 1900." In *The Riddle of Organismal Agency: New Historical and Philosophical Reflections*, edited by Alejandro Fábregas-Tejeda, Jan Baedke, Guido I. Prieto, and Gregory Radick, 97–113. Abingdon: Routledge.

Raerinne, Jani, and Jan Baedke. 2015. "Exclusions, Explanations, and Exceptions: On the Causal and Lawlike Status of the Competitive Exclusion Principle." *Philosophy & Theory in Biology* 7: e602. http://dx.doi.org/10.3998/ptb.6959004.0007.002.

Rama, Tiago. 2024. "The Historical Transformation of Individual Concepts into Populational Ones: An Explanatory Shift in the Gestation of the Modern Synthesis." *History and Philosophy of the Life Sciences* 46 (4): 35. https://doi.org/10.1007/s40656-024-00638-2.

Ratcliffe, Matthew. 2001. "A Kantian Stance on the Intentional Stance." *Biology and Philosophy* 16 (1): 29–52.

Rattasepp, Silver. 2010. "The Idea of Extended Organism in 20th Century Thought." *Hortus Semioticus* 6: 31–39.

Raynaud, Dominique. 1998. "La controverse entre organicisme et vitalisme: étude de sociologie des sciences." *Revue Française de Sociologie* 39: 721–750.

Reed, Edward S., and Rebecca K. Jones. 1977. "Towards a Definition of Living Systems: A Theory of Ecological Support for Behavior." *Acta Biotheoretica* 26 (3): 153–163.

Reiß, Christian. 2007. "No Evolution, No Heredity, Just Development—Julius Schaxel and the End of the Evo–Devo Agenda in Jena, 1906–1933: A Case Study." *Theory in Biosciences* 126 (4): 155–164.

Reiß, Christian. 2022. "From Organismic Biology as History and Philosophy to the History and Philosophy of Biology: the Work of Hans-Jörg Rheinberger in the German Context." *Berichte zur Wissenschaftsgeschichte* 45 (3): 384–396.

Reiss, John. 2009. *Not by Design: Retiring Darwin's Watchmaker*. London: University of California Press.

Rendón, Constanza, and Guillermo Folguera. 2012. "La relación organismo-ambiente en la extensión de la síntesis biológica." *Epistemología e Historia de la Ciencia* 18: 492–499.

Renwick, Chris. 2009. "The Practice of Spencerian Science: Patrick Geddes's Biosocial Program, 1876–1889." *Isis* 100 (1): 36–57.

Reznick, David N. 2013. "A Critical Look at Reciprocity in Ecology and Evolution." *American Naturalist* 181 (S1): S1–S8.

Richards, Robert J. 2000. "Kant and Blumenbach on the Bildungstrieb: A Historical Misunderstanding." *Studies in History and Philosophy of Science Part C: Studies in History and Philosophy of Biological and Biomedical Sciences* 31 (1): 11–32.

Richards, Robert J. 2004. *The Romantic Conception of Life: Science and Philosophy in the Age of Goethe*. Chicago: University of Chicago Press.

Rieppel, Olivier. 2016. *Phylogenetic Systematics: Haeckel to Hennig*. Boca Raton, FL: CRC Press.

Rignano, Eugenio. 1930. *The Nature of Life*. London: Kegan Paul, Trench, Trubner.

Riley, Charles Valentine. 1888. "On the Causes of Variation in Organic Forms." *Science* 12 (293): 123–125.

Riskin, Jessica. 2016. *The Restless Clock: A History of the Centuries-Long Argument over What Makes Living Things Tick*. Chicago: University of Chicago Press.

Riskin, Jessica. 2020. "Biology's Mistress, a Brief History." *Interdisciplinary Science Reviews* 45 (3): 268–298.

Ritter, William Emerson. 1916. "The Place of Description, Definition and Classification in Philosophical Biology." *Scientific Monthly* 3 (5): 455–470.

Ritter, William Emerson. 1919. *The Unity of the Organism, or The Organismal Conception of Life*. Boston: Richard G. Badger.

Ritter, William Emerson. 1932. "Why Aristotle Invented the Word Entelecheia." *Quarterly Review of Biology* 7 (4): 377–404.

Ritter, William Emerson. 1933. "Historical and Contemporary Relationships of Physical and Biological Sciences. A Critical Summary of the Papers Presented at This Session." *Archeion* 14 (4): 497–515.

Rivera-Yoshida, Natsuko, Alejandra Hernández-Terán, Ana E. Escalante, and Mariana Benítez. 2020. "Laboratory Biases Hinder Eco-Evo-Devo Integration: Hints from the Microbial World." *Journal of Experimental Zoology Part B: Molecular and Developmental Evolution* 334 (1): 14–24.

Robertson, Alan. 1977. "Conrad Hal Waddington. 8 November 1905–26 September 1975." *Biographical Memoirs of Fellows of the Royal Society* 23: 575–622.

Robison, Shea K. 2018. *Epigenetics and Public Policy: The Tangled Web of Science and Politics*. Santa Barbara, CA: ABC-CLIO.

Rodríguez-Caso, Juan Manuel. 2022. "The Institutionalisation of Biology in the British Association for the Advancement of Science, 1866–1894." *História, Ciências, Saúde-Manguinhos* 29: 993–1011.

Roll-Hansen, Nils. 1984. "E. S. Russell and J. H. Woodger: The Failure of Two Twentieth-Century Opponents of Mechanistic Biology." *Journal of the History of Biology* 17 (3): 399–428.

Roll-Hansen, Nils. 2000. "The Application of Complementarity to Biology: From Niels Bohr to Max Delbrück." *Historical Studies in the Physical and Biological Sciences* 30 (2): 417–442.

Roux, Wilhelm. 1895. *Gesammelte Abhandlungen über Entwickelungsmechanik der Organismen*, vol. 2. Leipzig, Germany: Wilhelm Engelmann.

Roux, Wilhelm, Carl Correns, Alfred Fischel, and E. Küster. 1912. *Terminologie der Entwicklungsmechanik der Tiere und Pflanzen*. Leipzig, Germany: Wilhelm Engelmann.

Ruiz-Mirazo, Kepa, Arantza Etxeberria, Alvaro Moreno, and Jesús Ibáñez. 2000. "Organisms and Their Place in Biology." *Theory in Biosciences* 119 (3): 209–233.

Ruiz-Mirazo, Kepa, Juli Peretó, and Alvaro Moreno. 2004. "A Universal Definition of Life: Autonomy and Open-Ended Evolution." *Origins of Life and Evolution of the Biosphere* 34 (3): 323–346.

Rupik, Gregory. 2024. *Remapping Biology with Goethe, Schelling, and Herder: Romanticising Evolution*. London: Routledge.

Ruse, Michael. 2000. "Teleology: Yesterday, Today, and Tomorrow?" *Studies in History and Philosophy of Science Part C: Studies in History and Philosophy of Biological and Biomedical Sciences* 31 (1): 213–232.

Russell, Bertrand. 1903. *The Principles of Mathematics*. Cambridge, UK: Cambridge University Press.

Russell, Edward Stuart. 1904. "Notes on *Depastrum cyathiforme*, Gosse." *Annals and Magazine of Natural History* 13 (73): 62–65.

Russell, Edward Stuart. 1907. "The Atractylis Coccinea of T. S. Wright." *Annals and Magazine of Natural History* 20 (115): 52–55.

Russell, Edward Stuart. 1908. "LUDWIG PLATE- Selectionsprinzip und Probleme der Artbildung—Dritte vermehrte Auflage. S. 493. Leipzig, Engelmann." *Rivista di Scienza* 4: 175–179.

Russell, Edward Stuart. 1909a. "Preliminary Notice of the Cephalopoda Collected by the Fishery Cruiser 'Goldseeker,' 1903–1908." *Annals and Magazine of Natural History* 3 (17): 446–455.

Russell, Edward Stuart. 1909b. "The Evidence of Natural Selection." *Rivista di Scienza* 5: 67–85.

Russell, Edward Stuart. 1914. "Recensioni." *Scientia* 15: 109–116.

Russell, Edward Stuart. 1916. *Form and Function: A Contribution to the History of Animal Morphology*. London: John Murray.

Russell, Edward Stuart. 1919. "Note on the Righting Reaction in *Asterina gibbosa* Penn." *Proceedings of the Zoological Society of London* 89 (3–4): 423–432.

Russell, Edward Stuart. 1923. "Psychobiology." *Proceedings of the Aristotelian Society* 23 (1): 141–156.

Russell, Edward Stuart. 1924. *The Study of Living Things: Prolegomena to a Functional Biology*. London: Methuen.

Russell, Edward Stuart. 1930. *The Interpretation of Development and Heredity: A Study in Biological Method*. Oxford: Clarendon Press.

Russell, Edward Stuart. 1931. "Detour Experiments with Sticklebacks (*Gasterosteus aculeatus* L.)." *Journal of Experimental Biology* 8 (4): 393–410.

Russell, Edward Stuart. 1934. *The Behaviour of Animals: An Introduction to Its Study*. London: E. Arnold.

Russell, Edward Stuart. 1935. "Valence and Attention in Animal Behaviour." *Acta Biotheoretica* 1 (1): 91–99.

Russell, Edward Stuart. 1936. "Playing with a Dog." *Quarterly Review of Biology* 11 (1): 1–15.

Russell, Edward Stuart. 1937. "Fish Migrations." *Biological Reviews* 12 (3): 320–337.

Russell, Edward Stuart. 1945. *The Directiveness of Organic Activities*. Cambridge, UK: Cambridge University Press.

Salisbury, E. J. 1930. "Concerning the Study of Plants." *Science Progress in the Twentieth Century (1919–1933)* 25 (97): 50–63.

Sandeman, George. 1896. *Problems of Biology*. London: Swan Sonnenschein.

Sarkar, Sahotra. 1999. "From the *Reaktionsnorm* to the Adaptive Norm: The Norm of Reaction, 1909–1960." *Biology and Philosophy* 14 (2): 235–252.

Schaffner, Kenneth F. 1967. "Antireductionism and Molecular Biology." *Science* 157 (3789): 644–647.

Schanck, Richard Louis. 1954. *The Permanent Revolution in Science*. New York: Philosophical Library.

Schaxel, Julius. 1917. "Mechanismus, Vitalismus und kritische Biologie." *Biologisches Zentralblatt* 37: 188–196.

Schaxel, Julius. 1919. *Grundzuege der Theoriebildung in der Biologie*. Jena, Germany: G. Fischer.

Schiller, Claire H. 1957. "Note by the Translator." In *Instinctive Behavior: The Development of a Modern Concept*, edited by Claire H. Schiller. New York: International Universities Press.

Schlichting, Carl D., and Matthew A. Wund. 2014. "Phenotypic Plasticity and Epigenetic Marking: An Assessment of Evidence for Genetic Accommodation." *Evolution* 68 (3): 656–672.

Schmidt, Hartmut. 1986. *Die lebendige Sprache: Zur Entstehung des Organismuskonzepts*. Berlin: Akademie der Wissenschaften der DDR, Zentralinstitut für Sprachwissenschaft.

Schnödl, Gottfried, and Florian Sprenger. 2021. *Uexkülls Umgebungen: Umweltlehre und rechtes Denken*. Lüneburg: Meson Press.

Schürch, Caterina. 2019. "Understanding Past Research Practice: A Case for iHPS." In *The Past, Present, and Future of Integrated History and Philosophy of Science*, edited by Emily Herring, Kevin Matthew Jones, Konstantin S. Kiprijanov, and Laura M. Sellers, 38–60. London: Routledge.

Schwab, Daniel B., Sofia Casasa, and Armin P. Moczek. 2019. "On the Reciprocally Causal and Constructive Nature of Developmental Plasticity and Robustness." *Frontiers in Genetics* 9:735. https://www.frontiersin.org/articles/10.3389/fgene.2018.00735.

Selcer, Perrin. 2018. *The Postwar Origins of the Global Environment: How the United Nations Built Spaceship Earth*. New York: Columbia University Press.

Semper, Karl. 1881. *Animal Life as Affected by the Natural Conditions of Existence*. New York: Appleton.

Serres, Michel. 1982. *Hermes: Literature, Science, Philosophy*. Baltimore: John Hopkins University Press.

Shields, Christopher. 2017. "What Organisms Once Were and Might Yet Be." *Philosophy & Theory in Biology* 9. http://dx.doi.org/10.3998/ptb.6959004.0009.007.

Simon, Herbert A. 1982. *Models of Bounded Rationality*. 2 vols. Cambridge, MA: MIT Press.

Simon, Herbert A. 1956. "Rational Choice and the Structure of the Environment." *Psychological Review* 63 (2): 129–138.

Simpson, George Gaylord. 1944. *Tempo and Mode in Evolution*. New York: Columbia University Press.

Simpson, George Gaylord. 1953. "The Baldwin Effect." *Evolution* 7 (2): 110–117.

Simpson, George Gaylord. 1958. "Behavior and Evolution." In *Behavior and Evolution*, edited by Anne Roe and George Gaylord Simpson, 507–535. New Haven, CT: Yale University Press.

Simpson, George Gaylord. 1961. *Principles of Animal Taxonomy*. New York: Columbia University Press.

Sims, Matthew. 2021. "A Continuum of Intentionality: Linking the Biogenic and Anthropogenic Approaches to Cognition." *Biology & Philosophy* 36 (6): 51. https://doi.org/10.1007/s10539-021-09827-w.

Slater, Matthew H. 2015. "Natural Kindness." *British Journal for the Philosophy of Science* 66 (2): 375–411.

Sloan, Phillip R. 2012. "How Was Teleology Eliminated in Early Molecular Biology?" *Studies in History and Philosophy of Science Part C: Studies in History and Philosophy of Biological and Biomedical Sciences*, 43 (1): 140–151.

Smith, Barry, and Achille C. Varzi. 2001. "Environmental Metaphysics." In *Metaphysics in the Post-Metaphysical Age*, edited by Uwe Meixner and Peter Simons, 231–239. Vienna: Öbv & Hpt.

Smith, Barry, and Achille C. Varzi. 2002. "Surrounding Space: The Ontology of Organism-Environment Relations." *Theory in Biosciences* 121 (2): 139–162.

Smith, Justin E. H. 2011. *Divine Machines: Leibniz and the Sciences of Life*. Princeton, NJ: Princeton University Press.

Smith, Subrena E. 2017. "Organisms as Persisters." *Philosophy & Theory in Biology* 9. http://dx.doi.org/10.3998/ptb.6959004.0009.014.

Smocovitis, Vassiliki Betty. 1992. *Unifying Biology: The Evolutionary Synthesis and Evolutionary Biology*. Princeton, NJ: Princeton University Press.

Smocovitis, Vassiliki Betty. 1994. "Disciplining Evolutionary Biology: Ernst Mayr and the Founding of the Society for the Study of Evolution and Evolution (1939–1950)." *Evolution* 48 (1): 1–8.

Sober, Elliott. 1984. *The Nature of Selection: Evolutionary Theory in Philosophical Focus*. Cambridge, MA: MIT Press.

Sober, Elliott. 2000. *Philosophy of Biology*, 2nd ed. Boulder, CO: Westview Press

Solomon, M. E. 1959. "A Symposium on Population Ecology." *Ecology* 40 (2): 325–326.

Spencer, Herbert. 1864. *The Principles of Biology*, vol. I. London: Williams and Norgate.

Spencer, Herbert. 1887. *The Factors of Organic Evolution*. New York: D. Appelton.

Spencer, Herbert. 1855. *The Principles of Psychology*. London: Longman, Brown, Green, and Longmans.

Spitzer, Leo. 1942. "Milieu and Ambiance: An Essay in Historical Semantics." *Philosophy and Phenomenological Research* 3 (1): 1–42.

Sprenger, Florian. 2019. *Epistemologien des Umgebens: Zur Geschichte, Ökologie und Biopolitik künstlicher Environments*. Bielefeld, Germany: Transcript Verlag.

Sprenger, Florian. 2023. "Surrounding and Surrounded: Toward a Conceptual History of Environment." *Critical Inquiry* 49 (3): 406–427.

Stebbins, George Ledyard. 1972. "Research on the Evolution of Higher Plants: Problems and Prospects." *Canadian Journal of Genetics and Cytology* 14 (3): 453–462.

Stebbins, George Ledyard. 1974. *Flowering Plants: Evolution Above the Species Level*. Cambridge, MA: Belknap Press.

Steele, John H., Kenneth H. Brink, and Beth E. Scott. 2019. "Comparison of Marine and Terrestrial Ecosystems: Suggestions of an Evolutionary Perspective Influenced by Environmental Variation." *ICES Journal of Marine Science* 76 (1): 50–59.

Steffes, David M. 2007. "Panpsychic Organicism: Sewall Wright's Philosophy for Understanding Complex Genetic Systems." *Journal of the History of Biology* 40 (2): 327–361.

Steigerwald, Joan. 2010. "Natural Purposes and the Reflecting Power of Judgment: The Problem of the Organism in Kant's Critical Philosophy." *European Romantic Review* 21 (3): 291–308.

Steigerwald, Joan. 2019. *Experimenting at the Boundaries of Life: Organic Vitality in Germany around 1800*. Pittsburgh: University of Pittsburgh Press.

Sterelny, Kim, and Paul E. Griffiths. 1999. *Sex and Death: An Introduction to Philosophy of Biology*. Chicago: University of Chicago Press.

Stoffregen, Thomas A. 2003. "Affordances as Properties of the Animal-Environment System." *Ecological Psychology* 15 (2): 115–134.

Stotz, Karola. 2017. "Why Developmental Niche Construction Is Not Selective Niche Construction: And Why It Matters." *Interface Focus* 7 (5): 20160157. https://doi.org/10.1098/rsfs.2016.0157.

Strawson, Peter Frederick. 1959. *Individuals: An Essay in Descriptive Metaphysics.* London: Methuen.

Strick, James. 1999. "Darwinism and the Origin of Life: The Role of H. C. Bastian in the British Spontaneous Generation Debates, 1868–1873." *Journal of the History of Biology* 32 (1): 51–92.

Strick, James Edgar. 2000. *Sparks of Life: Darwinism and the Victorian Debates over Spontaneous Generation.* Cambridge, MA: Harvard University Press.

Stroll, Avrum. 1979. "Two Conceptions of Surfaces." *Midwest Studies in Philosophy* 4 (1): 277–291.

Strom, K. Munster. 1929. "The Study of Limnology." *Journal of Ecology* 17 (1): 106–111.

Sturdy, Steve. 1988. "Biology as Social Theory: John Scott Haldane and Physiological Regulation." *British Journal for the History of Science* 21 (3): 315–340.

Sturdy, Steve. 2011. "The Meanings of 'Life': Biology and Biography in the Work of J. S. Haldane (1860–1936)." *Transactions of the Royal Historical Society* 21: 171–191.

Suárez, Javier, and Vanessa Triviño. 2019. "A Metaphysical Approach to Holobiont Individuality: Holobionts as Emergent Individuals." *Quaderns de Filosofia* 6 (1): 59–76.

Suárez-Díaz, Edna. 2017. *Evolución y moléculas: La molecularización de la biología evolutiva en contexto.* Mexico City: Centro de Estudios Filosóficos, Políticos y Sociales Vicente Lombardo Toledano.

Sultan, Sonia E. 2003. "Phenotypic Plasticity in Plants: A Case Study in Ecological Development." *Evolution & Development* 5 (1): 25–33.

Sultan, Sonia E. 2007. "Development in Context: The Timely Emergence of Eco-Devo." *Trends in Ecology & Evolution* 22 (11): 575–582.

Sultan, Sonia E. 2015. *Organism and Environment: Ecological Development, Niche Construction, and Adaptation.* Oxford: Oxford University Press.

Sultan, Sonia E. 2017. "Developmental Plasticity: Re-Conceiving the Genotype." *Interface Focus* 7 (5): 20170009. https://doi.org/10.1098/rsfs.2017.0009.

Sultan, Sonia E. 2021. "Eco-Evo-Devo." In *Evolutionary Developmental Biology: A Reference Guide*, edited by Laura Nuño de la Rosa and Gerd B. Müller, 1165–1177. Cham, Switzerland: Springer International Publishing.

Sultan, Sonia E., Armin P. Moczek, and Denis Walsh. 2022. "Bridging the Explanatory Gaps: What Can We Learn from a Biological Agency Perspective?" *BioEssays* 44 (1): 2100185. https://doi.org/10.1002/bies.202100185.

Sumner, Francis B. 1910. "The Reappearance in the Offspring of Artificially Produced Parental Modifications." *American Naturalist* 44 (517): 5–18.

Sumner, Francis B. 1919a. "Adaptation and the Problem of 'Organic Purposefulness.'" *American Naturalist* 53 (626): 193–217.

Sumner, Francis B. 1919b. "Adaptation and the Problem of 'Organic Purposefulness.' II." *American Naturalist* 53 (627): 338–369.

Sumner, Francis B. 1922. "The Organism and Its Environment." *Scientific Monthly* 14 (3): 223–233.

Sumner, Francis B. 1924. "The Stability of Subspecific Characters Under Changed Conditions of Environment." *American Naturalist* 58 (659): 481–505.

Surman, Jan, Katalin Stráner, and Peter Haslinger. 2014. "Nomadic Concepts in the History of Biology." *Studies in History and Philosophy of Science Part C: Studies in History and Philosophy of Biological and Biomedical Sciences* 48: 127–129.

Svensson, Erik I. 2018. "On Reciprocal Causation in the Evolutionary Process." *Evolutionary Biology* 45 (1): 1–14.

Tabery, James. 2008. "R. A. Fisher, Lancelot Hogben, and the Origin(s) of Genotype-Environment Interaction." *Journal of the History of Biology* 41 (4): 717–761.

Tamborini, Marco. 2018. "Expanding the History of Natural History." *Historical Studies in the Natural Sciences* 48 (3): 390–401.

Tamborini, Marco. 2022. "Organic Form and Evolution: The Morphological Problem in Twentieth-Century Italian Biology." *History and Philosophy of the Life Sciences* 44 (4): 54. https://doi.org/10.1007/s40656-022-00534-7.

Tamborini, Marco. 2023. *The Architecture of Evolution: The Science of Form in Twentieth-Century Evolutionary Biology*. Pittsburgh: University of Pittsburgh Press.

Tanaka, Mark M., Peter Godfrey-Smith, and Benjamin Kerr. 2020. "The Dual Landscape Model of Adaptation and Niche Construction." *Philosophy of Science* 87 (3): 478–498.

Tansley, Arthur George. 1920. "The Classification of Vegetation and the Concept of Development." *Journal of Ecology* 8 (2): 118–149.

Tarrant, Ann M. 2019. "Ecology and Physiology of Dormancy in a Changing World: Introduction to a Virtual Symposium in The Biological Bulletin." *Biological Bulletin* 237 (2): 73–75.

Taschwer, Klaus. 2014. "Expelled, Burnt, Sold, Forgotten, and Suppressed: The Permanent Destruction of the Institute for Experimental Biology and Its Academic Staff." In *The Academy of Sciences in Vienna: 1930 to 1945*, edited by Johannes Feichtinger, Herbert Matis, Stefan Sienell, and Heidemarie Uhl, 101–111. Vienna: Austrian Academy of Sciences Press.

Taschwer, Klaus. 2020. "Destroyed Research in Nazi Vienna: The Tragic Fate of the Institute for Experimental Biology in Austria." *Mètode Science Studies Journal* 10: 139–145.

Taschwer, Klaus, Johannes Feichtinger, Stefan Sienell, and Heidemarie Uhl, eds. 2016. *Experimental Biology in the Vienna Prater: On the History of the Institute for Experimental Biology 1902 to 1945*, 1st ed. Vienna: Austrian Academy of Sciences Press.

Taylan, Ferhat. 2018. *Mésopolitique: Connaître, théoriser et gouverner les milieux de vie (1750–1900)*. Paris: Éditions de la Sorbonne.

Taylan, Ferhat. 2021. "Mesology and Ecology: Two Alternative Views of the Environment and Their Political Implications in the Nineteenth Century." In *Hybrid Ecologies*, edited by Susanne Witzgall, Marietta Kesting, Maria Muhle, and Jenny Nachtigall, 31–41. Zurich: Diaphanes.

Taylan, Ferhat. 2022. "The Rise of the 'Environment': Lamarckian Environmentalism Between Life Sciences and Social Philosophy." *Biological Theory* 17 (1): 4–19.

Thompson, D'Arcy Wentworth. 1884. "The Regeneration of Lost Parts in Animals." *Mind* 9 (35): 415–420.

Thompson, D'Arcy Wentworth. 1913. *On Aristotle as a Biologist*. Oxford: Clarendon Press.

Thompson, John N. 1998. "Rapid Evolution as an Ecological Process." *Trends in Ecology & Evolution* 13 (8): 329–332.

Thompson, William Robin. 1928. "A Contribution to the Study of Biological Control and Parasite Introduction in Continental Areas." *Parasitology* 20 (1): 90–112.

Thompson, William Robin. 1929. "A Contribution to the Study of Morphogenesis in the Muscoid Diptera." *Transactions of the Royal Entomological Society of London* 77 (2): 195–244.

Thompson, William Robin. 1939. "Biological Control and the Theories of the Interactions of Populations." *Parasitology* 31 (3): 299–388.

Thompson, William Robin. 1956. "The Fundamental Theory of Natural and Biological Control." *Annual Review of Entomology* 1: 379–402.

Thorpe, Malcolm Rutherford. 1924. "Introduction." In *Organic Adaptation to Environment*, edited by Malcolm Rutherford Thorpe, xiii–xviii. New Haven, CT: Yale University Press.

Thorpe, William Homan. 1945. "The Evolutionary Significance of Habitat Selection." *Journal of Animal Ecology* 14 (2): 67–70.

Thorpe, William Homan. 1978. *Purpose in a World of Chance: A Biologist's View*. Oxford: Oxford University Press.

Tinbergen, Nikolaas. 1963. "On Aims and Methods of Ethology." *Zeitschrift für Tierpsychologie* 20 (4): 410–433.

Tincker, M. A. H. 1924. "Effect of Length of Day on Flowering and Growth." *Nature* 114 (2862): 350–351.

Toepfer, Georg. 2004. *Zweckbegriff und Organismus: Über die teleologische Beurteilung biologischer Systeme*. Würzburg, Germany: Königshausen & Neumann.

Toepfer, Georg. 2011a. *Historisches Wörterbuch der Biologie. Geschichte und Theorie der biologischen Grundbegriffe. Band 3: Parasitismus–Zweckmäßigkeit*. Stuttgart, Germany: J. B. Metzler.

Toepfer, Georg. 2011b. "Kant's Teleology, the Concept of the Organism, and the Context of Contemporary Biology." In *Final Causes and Teleological Explanation*, edited by Dominik Perler and Stephan Schmid, 107–124. Paderborn, Germany: Mentis.

Toepfer, Georg. 2011c. "Organismus." In *Historisches Wörterbuch der Biologie. Geschichte und Theorie der biologischen Grundbegriffe. Band 2: Gefühl–Organismus*, edited by Georg Toepfer, 777–842. Stuttgart, Germany: J. B. Metzler.

Toepfer, Georg. 2012. "Teleology and Its Constitutive Role for Biology as the Science of Organized Systems in Nature." *Studies in History and Philosophy of Science Part C: Studies in History and Philosophy of Biological and Biomedical Sciences* 43 (1): 113–119.

Tomasello, Michael. 2022. *The Evolution of Agency: Behavioral Organization from Lizards to Humans*. Cambridge, MA: MIT Press.

Torres, J.-L., O. Pérez-Maqueo, M. Equihua, and L. Torres. 2009. "Quantitative Assessment of Organism–Environment Couplings." *Biology & Philosophy* 24 (1): 107–117.

Torrey, Harry Beal. 1907. "The Method of Trial and the Tropism Hypothesis." *Science* 26 (662): 313–323.

Trappes, Rose. 2021. "Defining the Niche for Niche Construction: Evolutionary and Ecological Niches." *Biology & Philosophy* 36 (3): 31. https://doi.org/10.1007/s10539-021-09805-2.

Trappes, Rose, Behzad Nematipour, Marie I. Kaiser, Ulrich Krohs, Koen J. van Benthem, Ulrich R. Ernst, Jürgen Gadau, et al. 2022. "How Individualized Niches Arise: Defining Mechanisms of Niche Construction, Niche Choice, and Niche Conformance." *BioScience* 72 (6): 538–548.

Trestman, Michael. 2012. "The Environment, from a Behavioral Perspective." In *The Environment: Philosophy, Science, and Ethics*, edited by William P. Kabasenche, Michael O'Rourke, and Matthew H. Slater, 57–72. Cambridge, MA: MIT Press.

Trienes, Rudie. 1992. "Holism and Kantian Teleology in C. J. Van De Klaauw's Structuralization of Oecology." *Acta Biotheoretica* 40 (1): 11–22.

Triviño, Vanesa. 2022. "Towards a Characterization of Metaphysics of Biology: Metaphysics for and Metaphysics in Biology." *Synthese* 200 (5): 428. https://doi.org/10.1007/s11229-022-03897-3.

Triviño, Vanessa, and María Cerezo. 2015. "The Metaphysical Equivalence Between 3D and 4D Theories of Species." *Revista Portuguesa de Filosofía* 71 (4): 783–808.

Tucker, Richard. 1953. "Studies in Enology. I. The Concept of Enology, Its Method and Its Use in Teratological Investigations." *Acta Anatomica* 19 (1): 51–60.

Tuma, Julio R. 2009. "Biological Boundaries and the Vertebrate Immune System." *Biological Theory* 4 (3): 287–293.

Tuomi, Juha. 1981. "Structure and Dynamics of Darwinian Evolutionary Theory." *Systematic Zoology* 30 (1): 22–31.

Turbil, Cristiano. 2018. "Making Heredity Matter: Samuel Butler's Idea of Unconscious Memory." *Journal of the History of Biology* 51 (1): 7–29.

Turner, J. Scott. 2002. *The Extended Organism: The Physiology of Animal-Built Structures*. Cambridge, MA: Harvard University Press.

Turner, J. Scott. 2016. "Homeostasis and the Physiological Dimension of Niche Construction Theory in Ecology and Evolution." *Evolutionary Ecology* 30 (2): 203–219.

von Uexküll, Jakob. 1922. "Technische und mechanische Biologie." *Ergebnisse der Physiologie* 20 (1): 129–161.

von Uexküll, Jakob. 1926. *Theoretical Biology*. New York: Harcourt, Brace.

von Uexküll, Jakob. 1945. *Cartas biológicas a una dama*. Madrid: Revista de Occidente.

Uller, Tobias. 2023. "Agency, Goal-Orientation and Evolutionary Explanations." In *Evolution "on Purpose": Teleonomy in Living Systems*, edited by Peter Corning, Stuart A. Kauffman, Denis Noble, J. A. Shapiro, R. I. Vane-Wright, and A. Pross, 325–39. Cambridge, MA: MIT Press.

Uller, Tobias, and Heikki Helanterä. 2019. "Niche Construction and Conceptual Change in Evolutionary Biology." *British Journal for the Philosophy of Science* 70 (2): 351–375.

Uller, Tobias, and Kevin N. Lala, eds. 2019. *Evolutionary Causation: Biological and Philosophical Reflections*. Cambridge, MA: MIT Press.

Ungerer, Emil. 1919. *Die Regulationen der Pflanzen. Ein System der Teleologischen Begriffe in der Botanik*. Berlin: Julius Springer.

Vagelli, Matteo. 2019. "Historical Epistemology and the 'Marriage' Between History and Philosophy of Science." In *The Past, Present, and Future of Integrated History and Philosophy of Science*, edited by Emily Herring, Kevin Jones, Konstantin Kiprijanov, and Laura Sellers, 96–112. London: Routledge.

Van de Vijver, Gertrudis, Linda Van Speybroeck, Dani De Waele, Filip Kolen, and Helena De Preester. 2005. "Philosophy of Biology: Outline of a Transcendental Project." *Acta Biotheoretica* 53 (2): 57–75.

van der Klaauw, Cornelis Jakob. 1935. "Die Bedeutung der Teleologie Kants für die Logik der Ökologie." *Südhoffs Archiv für Geschichte der Medizin und der Wissenschaften* 27: 516–588.

Van Valen, Leigh. 1971. "Adaptive Zones and the Orders of Mammals." *Evolution* 25 (2): 420–428.

Van Valen, Leigh. 1973. "A New Evolutionary Law." *Evolutionary Theory* 1: 1–30.

Varela, Francisco, and Humberto Maturana. 1972. "Mechanism and Biological Explanation." *Philosophy of Science* 39 (3): 378–382.

Varela, Francisco G., Humberto R. Maturana, and R. Uribe. 1974. "Autopoiesis: The Organization of Living Systems, Its Characterization and a Model." *Biosystems* 5 (4): 187–196.

Varela, Francisco J. 1979. *Principles of Biological Autonomy*. New York: Elsevier North Holland.

Vernon, Horace Middleton. 1894. "The Effect of Environment on the Development of Echinoderm Larvae: An Experimental Inquiry into the Causes of Variation." *Proceedings of the Royal Society of London* 57: 382–385.

Vernon, Horace Middleton. 1900. "Certain Laws of Variation. I. The Reaction of Developing Organisms to Environment." *Proceedings of the Royal Society of London* 67 (435–441): 85–101.

Vigne, Paul, Clotilde Gimond, Céline Ferrari, Anne Vielle, Johan Hallin, Ania Pino-Querido, Sonia El Mouridi, et al. 2021. "A Single-Nucleotide Change Underlies the Genetic Assimilation of a Plastic Trait." *Science Advances* 7 (6): eabd9941. https://doi.org/10.1126/sciadv.abd9941.

Villegas, Cristina, and Vanessa Triviño. 2023. "Typology and Organismal Dispositions in Evo-Devo: A Metaphysical Approach." *ArtefaCToS. Revista de estudios sobre la ciencia y la tecnología* 12 (1): 79–102.

Virenque, Louis, and Matteo Mossio. 2023. "What Is Agency? A View from Autonomy Theory." *Biological Theory* 19 (1): 11–15. https://doi.org/10.1007/s13752-023-00441-5.

Virtanen, Reino. 1960. *Claude Bernard and His Place in the History of Ideas*. Lincoln, NE: University of Nebraska Press.

Volterra, Vito. 1926. "Fluctuations in the Abundance of a Species Considered Mathematically." *Nature* 118 (2972): 558–560.

von Dassow, George, Richard B. Emlet, and Svetlana A. Maslakova. 2013. "How the Pilidium Larva Feeds." *Frontiers in Zoology* 10 (1): 47. https://doi.org/10.1186/1742-9994-10-47

Waddington, Conrad Hal. 1942. "Canalization of Development and the Inheritance of Acquired Characters." *Nature* 150 (3811): 563–565.

Waddington, Conrad Hal. 1945. "The Directiveness of Organic Activities." *Nature* 156 (3973): 731–732.

Waddington, Conrad Hal. 1953. "Genetic Assimilation of an Acquired Character." *Evolution* 7 (2): 118–126.

Waddington, Conrad Hal. 1956. "Genetic Assimilation of the Bithorax Phenotype." *Evolution* 10 (1): 1–13.

Waddington, Conrad Hal. 1957. *The Strategy of the Genes: A Discussion of Some Aspects of Theoretical Biology*. London: Allen & Unwin.

Waddington, Conrad Hal. 1959a. "Evolutionary Systems: Animal and Human." *Nature* 183 (4676): 1634–1638.

Waddington, Conrad Hal. 1959b. "Evolutionary Adaptation." In *Evolution After Darwin, The University of Chicago Centennial*, edited by Sol Tax, 381–402. Chicago: University of Chicago Press.

Waddington, Conrad Hal. 1961. *The Nature of Life*. London: Allen & Unwin.

Waddington, Conrad Hal. 1962. *Biology for the Modern World*. New York: Barnes & Noble.

Waddington, Conrad Hal. 1965. "The Regulation of Population Density by Animals." *Ekistics* 20 (119): 185–190.

Waddington, Conrad Hal. 1975. *The Evolution of an Evolutionist*. Edinburgh: Edinburgh University Press.

Waddington, Conrad Hal, Joseph Needham, and Dorothy M. Needham. 1933. "Physico-Chemical Experiments on the Amphibian Organiser." *Nature* 132 (3328): 239–239.

Waddington, Conrad Hal, B. Woolf, and M. M. Perry. 1954. "Environment Selection by *Drosophila* Mutants." *Evolution* 8 (2): 89–96.

Wallace, Alfred Russell. 1913. *Social Environment and Moral Progress*. New York: Cassell.

Walsh, Denis M. 2012. "Mechanism and Purpose: A Case for Natural Teleology." *Studies in History and Philosophy of Science Part C: Studies in History and Philosophy of Biological and Biomedical Sciences* 43 (1): 173–181.

Walsh, Denis M. 2019. "The Paradox of Population Thinking: First Order Causes and High Order Effects." In *Evolutionary Causation: Biological and Philosophical Reflections*, edited by Tobias Uller and Kevin Lala, 227–246. Cambridge, MA: MIT Press.

Walsh, Denis M. 2021. "Aristotle and Contemporary Biology." In *The Cambridge Companion to Aristotle's Biology*, edited by Sophia M. Connell, 280–297. Cambridge, UK: Cambridge University Press.

Walsh, Denis M. 2022. "Environment as Abstraction." *Biological Theory* 17 (1): 68–79.

Walsh, Denis M. 2006. "Organisms as Natural Purposes: The Contemporary Evolutionary Perspective." *Studies in History and Philosophy of Science Part C: Studies in History and Philosophy of Biological and Biomedical Sciences* 37 (4): 771–791.

Walsh, Denis M. 2007. "The Pomp of Superfluous Causes: The Interpretation of Evolutionary Theory." *Philosophy of Science* 74 (3): 281–303.

Walsh, Denis M. 2008. "Teleology." In *Oxford Handbook of the Philosophy of Biology*, edited by Michael Ruse, 113–137. Oxford: Oxford University Press.

Walsh, Denis M. 2010. "Two Neo-Darwinisms." *History and Philosophy of the Life Sciences* 32 (2–3): 317–339.

Walsh, Denis M. 2012. "Situated Adaptationism." In *The Environment: Philosophy, Science, and Ethics*, edited by William P. Kabasenche, Michael O'Rourke, and Matthew H. Slater, 89–116. Cambridge, MA: The MIT Press.

Walsh, Denis M. 2015. *Organisms, Agency, and Evolution*. Cambridge, UK: Cambridge University Press.

Walsh, Denis M. 2017. "Chance Caught on the Wing: Metaphysical Commitment or Methodological Artifact?" In *Challenging the Synthesis: Adaptation, Development, Inheritance*, edited by Philippe Huneman and Denis M. Walsh, 239–261. Oxford: Oxford University Press.

Walsh, Denis M. 2018. "Objectcy and Agency: Towards a Methodological Vitalism." In *Everything Flows: Towards a Processual Philosophy of Biology*, edited by Daniel J. Nicholson and John Dupré, 167–185. Oxford: Oxford University Press.

Walsh, Denis M., André Ariew, and Mohan Matthen. 2017. "Four Pillars of Statisticalism." *Philosophy & Theory in Biology* 9 (1): 1–18. http://dx.doi.org/10.3998/ptb.6959004.0009.001.

Walsh, Denis M., and Gregory Rupik. 2023. "The Agential Perspective: Countermapping the Modern Synthesis." *Evolution & Development* 25 (6): 335–352.

Walsh, Denis M., and Sonia E. Sultan. 2024. "The Higher-Order Norm of Reaction: Biological Agency and Adaptive Phenotypic Response." In *The Riddle of Organismal Agency: New Historical and Philosophical Reflections*, edited by Alejandro

Fábregas-Tejeda, Jan Baedke, Guido I. Prieto, and Gregory Radick, 114–130. Abingdon: Routledge.

Walsh, James Joseph. 1900. "A Half-Century in Biology." *Catholic World* LXX: 466–479.

Warde, Paul, Libby Robin, and Sverker Sörlin. 2018. *The Environment: A History of the Idea*. Baltimore: John Hopkins University Press.

Ware, Ian M., Connor R. Fitzpatrick, Athmanathan Senthilnathan, Shannon L. J. Bayliss, Kendall K. Beals, Liam O. Mueller, Jennifer L. Summers, et al. 2019. "Feedbacks Link Ecosystem Ecology and Evolution Across Spatial and Temporal Scales: Empirical Evidence and Future Directions." *Functional Ecology* 33 (1): 31–42.

Wasserman, Ryan. 2004. "The Constitution Question." *Noûs* 38 (4): 693–710.

Watts, Elizabeth, Uwe Hoßfeld, and Georgy S. Levit. 2019. "Ecology and Evolution: Haeckel's Darwinian Paradigm." *Trends in Ecology & Evolution* 34 (8): 681–683.

Weaver, John E., and Frederic E. Clements. 1929. *Plant Ecology*. New York: McGraw-Hill.

Weber, Andreas, and Francisco J. Varela. 2002. "Life After Kant: Natural Purposes and the Autopoietic Foundations of Biological Individuality." *Phenomenology and the Cognitive Sciences* 1 (2): 97–125.

Weber, Bruce H., and David J. Depew, eds. 2003. *Evolution and Learning: The Baldwin Effect Reconsidered*. Cambridge, MA: MIT Press.

Weismann, August. 1892. *Das Keimplasma: Eine Theorie der Vererbung*. Jena, Germany: Gustav Fischer.

Weiss, Linda C. 2019. "Sensory Ecology of Predator-Induced Phenotypic Plasticity." *Frontiers in Behavioral Neuroscience* 12: 330. https://www.frontiersin.org/articles/10.3389/fnbeh.2018.00330.

Weldon, W. F. R. 1894. "The Study of Animal Variation." *Nature* 50 (1280): 25–26.

Werner, Konrad. 2021. "Structural Coupling and the Puzzle of Surfaces: Ontology of Boundaries from the Minimally Cognitive Perspective." *Adaptive Behavior* 29 (6): 601–615.

Wessely, Christina, and Nathan Stobaugh. 2019. "Watery Milieus: Marine Biology, Aquariums, and the Limits of Ecological Knowledge Circa 1900." *Grey Room* 75: 36–59.

West-Eberhard, Mary Jane. 2003. *Developmental Plasticity and Evolution*. New York: Oxford University Press.

West-Eberhard, Mary Jane. 2005. "Developmental Plasticity and the Origin of Species Differences." *Proceedings of the National Academy of Sciences* 102 (suppl_1): 6543–6549.

Wheeler, William Morton. 1905. "Ethology and the Mutation Theory." *Science* 21 (536): 535–540.

Wheeler, William Morton. 1926. "Emergent Evolution and the Social." *Science* 64 (1662): 433–440.

Wheeler, William Morton. 1929. "Present Tendencies in Biological Theory." *Scientific Monthly* 28 (2): 97–109.

Whitehead, Alfred North. 1929. *Science and the Modern World.* Cambridge, UK: Cambridge University Press.

Whitehead, Alfred North, and Bertrand Russell. 1910. *Principia Mathematica,* vol. 1. Cambridge, UK: Cambridge University Press.

Whyte, Lancelot Law. 1949. *The Unitary Principle in Physics and Biology.* London: Cresset Press.

Wiggins, David. 1968. "On Being in the Same Place at the Same Time." *Philosophical Review* 77 (1): 90–95.

Wiggins, David. 2001. *Sameness and Substance Renewed.* Cambridge, UK: Cambridge University Press.

Wigglesworth, Vincent Brian. 1961. "Insect Polymorphism: A Tentative Synthesis." In *Insect Polymorphism,* edited by J. S. Kennedy, 103–113. London: Bartholomew Press.

Wilbert, Hubert. 1961. "Über Festlegung und Einhaltung der mittleren Dichte von Insektenpopulationen." *Zeitschrift für Morphologie und Ökologie der Tiere* 50 (5): 576–615.

Wilbert, Hubert. 1970. "Cybernetic Concepts in Population Dynamics." *Acta Biotheoretica* 19 (2): 54–81.

Williams, George C. 1966. *Adaptation and Natural Selection: A Critique of Some Current Evolutionary Thought.* Princeton, NJ: Princeton University Press.

Williams, George C. 1992. "Gaia, Nature Worship and Biocentric Fallacies." *Quarterly Review of Biology* 67 (4): 479–486.

Willson, Mary F. 1981. "On the Evolution of Complex Life Cycles in Plants: A Review and an Ecological Perspective." *Annals of the Missouri Botanical Garden* 68 (2): 275–300.

Wilson, Edward Beecher. 1925. *The Cell in Development and Heredity,* 3rd ed. New York: Macmillan.

Wilson, Robert A. 2005. *Genes and the Agents of Life: The Individual in the Fragile Sciences Biology.* Cambridge, UK: Cambridge University Press.

Wilsterman, Kathryn, Mallory A. Ballinger, and Caroline M. Williams. 2021. "A Unifying, Eco-Physiological Framework for Animal Dormancy." *Functional Ecology* 35 (1): 11–31.

Wimsatt, William C. 1972. "Teleology and the Logical Structure of Function Statements." *Studies in History and Philosophy of Science Part A* 3 (1): 1–80.

Wingfield, John C., J. Patrick Kelley, Fréderic Angelier, Olivier Chastel, Fumin Lei, Sharon E. Lynn, Brooks Miner, Jason E. Davis, Dongming Li, and Gang Wang. 2011. "Organism–Environment Interactions in a Changing World: A Mechanistic Approach." *Journal of Ornithology* 152 (1): 279–288.

Winsor, Mary P. 2006. "The Creation of the Essentialism Story: An Exercise in Metahistory." *History and Philosophy of the Life Sciences* 28 (2): 149–174.

Winther, Rasmus G. 2020. *When Maps Become the World*. Chicago: University of Chicago Press.

Wise, M. Norton. 1994. "Pascual Jordan: Quantum Mechanics, Psychology, and National Socialism." In *Science, Technology, and National Socialism*, edited by Monika Renneberg and Mark Walker, 224–254. Cambridge, UK: Cambridge University Press.

Witteveen, Joeri. 2015. "'A Temporary Oversimplification': Mayr, Simpson, Dobzhansky, and the Origins of the Typology/Population Dichotomy (Part 1 of 2)." *Studies in History and Philosophy of Science Part C: Studies in History and Philosophy of Biological and Biomedical Sciences* 54: 20–33.

Witteveen, Joeri. 2016. "'A Temporary Oversimplification': Mayr, Simpson, Dobzhansky, and the Origins of the Typology/Population Dichotomy (Part 2 of 2)." *Studies in History and Philosophy of Science Part C: Studies in History and Philosophy of Biological and Biomedical Sciences* 57: 96–105.

Wolfe, Charles T. 2009. "Organisation ou organisme? L'individuation organique selon le vitalisme montpellierain." *Dix-huitième siècle* 41 (1): 99–119.

Wolfe, Charles T. 2011a. "From Substantival to Functional Vitalism and Beyond: Animas, Organisms and Attitudes." *Eidos: Revista de filosofía de la Universidad del Norte* 14: 212–235.

Wolfe, Charles T. 2011b. "Why Was There No Controversy over Life in the Scientific Revolution?" In *Controversies in the Scientific Revolution*, edited by Victor Boantza and Marcelo Dascal, 187–219. Amsterdam: John Benjamins.

Wolfe, Charles T. 2014a. "Holism, Organicism and the Risk of Biochauvinism." *Verifiche: Rivista Trimestrale Di Scienze Umane* 43 (1–3): 39–57.

Wolfe, Charles T. 2014b. "The Organism as Ontological Go-between: Hybridity, Boundaries and Degrees of Reality in Its Conceptual History." *Studies in History and Philosophy of Science Part C: Studies in History and Philosophy of Biological and Biomedical Sciences* 48: 151–161.

Wolfe, Charles T. 2024. "Varieties of Organicism: A Critical Analysis." In *Organization in Biology*, edited by Matteo Mossio, 41–58. Cham, Switzerland: Springer International Publishing.

Woltereck, Richard. 1909. "Weitere experimentelle Untersuchungen über Artveränderung, speziell über das Wesen quantitativer Artunterschiede bei Daphniden." *Verhandlungen der deutschen zoologischen Gesellschaft* 19: 110–173.

Woodger, Joseph Henry. 1925. "The Present Position of Biology Among the Sciences." *Science Progress in the Twentieth Century* 19 (76): 675–679.

Woodger, Joseph Henry. 1929. *Biological Principles: A Critical Study*. London: Kegan Paul.

Woodruff, Lorande Loss. 1908. "The Life Cycle of Paramecium when Subjected to a Varied Environment." *American Naturalist* 42 (500): 520–526.

Woodward, James. 2003. *Making Things Happen: A Theory of Causal Explanation*. Oxford: Oxford University Press.

Woodward, James. 2010. "Causation in Biology: Stability, Specificity, and the Choice of Levels of Explanation." *Biology & Philosophy* 25 (3): 287–318.

Woodward, Joseph Janvier. 1884. "On the Modern Philosophical Conceptions of Life." *Annals and Magazine of Natural History* 13 (76): 233–264.

Wright, Sewall. 1960. "The Treatment of Reciprocal Interaction, with or without Lag, in Path Analysis." *Biometrics* 16 (3): 423–445.

Wright, Sewall. 1966. *Evolution and the Genetics of Populations*, vol. 2: *Theory of Gene Frequencies*. Chicago: University of Chicago Press.

Yablo, Stephen. 1992. "Mental Causation." *Philosophical Review* 101 (2): 245–280.

Ylikoski, Petri. 2013. "Causal and Constitutive Explanation Compared." *Erkenntnis* 78 (2): 277–297.

Ylikoski, Petri, and Jaakko Kuorikoski. 2010. "Dissecting Explanatory Power." *Philosophical Studies: An International Journal for Philosophy in the Analytic Tradition* 148 (2): 201–219.

Zammito, John H. 2012. "The Lenoir Thesis Revisited: Blumenbach and Kant." *Studies in History and Philosophy of Science Part C: Studies in History and Philosophy of Biological and Biomedical Sciences* 43 (1): 120–132.

Zammito, John H. 2018. *The Gestation of German Biology: Philosophy and Physiology from Stahl to Schelling*. Chicago: University of Chicago Press.

Zeleny, Charles. 1905. "Compensatory Regulation." *Journal of Experimental Zoology* 2 (1): 1–102.

Zeman, J. Jay. 1982. "Peirce on Abstraction." *The Monist* 65 (2): 211–229.

Index

Note to readers: This index comprises only the names of scholars—both historical and contemporary—cited in the running text of the book and in the chapter endnotes; entries appearing exclusively in the references are not included. Endnotes are identified by the page number followed by "n" and the corresponding note number. Page numbers set in italics denote figure captions.

Publisher contact:
The MIT Press
Massachusetts Institute of Technology
77 Massachusetts Avenue, Cambridge, MA 02139
mitpress.mit.edu

EU Authorised Representative:
Easy Access System Europe, Mustamäe tee 50,
10621 Tallinn, Estonia
gpsr.requests@easproject.com

Printed by Integrated Books International,
United States of America